ENERGY OF MATTER

ENERGY OF MATTER

Editor

Nino Albini

scitus
academics

Energy of Matter

Edited by **Nino Albini**

This Book is Printed on Acid-Free Paper. ♾

ISBN: 9781681178523

Library of Congress Control Number: 2017948879

SCITUS Academics LLC,

616, Corporate Way, Suite 2, 4766,

Valley Cottage, NY 10989

www.scitusacademics.com

Notice

Printed in United States of America.

Preface

Physics is the study of matter, energy and the way they interact. Matter is the substance of which all material is made. That means objects which have mass. Energy is used in science to describe how much potential a physical system has to change. In physics, energy is a property of matter. It can be transferred between objects, and converted in form. It cannot be created or destroyed. As Einstein showed us, light and matter and just aspects of the same thing. Albert Einstein's most famous equation says that energy and matter are two sides of the same coin. It suggests that the concept of mass is indeed, less basic than what can be believed from everyday experiences with massive bodies. In fact, energy can be transformed into massive particles, and mass can be transformed into energy. Energy in all its different appearances is a key concept in physics. The study of matter is important because, without matter it is very hard to classify things. Matter is important because it helps a lot in our everyday lives which makes our life to be more convenient and easier.

This book, provides easy access to scientific knowledge, describes how matter is related to energy, which the human body needs and uses to do just about anything. The book will be of important not only to students and scientists but engineers as well (and anybody who is working with matter).

Table of Contents

CHAPTER 1

On the Real Einstein Beauty $E = Kmc^2$

A. J. Babchin[1], M. S. El Naschie[2*]

[1]*Alberta Research Council (ARC), Edmonton, Canada*

[2]*Department of Physics, University of Alexandria, Alexandria, Egypt*

ABSTRACT

The paper suggests that $E = mc^2$ may be open to misinterpretation and that in this form it is not what Einstein advanced first. It is further suggested to return to the slightly less compact formula $E = Kmc^2$ where $a < K < 1$ which has the merit of accounting for the measured ordinary energy density of the cosmos ($K = 1/22$) and the conjectured missing dark energy density of the universe ($K = 21/22$) from the view point of economical notation.

Keywords: Special Relativity, Nuclear Physics, Quantum Relativity, Quantum Entanglement, $E = Kmc^2$, Dark Energy, Ordinary Energy, Pair Annihilation, Pair Creation, Dvoretzky's Theorem, Bohm Guiding Wave

1. INTRODUCTION

The present paper started by the following innocent but for at least historical reasons, a justified question [1] as we will see:

DOES THE FORMULA E = mC^2 BELONGS TO EINSTEIN?

Now those of us who are passionate readers of the classical work of Einstein [1] - [12] , will not find this Formula as his work as a result [1] - [5] [10] - [12] . Instead one will notice that Einstein came to the result that l = E/C^2, where (l) is the famous mass defect due to the fission of Radium salts, E is the energy generated by fission and C is the speed of light. Thus, being the most basic equation of nuclear physics, Einstein result did not claim that mass of arbitrary chemistry can be entirely transformed to energy due to nuclear reaction. Such a leap is justified under certain circumstances in high energy physics and astrophysics but as certain, not in chemistry.

It's hard to nail down those responsible for transforming the Einstein original but they probably proceeded in the following way. As l is a very small mass (defect of mass in reality), it was suggested to rename l = dm, and the small energy, associated with dm was taken as dE. As a result the following formula appeared:

$$\mathrm{d}E = \mathrm{d}m \cdot C^2$$

There is nothing wrong with this formula as it corresponds to Einstein's result in a slightly different notation. The misleading error occurred upon integration of dm from 0 to entire mass m of the body of arbitrary Chemistry, and dE is integrated to give the energy E, corresponding to the mass of the entire body. The fact is however this formula cannot be integrated, unless dm is treated as small portion of the defect of mass, then the integration limits are from 0 to l, resulting in the initial Einstein formula.

Unfortunately, this practically incorrect result E = mC^2 [1] - [5] can be found in almost any textbook on General Physics [2] , usually without derivation, and is ascribed to Einstein, who really does not bear any responsibility for this widespread error of interpretation.

1.1. About the Present Article

The present article has grown out of an extensive open discussion invited by a question posed by the first author (A.J.B.) on Research Gate which is nowadays a leading and popular forum for research in general science, engineering, medicine and technology as well as humanities and social sciences [1] . The debate included at the time of writing 12 scientists and filled about 14 pages when downloaded. Since the subject is probably the most famous equation in the history of physics, in fact arguably in the history of science, the present authors who were able to come to a general consensus about the merit of the different arguments, decided to go ahead and publish the outcome as a formal paper in the reputable peer reviewed journal. We do not presume that this paper is the last word on a vast subject [1] - [18] but we are confident that it is a non-trivial and fundamental contribution to a pillar of modern physics [1] - [19] .

1.2. About $E = mc^2$

L.B. Okun, a world renowned expert on Einstein relativity and particularly on $E = mc^2$, put his misgivings plainly in a formal scientific paper by entitling it "The Einstein formula; $E_o = mc^2$. Isn't the Lord laughing?" published in 2008 in a leading Russian physics journal [2] . The controversy about $E = mc^2$ has a long history and an even longer list of published papers [1] - [14] . However in our present work we advance other arguments showing that the highly popular but open to misunderstanding interpretation for converting mass to energy, i.e. $E = mc^2$ is first not due to Einstein and second Albert Einstein is not responsible for all these misunderstandings in the first place. Subsequently we show that writing $E = Kmc^2$ is more general and accounts for the deeper meaning of Einstein's intention as well as for ordinary energy and dark energy by setting $K = 1/22$ for the first and $K = 21/22$ for the second so that at the end we find $E = (mc^2/22) + mc^2(21/22) = mc^2$ as should be [6] and in full agreement with cosmological measurements and observations which led twice to the Nobel Prize, once for COBE and the second in 2011 for Type 1a supernova and WMAP [6] [8] [13] .

2. THE CONVENTIONAL WIDESPREAD DERIVATION OF $E = MC^2$

The most common way used in deriving $E = mc^2$ is via the following example of electron-positron production and annihilation:

If a photon enters the coulomb field of a nucleus or another charged particle, this photon can disappear by so called pair production if the energy of the photon is above an energetic threshold, the pair production minimal energy. For the production of an electron-positron pair this minimal energy is 1022 keV plus a small rest of recoil energy for the coulomb field spending nucleus or another particle like atomic shell electron or free unbound electrons. The value of m_e/M_n is typically about 10^{-6}, the threshold formula (Figure 1)

$$E_{\gamma,\min} = 2m_e c^2 \left(1 + m_e / M_n\right)$$

The symbolic formula for the process is

$$\gamma = e^{+} + e^{-}$$

This is a well known elementary particle process. Its probability depends on the square of the atomic number of the field producing nucleus and increases with about the logarithm of the photon energy above the threshold.

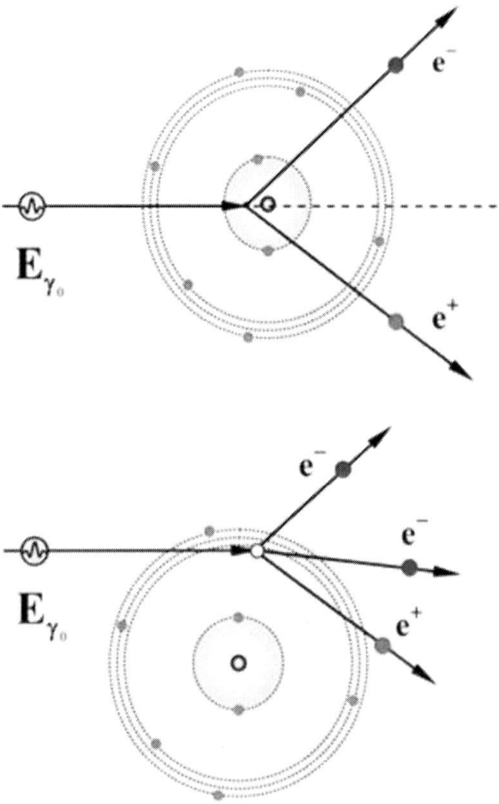

Figure 1. Illustration for electron-positron production.

The energy of the photon is changed into particle masses, and if then a difference of energy remains as kinetic energy of the two particles. From conservation laws this particle pair must consist of a matter-antimatter pair to keep the particle numbers etc. at zero.

$$E_{kin} = E_\gamma - 2 \cdot m_0 \cdot c^2 = E_\gamma - 1022 \text{ keV}$$

In the case of zero remainingkineticenergy,wefinddirectly the famous Einstein mass- energy relation

$$E = mc^2$$

which means the direct equivalence of energy and mass.

The revised process of pair production is pair annihilation. It happens if a particle and the corresponding antiparticle collide. The example here is the electron-positron annihilation.

If this process happens with zero rest kinetic energy of the two colliding leptons, we immediately get the equation

$$e^+ + e^- = 2m_0c^2 = 2 \times 511 \text{ keV}$$

The two annihilation photons are emitted in a typical 180 degree angle, which is applied in nuclear medicine imaging as PET technology. Or in a more symbolic way you can write again

$$mc^2 = E$$

the direct equivalence again of mass and energy (Figure 2).

If the annihilation takes place with remaining energy of the electron or the positron, the two annihilation photons are emitted under a sharp labor angle smaller than 180 degree.

3. DISSECTING E = MC² INTO ITS TWO QUANTUM RELATIVITY COMPONENTS

Let us start from the generally accepted De Broglie-Bohm pilot wave picture, which is probably the only quantum theory which Einstein was ready to at least partially “swallow” to put it mildly. We take this picture because it agrees completely with von Neumann-Connes’ pointless noncommutative geometry as our spacetime and consequently also with the fundamental equation [8] [9]

$$D = a + b\phi,\ a, b \in Z \text{ and } \phi = \left(\sqrt{5} - 1\right)/2$$

which shows that the "pilot" quantum wave is the cobordism, i.e. "surface" of the quantum particle. In particular the "pilot" pre-quantum wave would be fixed by the empty set bi-dimension

$$D(\text{empty}) = \left(-1, \phi^2\right)$$

while the pre-quantum particle would be fixed by the zero set

$$D(\text{zero}) = (0, \phi).$$

Let us calculate the quasi volume of the pre-quantum particle and the quasi area of its surface, i.e. the surface constituting the "pilot" wave. Noting that the basic dimensionality of spacetime is that of Kaluza-Klein five dimensions [14] then it follows that the volume of the quantum particle must be the correlated multiplicative value

$$\text{vol}_5(5) = (5)(\phi)^2$$

where ϕ and ϕ^2 are the corresponding Hausdorff dimensions given by $D = a + b\phi$ and interpreted as a Hausdorff volume measure. Now taking Newton's kinetic energy as our canonical equation, we see easily that the energy density corresponding to the above measures must be [8] [9]

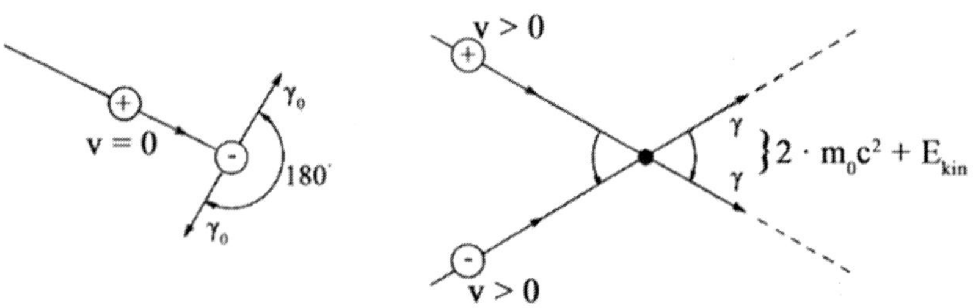

Figure 2. Illustration for pair annihilation process.

$$E(O)=\frac{1}{2}m(v\to c)^2(\phi^5)=(\phi^5/2)(mc^2)\simeq mc^2/22$$

and

$$E(D)=\frac{1}{2}m(v\to c)^2(5\phi^2)=(5\phi^2/2)(mc^2)\simeq mc^2(21/22)$$

Adding ordinary energy E(O) to dark E(D) it is delightful to see that we retrieve Einstein's correct formulas [6] [8] [9]

$$E=(\phi^5/2)(mc^2)+(5\phi^2/2)(mc^2)=[(\phi^5/2)+(5\phi^2/2)]mc^2$$
$$=[k_1+k_2]mc^2=Kmc^2=mc^2$$

where $K=k_1+k_2=(\phi^5+5\phi^2)/2=2/2=1$. This remarkable result could be obtained more directly from the pure mathematics of measure theory of Banach spaces [15] - [19] . To do that we simply resort to Dvoretzky's theorem and regard particle and wave as a high dimensional sphere. Consequently such a sphere would have 96 percent of its volume concentrated on the higher dimensional surface while 4 percent constitutes the deceptive internal bulk of the sphere. This is practically the same previous result. In conclusion we must say that there is hardly any doubt about this result because it is in astounding nearly surreal agreement with unprecedented high precision cosmic measurements and observations connected to COBE and Type 1a supernova which we repeatedly refer to.

4. DISCUSSION

It is not always easy to set an exact limit for where we become relativistic or render quantum starting from classical physics. Consequently we should not be astonished when we see that at the edge of the universe at hyperbolic infinity of the boundary of the holographic boundary we have an accumulation of dark energy which accounts for 95.5% of the total energy density of the universe given by $E\cong mc^2(21/22)$, i.e. $K=(21/22)$ in the formula E = Kmc^2 which we propose to replace with E = mc^2. However for K = 1 we have a very special situation indeed for which by chance or providence we found E(max) = (1)(mc^2) and that by a totally ingenious leap of faith rather than a systematic derivation as noted by the grand connoisseur of relativity, Professor Wolfgang Rindler [7] . All of the present derivation follows actually smoothly and almost effortlessly from the Rindler wedge of the Rindler spacetime geometry and topology where everything was encoded in the hyperbolic geometry without us noticing, nor in fact Rindler himself noticing that it is telling us all about $E = Kmc^2$, ordinary energy and dark energy density of the universe but in a very subtle language and quite different from the language used in the present work. Never the less, $E = Kmc^2$ is the physical incarnation of the old proverb which says that all roads lead to Rome or $E = Kmc^2$.

5. CONCLUSION

The celebrated formula $E = mc^2$ has a long history dating back to before Einstein, for instance Poincare. Not only that but it is also somewhat misleading and better replaced by $E = Kmc^2$. In this second form E becomes the sum of two types of energy, the measurable ordinary energy $E(O) = mc^2/22$ and the supposedly missing dark energy $E(D) = mc^2(21/22)$ [19] . This reveals hitherto unsuspected quantum roots for $E = mc^2$. In fact classical relativity already implies the total unity of Einstein's relativity, classical mechanics and quantum mechanics. This is so because it is the relative size of what we are measuring vis-à-vis our own size and the size of our experimental device is what makes things relatively very large where relativity is applicable or relatively very small where quantum mechanical description takes over. No wonder then that the quantum relativity formula $E(O) = mc^2/22$ when added to the quantum cosmology formula $E(D) = mc^2(21/22)$ gives us the famous formula $E = mc^2$ which is only a scaling of Newton's $E = 1/2mv^2$sending $1/2v^2$ to c^2. Nothing could possibly be more astoundingly simple than that.

ACKNOWLEDGEMENTS

The authors are deeply indebted to all who participated in the relevant discussion on Research Gate questions and answers of Ref. [1] .

REFERENCES

1. Babchin, A.J. (2015) Does the Formula E = mc^2 Belong to Einstein? Research Gate Questions and Answers. https://www.researchgate.net/post/Does_the_formula_ E_mC2_ belong_to_Einstein
2. Okun. L.B. (2008) The Einstein Formula Eo = mc2. Isn't the Lord Laughing? Physics-Uspekhi, 51, 513-527.

3. Schwartz, H.M. (1977) Einstein's First Paper on Relativity. American Journal of Physics, 45, 18-25.
4. Panos, P. (2008) The Mass Energy Relation of Einstein = mc2 Is Not Generally Valid. http://www.panospappas.gr/kmc2eng.htm
5. Alekseevich, U.N. (1950) Izbrannyyeesochineniya. M. 1, 1873.
6. El Naschie, M.S. (2014) From E = mc2 to E = mc2/22—A Short Account of the Most Famous Equation in Physics and Its Hidden Quantum Entangled Origin. Journal of Quantum Information Science, 4, 284-291.
7. Rindler, W. (2006) Relativity (Special, General and Cosmological). Oxford University Press, Oxford.
8. El Naschie, M.S. (2015) An Exact Mathematical Picture of Quantum Spacetime. Advances in Pure Mathematics, 5, 560-570.
9. El Naschie, M.S. (2015) If Quantum "Wave" of the Universe Then Quantum "Particle" of the Universe: A Resolution of the Dark Energy Question and the Black Hole Information Paradox. International Journal of Astronomy & Astrophysics, 5, 243-247.
10. Einstein, A. (1905) Ist die Trägheit eines Körpers von seinem Energieinhalt abhängig? (German) [Does the Inertia of a Body Depend on Its Energy Content?]. Annalen der Physik, 323, 639-641. (English Translations in [4, 67-71]).
11. Einstein, A. (1905) Zur Elektrodynamik bewegter Körper. (German) [On the Electrodynamics of Moving Bodies]. Annalen der Physik, 322, 891-921. (English Translations in [4, 35-36 and 7]).
12. Lorentz, H.A., Einstein, A., Minkowski, H. and Weyl, H. (1952) The Principle of Relativity: A Collection of Original Memoirs on the Special and General Theory of Relativity. With Notes by A. Sommerfeld. Translated by W. Perrett and G.B. Jeffery. Dover Publications, Inc., New York.
13. El Naschie, M.S. (2014) The Measure Concentration of Convex Geometry in a Quasi Banach Spacetime behind the Supposedly Missing Dark Energy of the Cosmos. American Journal of Astronomy & Astrophysics, 2, 72-77.
14. Wesson, P.S. (2006) Five-Dimensional Physics. World Scientific, Singapore.
15. El Naschie, M.S. (2015) The Counterintuitive Increase of Information Due to Extra Spacetime Dimensions of a Black Hole and Dvoretzky's Theorem. Natural Science, 7, 483-487.
16. El Naschie, M.S. (2015) Application of Dvoretzky's Theorem of Measure Concentration in Physics and Cosmology. Open Journal of Microphysics, 5, 11-15.
17. El Naschie, M.S. (2015) A Resolution of the Black Hole Information Paradox via Transfinite Set Theory. World Journal of Condensed Matter Physics, 5, 249-260.
18. El Naschie, M.S. (2015) Casimir-Dark Energy Nano Reactor Design Proposal Based on Fractals. International Journal of Innovation in Science and Mathematics, 3, 187-194.
19. Marek-Crnjac, L. (2015) On El Naschie's Fractal Cantorian Spacetime and Dark Energy—A Tutorial Review. Natural Science, 7, 581-598.

CHAPTER 2

Dark Energy as a Property of Dark Matter

Jacques Leibovitz

independent author, San Jose, USA

ABSTRACT

A novel model of dark matter (DM), elastically compressible, can contribute to the acceleration of our Universe expansion. While each galaxy compresses its own DM within its gravitation field, the DM bordering neighboring galaxies, far from their centers, is pulled apart. It is shown that, although the DM pressure tends to zero at such locations, the DM compressibility tends to infinity. This allows the DM to expand between galaxies without gravitation hindrance. The model is consistent with the coupled distributions of baryonic and dark matters, with black hole formation at the centers of large galaxies, with galactic flat rotation curves, with a Tully-Fisher relation, and with Milgrom's MOND relation. Results are discussed.

Keywords: dark energy, dark matter

1. INTRODUCTION

A main problem in astrophysics is the discovery, about two decades ago, that our Universe expansion is accelerating, instead of slowing down as predicted by the Big Bang theory [1] Scientists hypothesize the existence of an anti-gravity energy field dubbed Dark Energy (DE). It has been claimed that the identification and understanding of that DE would be the greatest accomplishment of the century [2].

A less recognized problem is the Big Bang itself, the explosion of the biggest of black holes, which the present theory and model of black holes cannot accommodate. Instead, the Big Bang theory deals with the evolution of our Universe after the initial Bang [1]. Inclusion of the DE proposed here could destabilize black holes sufficiently to help resolve that problem.

An elastically compressible DM might contribute to the accelerated expansion of our Universe and to the destabilization of black holes. In the following, we show that such model of DM is consistent with the observed DM phenomena. Then, we derive a possible mechanism of acceleration of our Universe expansion.

1.1. Dark Matter or New Physics?

The prevailing scientific view today is that DM is constituted of weakly interacting massive particles (WIMPs). At this time, such view must be considered a hypothesis because, so far, a constitutive particle of DM has not been detected, and computer simulations based on the predicted microscopic properties of WIMPs have not succeeded in modeling the observed behavior of DM [1-15].

An alternative hypothesis is that DM is not a substance at all and that the DM-phenomena result from a "new physics." Some scientists proposed a modification of Newton dynamics (MOND) for weak acceleration as such alternative [16-23]. Milgrom's version of MOND [16-22] proposes at very low accelerations, corresponding to very large distances R, a relation that departs from Newton laws as follows: defining γ_o as a small constant (= about 10^{-8} cm·s^{-2}) of acceleration dimensions, then for $\gamma \ll \gamma_o$ MOND departs from Newton laws by postulating:

$$\gamma_{\rm small} = F/M = -\,(GM\gamma_o)^{1/2}R^{-1} = -(V_c)^2R^{-1}. \quad (1.1.1)$$

From equation (1.1.1):

$$V_c = (GM\gamma_o)^{1/4}, \quad (1.1.2)$$

at very low accelerations, corresponding to very large distances R_{large}.

Although MOND's relation does not define the dynamics in the intermediate range where $\gamma \approx \gamma_o$, it is tailored to produce the galactic rotation curve plateau in a form (equation (1.1.2)) fitting a Tully-Fisher relation (TFR). However, MOND relates the luminous mass of the galaxy with the limiting speed (equation (1.1.2)) while the TFR implies a relation with the maximum, or peak speed.

While very successful at modeling much of DM behavior, serious reasons preclude the scientific acceptance of MOND [16-21,24,25]. In particular, Milgrom's derivation of MOND, based on a departure from Newton laws of gravitation, though speeds are small compared with the speed of light, conflicts with general relativity and makes that derivation unacceptable. However, because some doubt that DM particles will be substantiated, it has been suggested to take a new look at MOND [26]. On the other hand, the recent observation of dissociation between DM and baryonic clouds during a collision of galactic clusters [27] tends to tip the scales, at this time, in favor of the existence of DM.

Here, we explore another "new physics" consisting of an elastically compressible DM, with a mass density distribution ρ. The exploration of such "new physics" appears fruitful, leading to the derivation of an equation of state (EoS) for such DM in a gravitation field, of an equation governing the coupled distributions of baryonic and dark matters around galaxies, of a possible mechanism of black hole formation at the centers of large galaxies, of galactic flat rotation curves, of a Tully-Fisher relation, of Milgrom's MOND relation, and of a possible mechanism for the accelerated expansion of our universe.

1.2. Alternative Model of Dark Matter

a) DM is a non-baryonic substance, consistent with the observation of a collision between galactic clusters [27].

b) DM distribution is spherically symmetric around some spiral galaxies, without flattening at the poles [3, 6,10,28].

c) Consistent with assumption (b), DM does not necessarily partake in galaxy rotation [1,3,4,10,28]. The concept that DM is constituted of particles conflicts with this assumption.

d) Dark matter has no centrifugal force to balance the centripetal gravitation pull, consistent with assumption (c).

e) Instead, we assume that DM is elastically compressible, without energy dissipation, and develops a DM-pressure P, which balances the gravitation pull on DM.

f) Because DM-WIMPs have no self interaction sufficient to develop DM-pressure high enough to balance the gravitation pull, assumption (e) excludes the possibility that DM be constituted of WIMPs, consistent also with the absence of success, so far, of efforts to prove the existence of DM-WIMPs [1-4,7,8,11-15].

g) Newton dynamics may be used as a good approximation of general relativity in the limits of speeds small relative to the speed of light and of weak gravitation.

2. COUPLED DISTRIBUTIONS OF BARYONIC AND DARK MATTERS

To simplify, we consider only systems of spherical symmetry.

Including DM mass, Newton's equations applied to a spherically symmetric galaxy yield:

$$F = -G(M + m)M'R^{-2} \tag{2.1}$$

where M and m (g) are the baryonic (mostly luminous) mass and the DM-mass, respectively, both within a radius R from the galaxy center, and F is the gravitation force applied by the sum of M and m on a gravitating small test mass M' at radial distance R from the galaxy center.

The gravitational acceleration γ of M' at R then is:

$$\gamma = -(V_c)^2/R = F/M' = -G(M + m)R^{-2}. \tag{2.2}$$

from which:

$$(V_c)^2 = -G(M + m)R^{-1} \tag{2.3}$$

and $V_c = [G(M + m)R^{-1}]^{1/2}.$ (2.4)

The mass m may be expressed as:

$$m = 4\pi\int \rho R^2 \mathrm{d}R \tag{2.5}$$

from which:

$$dm = 4\pi\rho R^2 dR \tag{2.6}$$

where ρ is the DM density at location R.

2.1. Equation of State of Dark Matter in a Gravitation Field

Staying with the simplicity of spherical symmetry, let us consider a spherically symmetric distribution of baryonic mass, its associated DM, and their gravitation field of center O, at infinite distance from any other mass.

Because DM does not necessarily rotate around galaxies (assumption c), it lacks the centrifugal force that might balance the gravitation force on DM (assumption d). Instead, the gravitational field elastically compresses DM to a pressure distribution P ($g \cdot cm^{-1} \cdot s^{-2}$), function of radial distance R, without energy dissipation (assumption e). Therefore, the pressure P is a function only of location R and of other functions of R. The DM pressure balances the gravitation force and may be expressed in differential form as the ratio of the gravitation differential force dF_m exerted by the sum of M and m on the differential mass dm of the DM confined in a spherical shell of radius R, and differential thickness dR, to the surface area S (= $4\pi R^2$) of that shell:

$$dP = -G(M+m)R^{-2}dm(4\pi)^{-1}R^{-2} = \gamma(4\pi)^{-1}R^{-2}dm = \rho dR < 0. \tag{2.1.1}$$

in which Equation (2.6) was used in the last step.

Equation (2.1.1) shows that the pressure P is expressible in terms of the location R, the DM mass density ρ at location R, and the gravitation acceleration γ (of a test mass M') at that location. The pressure might be expressed as:

$$P - P_{oo} = -A\rho\gamma R = A\rho(V_c). \tag{2.1.2}$$

where P_{oo} is the DM pressure at infinite distance, and A is a dimensionless constant. When the value of A is determined, equation (2.1.2) provides the EoS for DM in the spherically symmetric gravitation field.

To determine the value of the integration constant A, we derive an expression for the distribution of DM mass-density ρ around a galaxy and then compare it to the boundary condition at very large radii R_{large}. The procedure is as follows:

Form the ratio of equation (2.1.1) to equation (2.1.2), side by side:

$$\frac{dP}{P-P_\infty}=\frac{\rho\gamma dR}{-A\rho\gamma R}=-\frac{dR}{AR} \qquad (2.1.3)$$

and by integration:

$$\mathrm{Log}(P - P_{oo}) + A^{-1}\mathrm{Log}\ (R) = \mathrm{Log(constant)} \qquad (2.1.4)$$

or:

$$(P - P_{oo})R^{(1/A)} = -A\rho\gamma R^{[1+(1/A)]} = A\rho(V_c)^2R^{(1/A)} = \text{constant} = Ak \qquad (2.1.5)$$

where the integration constant is conveniently written as Ak to yield:

$-\rho\gamma R^{[1+(1/A)]} = \rho(V_c)^2R^{(1/A)} = k$, another constant specific to the galaxy. (2.1.6)

The constant k is specific to the galaxy because it is a function of the total luminous mass M and of its distribution in the galaxy, so that 2 galaxies of equal total luminous mass but with different distributions can have different constants k.

For a galactic flat rotation curve, V_c becomes constant at very large radial distances R_{large}, which requires that the DM mass density be distributed there as:

$\rho(R_{large})^2 = K$, another constant specific to the galaxy. (2.1.7)

From equation (2.1.6):

$$\rho R^{(1/A)} = k/(V_c)^2. \qquad (2.1.8)$$

Equation (2.1.8) shows that the necessary and sufficient condition for V_c to remain constant in any range of radial locations R, is that $\rho R^{(1/A)}$ remain constant in that range. Comparison with equation (2.1.7), shows that equation (2.1.8) has the required form, and that the constant A = 1/2.

With the determination of A = 1/2, equation (2.1.2) becomes the sought EoS:

$$P - P_{\infty} = -\rho \gamma R/2 = \rho V_c^2/2. \tag{2.1.9}$$

Because of the restriction to spherical symmetry, equation (2.1.9) is the EoS derived for DM outside any galaxy and inside any spherically symmetric galaxy, except perhaps at sufficiently high DM-pressures where DM might undergo phase changes with corresponding changes of properties. Such changes might occur sufficiently close to some galaxy centers.

However, being a material property of DM, the validity of the EoS (equation 2.1.9) is not limited to spherical symmetry and would be valid outside and inside any galaxy, except perhaps at sufficiently high DM pressures where DM might undergo phase changes with corresponding changes of DM properties. Such changes might occur sufficiently close to some galaxy centers.

2.2. Equation Governing the Coupled Distributions of Baryonic and of Dark Matter

With the determination of the constant A = 1/2, equation (2.1.8) becomes:

$$-\rho \gamma R^3 = \rho (V_c)^2 R^2 = k. \tag{2.2.1}$$

Because V_c depends on the distribution of both the baryonic mass and the DM mass, its participation in equation (2.2.1) means that this equation (2.2.1) governs both distributions.

Equation (2.2.1) satisfies the known trivial condition that for V_c to become constant at sufficiently large distances R_{large}, the expression ρR^2_{large} must become constant there (equation (2.1.7)).

Still, it remains to show that both V_c and ρR^2_{large} do become constant at sufficiently large distances R_{large}. We now derive that result and, therefore, the flat galactic rotation curves.

2.3. Derivation of Galactic Flat Rotation Curves

Combine equations (2.2 and 2.5) and rearrange to yield:

$$RV_c^2 - GM = Gm = 4\pi G\int_a^b \rho R^2 \mathrm{d}R \quad (2.3.1)$$

The validity of equation (2.3.1) is restricted to spherical symmetry.

Form the differential on both sides of equation (2.3.1), to obtain:

$$(V_c)^2\mathrm{d}R + R\mathrm{d}(V_c)^2 - G\mathrm{d}M = 4\pi G\rho R^2\mathrm{d}R. \quad (2.3.2)$$

Substitute for ρR^2 from equation (2.2.1) to yield:

$$(V_c)^2\mathrm{d}R + R\mathrm{d}(V_c)^2 - G\mathrm{d}M = 4\pi Gk[\mathrm{d}R/(V_c)^2]. \quad (2.3.3)$$

Express dM in terms of the luminous mass density ρ_b:

$$\mathrm{d}M = 4\pi\rho_b R^2\mathrm{d}R. \quad (2.3.4)$$

Substitute into equation (2.3.3), multiply throughout by $(V_c)^2/R$, and rearrange to yield:

$$(V_c)^4\mathrm{d}R/R + (V_c)^2\mathrm{d}(V_c)^2 - G4\pi\rho_b(V_c)^2R\mathrm{d}R = 4\pi Gk[\mathrm{d}R/R].$$

or:

$$G4\pi k(\rho_b/\rho)\mathrm{d}R/R + 4\pi Gk[\mathrm{d}R/R] - (V_c)^4\mathrm{d}R/R - (V_c)^2\mathrm{d}(V_c)^2 = 0. \quad (2.3.5)$$

Solution of equation (2.3.5) yields the flat galactic curves. Various approaches are possible. One approach is, for any particular galaxy, to determine ρ_b from luminosity measurements, then solve equations (2.2.1) and (2.3.5) simultaneously, for V_c and ρ as functions of R.

CHAPTER 3

Distribution of Mass and Energy in Five General Cosmic Models

Fadel A. Bukhari

Department of Astronomy, Faculty of Science, King Abdulaziz University, Jeddah, KSA

ABSTRACT

Distributions of the universe horizon distance and universe horizon volume were investigated in the light of five general cosmic models which were constructed in a previous study. Both distributions increase so slowly up to $t \approx 21.5444$ Myr, then they start raising very fast up to $t \approx 60$ Gyr. Afterwards, they increase again very slowly until $t \approx 124$ Gyr. Distributions of mass of radiation, matter and dark energy within the horizon volume of the universe were also studied in the five general cosmic models. The masses of both radiation and matter decrease gradually with time while the mass of dark energy increases. The mass of radiation prevailed in the early universe up to $t \approx 34627.5$ - 55916.2 yr, where it becomes equal to the mass of matter. Then the mass of matter dominated until $t \approx 9.4525$ - 10.0632 Gyr, where it becomes equal to the mass of dark energy. Thenceforward, the mass of dark energy prevails the universe. The cosmic space becomes approximately matter empty in the so far future of the universe.

Keywords: General Cosmic Models, Distribution of Mass and Energy

1. INTRODUCTION

In a previous study [1] the distribution of density parameters of radiation, matter and dark energy were investigated in details in five general cosmic models. Hence, it would be interesting to study the distributions of equivalent mass of radiation, mass of matter and equivalent mass of dark energy within the horizon volume of the universe in the general models.

Therefore, it is necessary to start this study by investigating the distributions of the horizon distance and horizon volume of the universe in the general models at different time intervals depending on the bases discussed in [2] . Description of methodology is given in Section 2 while algorithm would be illustrated in Section 3. Results and discussion are presented in Section 4. Conclusion is shown in Section 5.

2. METHODOLOGY

We have seen in [2] that the horizon distance and horizon volume of the universe at the present time are respectively

$$d_h(t_o)=\frac{c}{H_o}\int_0^1 \frac{1}{a}\left[1-\Omega_{\Lambda,o}\left(1-a^2\right)+\Omega_{m,o}\left(\frac{1}{a}-1\right)+\Omega_{r,o}\left(\frac{1}{a^2}-1\right)\right]^{-\frac{1}{2}} da. \tag{1}$$

$$V_h(t_o)=\frac{8\pi}{3}d_h^3(t_o). \tag{2}$$

where $c, H_o, \Omega_{r,o}, \Omega_{m,o}, \Omega_{\Lambda,o}$ are all defined as in [1] . Thus the horizon distance of the universe at any given time is

$$d_h(t)=\frac{c}{H_o}\int_0^a \frac{1}{a}\left[1-\Omega_{\Lambda,o}\left(1-a^2\right)+\Omega_{m,o}\left(\frac{1}{a}-1\right)+\Omega_{r,o}\left(\frac{1}{a^2}-1\right)\right]^{-\frac{1}{2}} da. \tag{3a}$$

Consequently the change in the horizon distance of the universe in the time interval between two instants of scale factors a_1, a_2 is written as

$$\Delta d_h(t) = \frac{c}{H_o}\int_{a_1}^{a_2} \frac{1}{a}\left[1-\Omega_{\Lambda,o}\left(1-a^2\right)+\Omega_{m,o}\left(\frac{1}{a}-1\right)+\Omega_{r,o}\left(\frac{1}{a^2}-1\right)\right]^{-\frac{1}{2}} da. \tag{3b}$$

The horizon volume of the universe at any given time is

$$V_h(t) = \frac{8\pi}{3} d_h^3(t). \tag{4}$$

It is also obvious from [2] that the total density of the universe is given by

$$\rho(t) = \rho_{c,t}\Omega(t). \tag{5}$$

where

$$\Omega(t) = \rho_{c,t}\left(\Omega_{m,t} + \Omega_{r,t} + \Omega_{\Lambda,t}\right). \tag{6}$$

$$\rho_{c,t} = \frac{3H^2}{8\pi G}. \tag{7}$$

$$\Omega_{m,t} = \frac{\rho_{m,t}}{\rho_{c,t}} = \left(\frac{H_0}{H}\right)^2 \frac{\Omega_{m,o}}{a^3}. \tag{8}$$

$$\Omega_{r,t} = \frac{\rho_{r,t}}{\rho_{c,t}} = \left(\frac{H_0}{H}\right)^2 \frac{\Omega_{r,o}}{a^4}. \tag{9}$$

$$\Omega_{\Lambda,t} = \frac{\rho_{\Lambda,t}}{\rho_{c,t}} = \left(\frac{H_0}{H}\right)^2 \Omega_{\Lambda,o}. \tag{10}$$

$$H(t) = \frac{H_o}{a}\left[1-\Omega_{\Lambda,o}\left(1-a^2\right)+\Omega_{m,o}\left(\frac{1}{a}-1\right)+\Omega_{r,o}\left(\frac{1}{a^2}-1\right)\right]^{\frac{1}{2}}. \tag{11}$$

Hence, the total mass within the horizon volume of the universe at any given time is expressed as

$$M_h(t) = V_h(t)\rho(t). \tag{12}$$

The mass of matter $M_{m,t}$, the equivalent mass of radiation $M_{r,t}$ and the equivalent mass of dark energy $M_{\Lambda,t}$ within the horizon volume of the universe at any given time are given by

$$M_{m,t} = M_h(t)\frac{\rho_{m,t}}{\rho(t)}$$

$$M_{m,t} = M_h(t)\frac{\Omega_{m,t}}{\Omega(t)}. \tag{13}$$

$$M_{r,t} = M_h(t)\frac{\Omega_{r,t}}{\Omega(t)}. \tag{14}$$

$$M_{\Lambda,t} = M_h(t)\frac{\Omega_{\Lambda,t}}{\Omega(t)}. \tag{15}$$

The cosmic time is given by Equation (16) in [1] as

$$t = \frac{1}{H_o}\int_0^a \left[1 - \Omega_{\Lambda,o}\left(1-a^2\right) + \Omega_{m,o}\left(\frac{1}{a}-1\right) + \Omega_{r,o}\left(\frac{1}{a^2}-1\right)\right]^{-\frac{1}{2}} da. \tag{16a}$$

Thus the time interval between two instants with scale factors a_1, a_2 during the universe expansion is expressed as

$$\Delta t = \frac{1}{H_o}\int_{a_1}^{a_2} \left[1 - \Omega_{\Lambda,o}\left(1-a^2\right) + \Omega_{m,o}\left(\frac{1}{a}-1\right) + \Omega_{r,o}\left(\frac{1}{a^2}-1\right)\right]^{-\frac{1}{2}} da. \tag{16b}$$

3. ALGORITHM

In determination of the distributions of $d_h(t), V_h(t), M_h(t), M_{m,t}, M_{r,t}$ and $M_{\Lambda,t}$ we use the following steps:

1) Set $t = 0, d_h = 0, K_1 = 1, K_2 = 2000,$ then insert the value of $a_{max} = 0.098$ for $t \le 0.5$ Gyr, $a_{max} = 10$ for $0.5 < t \le 50$ Gyr and $a_{max} = 1000$ for $50 < t \le 124$ Gyr.

2) Compute $DA = \frac{a_{max}}{DBLE(K_2)}$.

3) Start general DO loop $I = K_1, K_2$ which includes the following sub steps:

4) $a_1 = DA(I-1), a_2 = DAI.$

5) Calculate new value of cosmic time t numerically using(16-b), where $t = t + \Delta t$.

6) Determinate new value of the universe horizon distance d_h numerically using (3-b) where, $d_h = d_h + \Delta d_h$.

7) Obtain the corresponding values of $V_h, H, \rho_{c,t}, \Omega_{m,t}, \Omega_{r,t}, \Omega_{\Lambda,t}, \Omega(t), \rho(t), M_h(t), M_{m,t}, M_{r,t}, M_{\Lambda,t}$ using (4), (11), (7), (8), (9), (10), (6), (5), (12), (13), (14) and (15) respectively.

8) Continue the general DO loop.

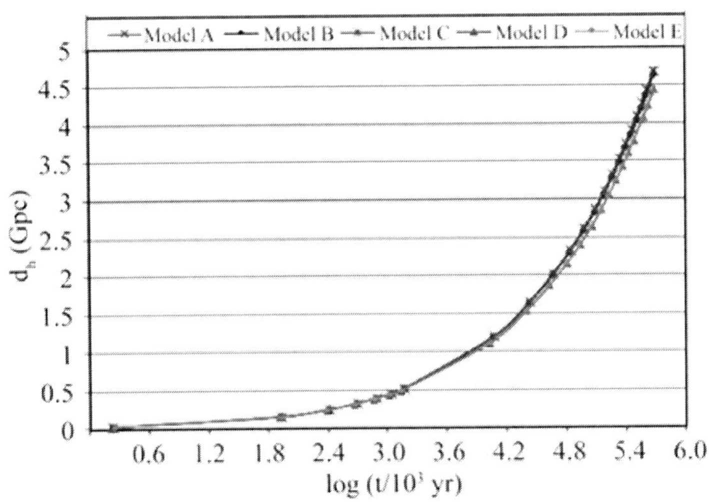

(a)

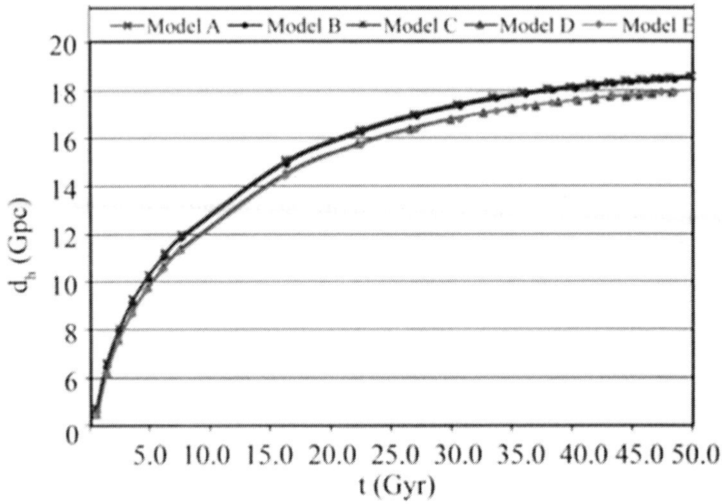

(b)

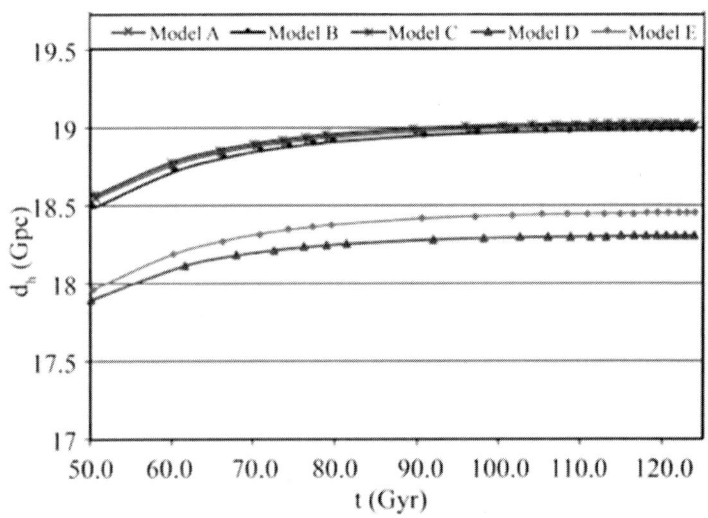

(c)

Figure 1. The distribution of the universe horizon distance in the general cosmic models (a) up to t = 0.5 Gyr; (b) in the range t = 0.5 - 50 Gyr; (c) in the range t = 50 - 124 Gyr.

4. RESULTS AND DISCUSSION

The distribution of the universe horizon distance in the general models up to $t = 0.5$ Gyr is shown in Figure 1(a). The distributions of all models coincide on each other until $t \approx 21.5444$ Myr. Then, the distributions of models A, B and C coincide on each other and get upper than the coincided distributions of models D and E. The universe horizon distance increases quite slowly with time in all general models up to $t \approx 21.5444$ Myr, hence it stats raising very fast. The distribution of the universe horizon distance in the general models in the range $t = 0.5$ - 50 Gyr is illustrated in Figure 1(b). The distributions of all models coincide on each other up to $t \approx 1.1111$ Gyr. Afterwards, the distributions of models A, B and C coincide on each other and become higher than the coincided distributions of models D and E. In all general models the universe horizon distance distributions increase very fast with time. The distribution of the universe horizon distance in the general models in the range $t = 50$ - 124 Gyr is presented in Figure 1(c). The distributions of models A, B and C are close to each other and lie upper than the distributions of models E and D. The increase in the universe horizon distance gets very small with increasing time in all general models.

Table 1 shows the universe horizon distances in the general models at special times. These times are the time of radiation-matter mass equivalence t_{rm}, the time of matter-dark energy mass equivalence $t_{m\Lambda}$, the present time $t_o = 13.7 \pm 0.2$ Gyr and the time $t_n = 124$ Gyr.

The results illustrated in Figures 1(a)-(c) are supported by those displayed in Figures 2(a)-(c) which show the distributions of the universe horizon volume in the general models in the ranges up to t = 0.5 Gyr, t = 0.5 - 50 Gyr and t = 50 - 124 Gyr respectively. Table 2 presents the universe horizon volumes in the general models at the special times $t_{rm}, t_{m\Lambda}, t_o$ and t_n.

The distribution of mass and energy within the universe horizon volume of the universe in any general model up to $t = 0.5$ Gyr is exhibited in Figure 3(a). The distributions of both radiation and matter decrease gradually with time and intersect at the time $t_{rm} = 34627.5$ - 55916.2 yr as shown in Table 3. On the other hand, the distribution of dark energy increases gradually until it intersects with the radiation distribution at the time $t_{r\Lambda} = 0.5166$ - 0.5839 Gyr as seen in Table 4.

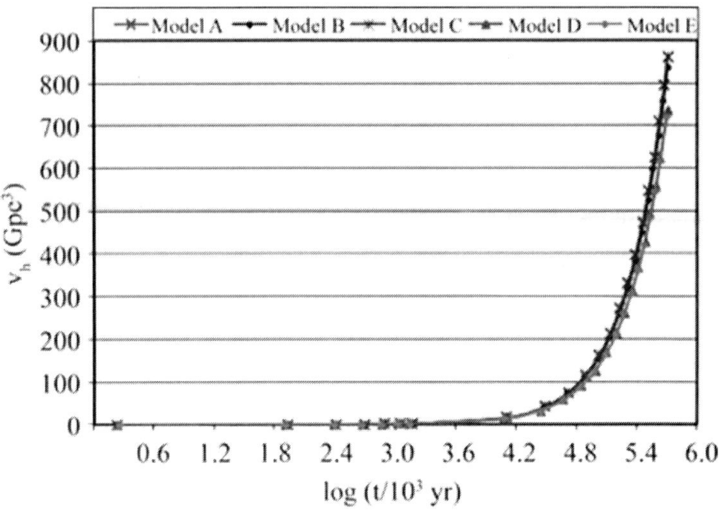
Model A
Model B
Model C
Model D
Model E
900
800
700
600
500
400
300
200
100
0
0.6
1.2
1.8
2.4
3.0
3.6
4.2
4.8
5.4
6.0
v_h (Gpc^3)
log (t/10^3 yr)

(a)

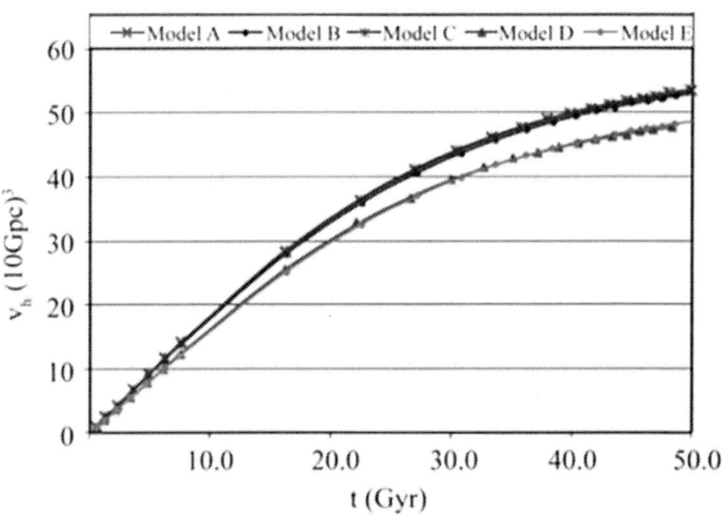
Model A
Model B
Model C
Model D
Model E
60
50
40
30
20
10
0
10.0
20.0
30.0
40.0
50.0
v_h (10Gpc)^3
t (Gyr)

(b)

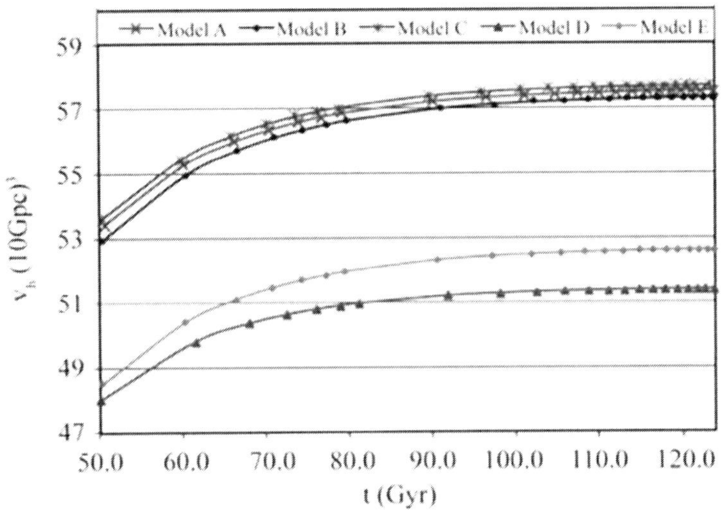

(c)

Figure 2. The distribution of the universe horizon volume in the general cosmic models (a) up to t = 0.5 Gyr; (b) in the range t = 0.5 - 50 Gyr; (c) in the range t = 50 - 124 Gyr.

Table 1. Horizon distances of the universe in the general cosmic models at special times.

Model	$d_h(t_{rm})$/Mpc	$d_h(t_{m\Lambda})$/Gpc	$d_h(t_0)$/Gpc	$d_h(t_s)$/Gpc
A	118.1520	12.7667	14.2969	19.0103
B	112.1480	12.7586	14.1588	18.9849
C	118.0740	12.7812	14.2780	19.0274
D	98.0130	12.3367	13.8666	18.3062
E	94.3170	12.4269	13.8070	18.4510

Table 2. Horizon volumes of the universe in the general cosmic models at special times.

Model	$V_h(t_{rm})$/Mpc^3	$V_h(t_{m\Lambda})$/(10 Gpc)^3	$V_h(t_0)$/(10 Gpc)^3	$V_h(t_s)$/(10 Gpc)^3
A	13820.0	17.4322	24.4819	57.5555
B	11820.0	17.3990	23.7792	57.3248
C	13790.0	17.4916	24.3849	57.7111
D	7890.0	15.7295	22.3372	51.3939
E	7030.0	16.0772	22.0504	52.6234

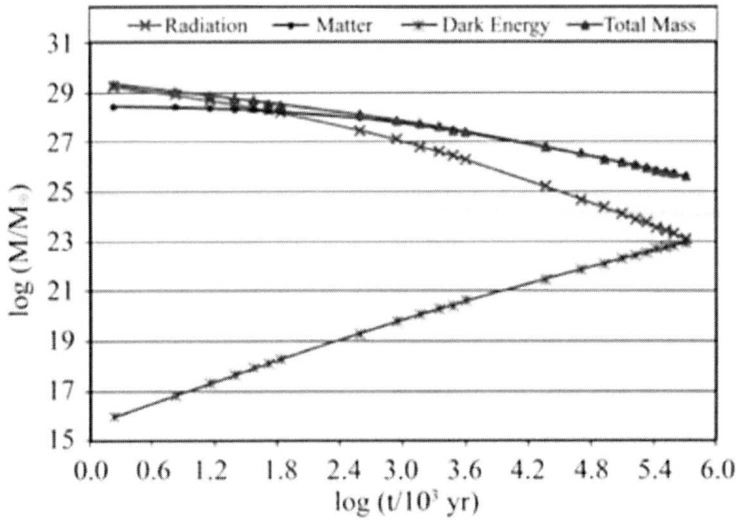
Radiation
Matter
Dark Energy
Total Mass
31
29
27
25
23
21
19
17
15
log (M/M☉)
0.0 0.6 1.2 1.8 2.4 3.0 3.6 4.2 4.8 5.4 6.0
log (t/10³ yr)

(a)

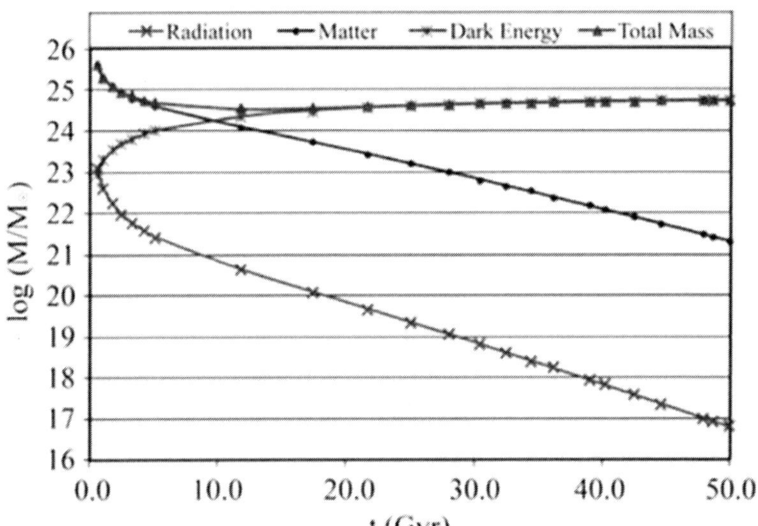
Radiation
Matter
Dark Energy
Total Mass
26
25
24
23
22
21
20
19
18
17
16
log (M/M☉)
0.0 10.0 20.0 30.0 40.0 50.0
t (Gyr)

(b)

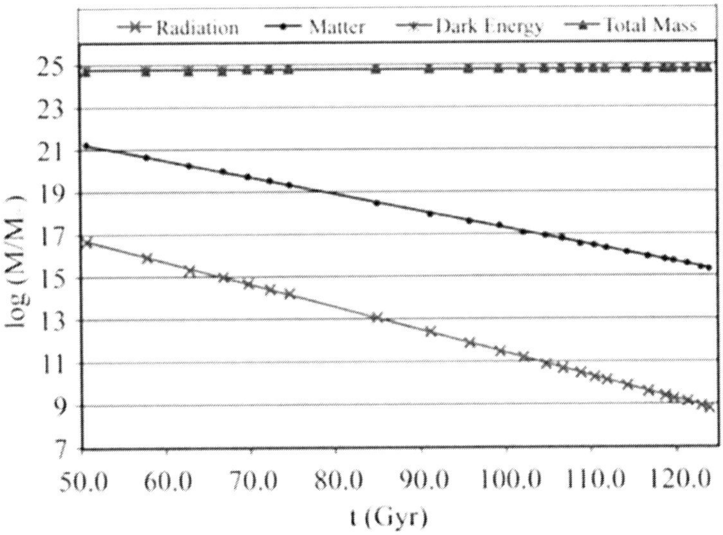

(c)

Figure 3. The distribution of mass and energy within the universe horizon volume in any general cosmic model (a) up to t = 0.5 Gyr; (b) in the range t = 0.5 - 50 Gyr; (c) in the range t = 50 - 124 Gyr.

Table 3. Cosmic times at which $M_r(t) = M_m(t)$ within the universe horizon volume in the general cosmic models.

Model	$t_{rm}/10^3$ yr	$\mathrm{Log}(M_{rm}/M_\odot)$	$\mathrm{Log}(M_\Lambda/M_\odot)$
A	55.9162	28.2526	18.1498
B	49.6665	28.2876	18.0712
C	55.8428	28.2529	18.1562
D	38.4783	28.3338	17.9566
E	34.6275	28.3753	17.8515

The distribution of total mass coincides with that of radiation up to $t \approx 5843.4141$ yr. Afterwards, the two distributions diverge from each other. However, the distribution of the total mass coincides on the distribution of matter from the time $t \approx 857695.9$ yr onwards.

The distribution of mass and energy within the universe horizon volume of the universe in any general model in the range $t = 0.5$ - 50 Gyr is displayed in Figure 3(b). It is obvious that the distributions of matter and radiation decrease gradually with time and the former lies above the later.

Table 4. Cosmic times at which $M_r(t)=M_\Lambda(t)$ within the universe horizon volume in the general cosmic models.

Model	$t_{r\Lambda}$ / Gyr	$\mathrm{Log}(M_{r\Lambda}/M_\odot)$	$\mathrm{Log}(M_m/M_\odot)$
A	0.5839	23.0153	25.5410
B	0.5736	22.9834	25.5375
C	0.5808	23.0202	25.5445
D	0.5166	22.9407	25.5350
E	0.5259	22.8906	25.5216

Table 5. Cosmic times at which $M_m(t)=M_\Lambda(t)$ within the universe horizon volume in the general cosmic models.

Model	$t_{m\Lambda}$ / Gyr	$\mathrm{Log}(M_{m\Lambda}/M_\odot)$	$\mathrm{Log}(M_r/M_\odot)$
A	9.4650	24.2507	20.8831
B	9.6930	24.2393	20.8339
C	9.4525	24.2594	20.8937
D	9.6289	24.2564	20.7974
E	10.0632	24.2109	20.7030

Table 6. Masses of radiation, matter and dark energy within the universe horizon volume in the general cosmic models at $t=t_o$.

Model	$\mathrm{Log}(M_r/M_\odot)$	$\mathrm{Log}(M_m/M_\odot)$	$\mathrm{Log}(M_\Lambda/M_\odot)$
A	20.4531	23.9625	24.3973
B	20.4324	23.9683	24.3785
C	20.4570	23.9663	24.4086
D	20.4035	23.9954	24.3981
E	20.3684	23.9921	24.3328

The distribution of dark energy increases with time and intersects with the distribution of matter at $t=9.4525$ - 10.0632 Gyr as illustrated in Table 5. The distribution of the total mass coincides on the distribution of the matter up to $t=4.5714$ Gyr, then they diverge from each other. Furthermore, the distribution of the total mass coincides on the distribution of the dark energy from $t = 18.2857$ Gyr onwards. Masses of radiation, matter and dark energy within the universe horizon volume in the general models at the present time are illustrated in Table 6.

Table 7. Masses of radiation, matter and dark energy within the universe horizon volume in the general cosmic models at $t = t_n$.

Model	$\mathrm{Log}(M_r/M_\odot)$	$\mathrm{Log}(M_m/M_\odot)$	$\mathrm{Log}(M_\Lambda/M_\odot)$
A	8.8400	15.3480	24.7694
B	8.9797	15.4781	24.7571
C	8.7546	15.2847	24.7779
D	8.1326	14.8863	24.7706
E	8.8615	15.4585	24.7258

Table 8. Equivalent number of the Coma-like clusters to the mass of matter within the universe horizon volume in the general cosmic models at special times.

Model	t_{rm}	$t_{m\Lambda}$	t_o	t_n
A	8.9448×10^{22}	8.9057×10^{8}	4.5864×10^{8}	1.1142
B	9.6955×10^{22}	8.6750×10^{8}	4.6480×10^{8}	1.5034
C	8.9510×10^{22}	9.0859×10^{8}	4.6267×10^{8}	0.9631
D	10.7838×10^{22}	9.0234×10^{8}	4.9473×10^{8}	0.3848
E	11.8651×10^{22}	8.1259×10^{8}	4.9099×10^{8}	1.4370

The distribution of mass and energy within the universe horizon volume in any general model in the range $t = 50 - 124$ Gyr is exhibited in Figure 3(c). Again the distribution of both matter mass and radiation mass decrease with time and the former is higher than the later.

The distributions of dark energy mass and total mass coincide on each other. Masses of radiation, matter and dark energy within the universe horizon volume in the general models at t_n are given in Table 7.

Table 8 shows the equivalent number of the Coma-like clusters to the mass of matter within the universe horizon volume $N_{COMA}(t)$ in the general models at the special times $t_{rm}, t_{m\Lambda}, t_o$ and t_n. It is obvious that this content of matter strongly decreases with time such that the cosmic space becomes almost matter empty in the far future of the universe.

5. CONCLUSION

In this article distributions of the universe horizon distance and universe horizon volume were determined in the five general cosmic models which were established previously. The two distributions were found increasing slowly up to $t \approx 21.5444$ Myr, hence they raise appreciably fast up to $t = 60$ Gyr, then they increase again so slowly until $t = 124$ Gyr. Distributions of mass of radiation, matter and dark energy within the universe horizon volume were also investigated in the five general models. The masses of radiation and matter are decreasing with time although the mass of dark energy is increasing. The mass of radiation was dominant in the early

Fadel A. Bukhari universe up to $t = 34627.5 - 55916.2$ yr, where it becomes equivalent to the mass of matter. Afterwards, the mass of matter prevailed until $t = 9.4525 - 10.0632$ Gyr, where it becomes equal to the mass of dark energy. From this time onwards the mass of dark energy dominates the universe. The cosmic space gets approximately matter empty in the very remote future of the universe.

REFERENCES

1. Bukhari, F.A. (2013) Five General Cosmic Models. Journal of King Abdulaziz University: Science, 25.
2. Bukhari, F.A. (2013) Cosmological Distances in Five General Cosmic Models. International Journal of Astronomy and Astrophysics, 3, 183-188.

CHAPTER 4

The Nature of Dark Matter and of Dark Energy

Jacob Schaf

Instituto de Física, Universidade Federal do Rio Grande do Sul (UFRGS), Porto Alegre, Brazil

ABSTRACT

The aim of the present work is showing that the observed non-Keplerian rotation of the galaxies is predicted by the space dynamic gravitational mechanism without dark matter and that the current explanation of the galactic gravitational dynamics in terms of dark matter is hypothetical. Visibly, the need of dark matter is an emergency created by a mistaken view about the nature of space and gravitation. Furthermore, in the view of the present work, the accelerated expansion of the universe is expansion of the Higgs quantum space itself, driven by minimization of its total energy. Spontaneous breaking of the U(1) gauge symmetry and Bose-Einstein condensation of the Higgs bosons creates a potential energy well in the form of the Mexican sombrero. If the density $\phi^{*}\phi$ of the condensate is larger than the equilibrium density, then the Higgs quantum space can minimize its energy by volumetric expansion. The idea that the accelerated expansion of the universe is driven by a positive potential energy, dubbed dark energy, needs to be unveiled.

Keywords: Dark Matter, Dark Energy, Gravity, Gravitational Physics, Gravitational Fields

1. INTRODUCTION

It is well known since decades that the orbital velocity of the stars round the galactic nucleus falls not off with distance as predicted by the current theories of gravity. Many published papers [1] -[3] display observed velocity profiles of the orbital velocity of the stars in the Andromeda galaxy, of our Milky Way galaxy as well as for other galaxies as a function of the distance from the galactic center, determined by spectroscopic methods. Figure 1 displays the observed velocity profile of our Milky Way Galaxy together with a typical Keplerian rotation. From the viewpoint of the current theories, the rotation rate of galaxies is much too fast to be accounted for by the gravitational force generated by their content and distribution of visible matter.

Instead of mistrusting the current theoretical view about space and gravitation, people preferred to mistrust the actual instruments of observation. Accordingly, these observations are alleged to see only a small part of the matter content of the galaxy and that a huge amount of invisible gravitating dark matter must be acting there. The nature of dark matter is really very exotic. It acts by gravity, but does not scatter, absorb or emit electromagnetic radiation. Actually a variety of very massive hypothetical non-hadronic particles (MACHOS, WIMPS etc.) are being proposed as possible candidates for dark matter. Challengingly, the amount of dark matter, necessary to account for the observed galactic rotation rate, must be about 5 times larger than that of the whole visible matter. To present date nobody has idea where such a huge amount of dark matter could be hided and what its nature is. Some authors assert that an external halo of dark matter round the galaxies may explain the observed galactic rotation. However, if dark matter interacts by gravity, why then is it not concentrated within stars and the center of galaxies? Besides its very exotic nature, its conduct is even much more exotic. The abnormal galactic rotation rate is actually the most challenging impasse in the current theories of space and gravitation.

The distribution of mass within galaxies is neither homogeneous nor spherically symmetric. Galaxies have a central bulge, formed by elliptically orbiting stars that normally harbors within it a super-massive black hole. Beyond the bulge, up from about 10^4 light years, galaxies are formed by a swarm of stars that orbit round the galactic center along nearly equatorial circular orbits within a thin disk that extends up to about 5×10^4 light years. Almost the whole mass of a galaxy is in the form of stars that are orbiting round the galactic nucleus under the collective self-consistent gravitational field created by them. The stars are separated from each other by large distances, in the order of light years, so that most of the space within a galaxy is empty. The density of the galactic matter undergoes weak variation with distance from the galactic center, due to their spiraled structure. Our Milky Way Galaxy is an old barred spiral galaxy having a central bulge of about 10^4 light years across, that is surrounded by a disk-shaped swarm of hundreds of billions of stars orbiting along closely circular orbits that extend out to about 4.5×10^4 light years. The bulge contains a central region of lower mass density having a diameter of some thousand light years wherein it exhibits a bar shaped halo of hot and dense flowing gases and within the bar it harbors a super-massive black hole. Almost the whole mass of the galaxy is orbiting round the galactic center within the bulge and the disk in the form of dense stars under their self-consistent gravitational field.

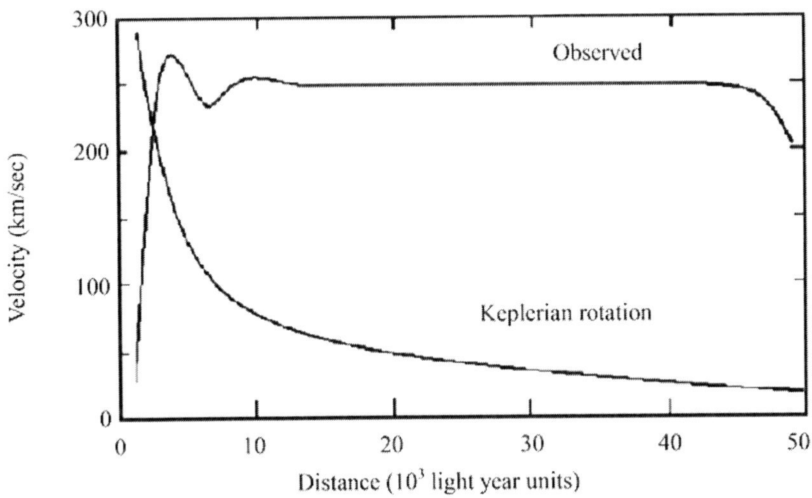

Figure 1. The figure displays the profile of the observed rotation rate of the Milky Way galaxy as a function of distance from the galactic center and the Keplerian rotation rate.

The challenging point in the galactic gravitational dynamics is that within the galactic disk the orbital velocity of the stars round the galactic center does not fall off with distance from the galactic center, as predicted by the current theories of gravitation. In some galaxies the velocity even increases with distance. However, at the outer border of the galaxy the galactic rotation rate falls off rather steeply. None of the current theories of gravity can explain this galactic gravitational dynamics. Explanations have been attempted introducing a preposterous amount of dark matter. In order to be able to explain the galactic gravitational dynamics the dark matter must form an external halo of matter round the galaxy. However if this dark matter interacts by gravity, why then is it not concentrated within the stars and in the galactic center like usual matter? In the view of the present work, the idea of dark matter is an emergency idea enforced by wrong concepts about the nature of space and the gravitational physics. In contrast, the physical mechanism of gravity, proposed by the theory of space dynamics, predicts the observed galactic gravitational dynamics without the need of dark matter.

2. BRIEF OUTLINE OF THE SPACE DYNAMIC GRAVITATIONAL MECHANISM

According to the new theory of space and gravitation, named space dynamics, [4] -[9] the real space, ruling the inertial motion of matter and the propagation of light, is moving in the ordinary three dimensions round each astronomical body throughout the universe according to a Keplerian velocity field, consistent with the local main astronomical motions (please see Equation (1) and Figure 2). With respect to this real space earth, the other planets and satellites of the solar system as well as all the astronomical bodies in general throughout the universe very nearly are stationary. This directly explains the null results of the Michelson light anisotropy experiments. Such real space, ruling the inertial motion of matter and the propagation of light, needs not to be invented. The Quantum Field Theory (QFT) underlying the Standard Elementary Particle Model (SEPM), that now is fairly well established, entails the idea that ordinary 3D-space is filled up with the Higgs condensate (HC). This HC is a very powerful quantum space (QS) stable up to 10^{15} degrees Kelvin, that is formed by Bose-Einstein condensation of the zero spin Higgs bosons by spontaneously breaking the universal U(1) gauge symmetry. This HC is a quantum fluid, totally analogous to the superconducting condensate (SCC). The HC is to the weak and strong nuclear fields (elementary particle fields) what the SCC is to photons. Quantum fluids are well known to intrinsically quantize all their cyclic excitations. Likewise superconductors develop screening currents (velocity fields of the SCC) and a Lorentz force field in the presence of magnetic fields, confining and compressing these fields by the Meissner effect and making them short-range, thereby giving inertial mass to the photons within the superconductor, the HC develops velocity fields of the HC round matter fields, confining them by the Higgs mechanism, giving inertial mass to the elementary particles, that is, giving them mechanical properties. This lets clear that the HC is the QS that rules the inertial motion of matter and the propagation of light and hence is the locally ultimate but moving reference for rest and for motions of matter and light. As the HC rules the inertial motion of matter, its motion according to a non-uniform velocity field, generates a field of inertial (gravitational) forces compressing matter into dense matter bodies.

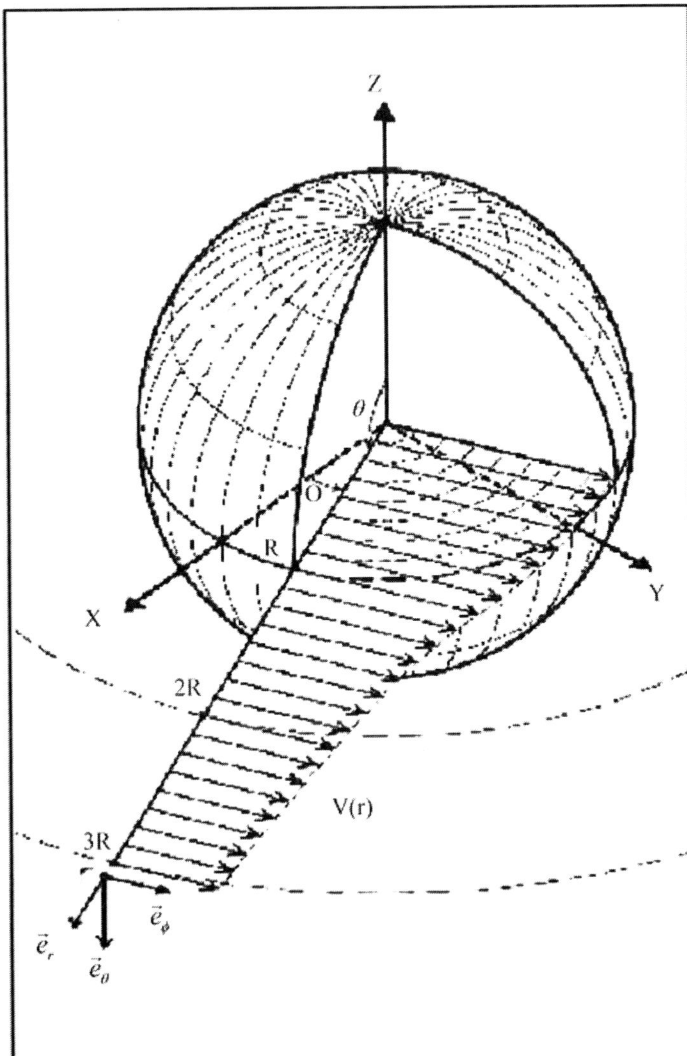

Figure 2. The velocity of the Keplerian velocity field $V(r)$ of the QS, due to a spherically symmetric body of radius R, with a homogeneous density of mass. The figure displays the velocity as a functions of r, from the origin up to $r = 3R$. This profile of the velocities is exactly the same along any direction, for all ϕ and all θ from the equator to the poles.

Recent experimental observations, achieved with the help of the tightly synchronized clocks of the Global Positioning System (GPS) [10] -[12] show that this QS-dynamics is not simply a guess. They show very clearly that the QS, ruling the propagation of light and the rate of clocks, is moving round earth according to a Keplerian velocity field consistently with the Moon's orbital motion and round the Sun consistently with the earth's orbital motion. [4] -[6] In order to describe this velocity field let us define non-rotating systems of orthogonal coordinate axes (XYZ), with origins fixed respectively to the center of earth, to the center of the sun and to the galactic center, the direction of these axes being fixed with respect to very distant astronomical objects and the Z axis along to rotation axis of earth, of the solar system and of the galaxy. Let (r,θ,ϕ) be the usual sperical coordinates and (e_r,e_θ,e_ϕ) unit vectors pointing along the positive direction of these sperical coordinates.

In terms of these usual spherical coordinates the Keplerian velocity field associated with a body of radius R and mass M for $r > R$ is:

$$V(r) = [\gamma M/r]^{1/2}\, e_\phi \quad (r > R) \tag{1}$$

where γ is the gravitational constant (please see Figure 2). Within the homogeneous and spherically symmetric matter body $r < R$ the velocity field of the QS is given by:

$$V(r) = V(R)\left[\left(3 - r^2/R^2\right)/2\right]^{1/2} e_\phi \quad (r < R) \tag{2}$$

where $V(R)$ is the velocity of the QS at the surface of the body. It is important to observe that within the gravitational source the velocity still increases toward the gravitational center, but becomes completely flat near to the center (see Figure 2).

In Refs. [6] [7] it is shown that this Keplerian velocity field of the QS Equation (1) straightforwardly creates the observed gravitational dynamics on earth and within the solar system. It also is shown to correctly generate all the observed effects of the gravitational fields on the propagation of light and the rate of clocks, there including all the new effects recently discovered with the help of the GPS. Within this Keplerian velocity field the circular orbital motions of the planets round the Sun and of the natural satellites round the planets trace out very closely the circulation field of the QS and hence closely rest with respect to this QS. These bodies are essentially carried around by the moving QS. In fact the planets have only very low velocities with respect to the local moving QS. In particular, the small orbital eccentricity of the planet Earth shows that it moves with only about 100 m/sec with respect to the QS in the solar Keplerian velocity field. The effects of such low velocities are much too small to be detected even by the actual most sensitive instruments.

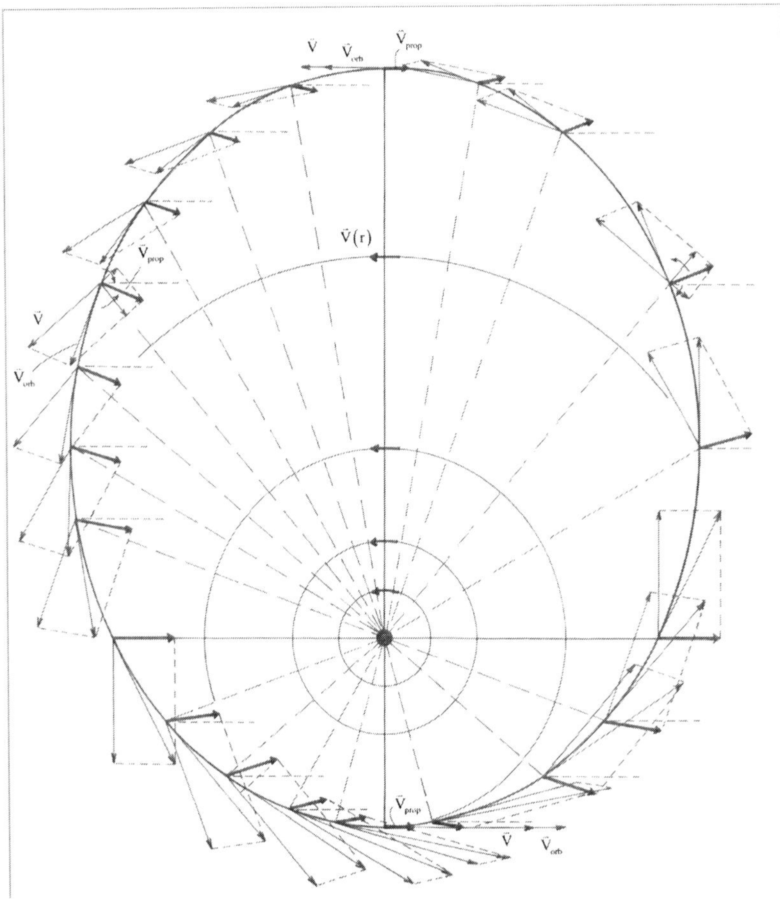

Figure 3. The figure is a very precise graphical representation of the orbital velocity v_{orb}, the velocity with respect to the QS v_{prop} and the local velocity of the QS in the Keplerian velocity field $V(r)$ for a large number of points along an elliptical orbit with eccentricity $\grave{o}=0.5$. The rotation rate of v_{prop} can be read precisely enough in the figure. Specifically the rotation rate of the ϕ velocity component can be read at the top of the figure, while that of the r velocity component can be read at the left hand side (please see Equation (3)).

Within an astronomical scale of velocities, the planets do very nearly rest with respect to the QS. This explains the null results of the Michelson experiments, searching for light anisotropy due to the orbital and cosmic motion of earth.

The very slow motion of the planets with respect to the QS is ruled by the usual principle of inertia. Due to the differential velocity of the QS in the Keplerian velocity field, the velocity vector of the free bodies with respect to the QS continuously changes direction and size. These variations involve effects of refraction and wavelength stratching-compression. Figure 3 is a highly precise graphical display, showing how the direction and length of the velocity vector of a free body with respect to the local QS changes along an equatorial excentric orbit. From this variation of the velocity vector the effective time rate of the opposite refraction of the ϕ velocity component (top) and the r valocity component (left) can be read. These rotation rates are characteristic of the Keplerian velocity fields of the QS. Please observe the 4π periodicity in this variation.

Equations (3) show the result of this lecture:

$$\boldsymbol{W}_r(r) = -\frac{1}{2}\left[\gamma M/r^3\right]^{1/2} \boldsymbol{e}_\theta \tag{3a}$$

$$\boldsymbol{W}_\phi(r) = \left[\gamma M/r^3\right]^{1/2} \boldsymbol{e}_\theta \tag{3b}$$

Using Equations (3a) and (3b) the hyperbolic rotation matrix can be defined:

$$\boldsymbol{A} = \begin{pmatrix} 0 & W_\phi \mathrm{d}t \\ -W_r \mathrm{d}t & 0 \end{pmatrix} = \begin{pmatrix} 0 & W\mathrm{d}t \\ \frac{1}{u} W\mathrm{d}t & 0 \end{pmatrix} \tag{4}$$

The collumn matrix of the r and ϕ velocity components of the body with respect to the QS, on which the hyperbolic rotation matrix acts is:

$$\boldsymbol{v}(t) = \begin{pmatrix} v_r(t) \\ -v_\phi(t) \end{pmatrix} \tag{5}$$

In terms of the matrices Equations (4) and (5) the equation ruling the vertical free-fall can be written as:

$$\frac{\mathrm{d}\boldsymbol{v}}{\mathrm{d}t} = A\boldsymbol{v} \tag{6}$$

This equation develops into:

$$\boldsymbol{v}(t) = \boldsymbol{v}_0 \mathrm{Exp}\left[\int_0^t A(r(t'))\mathrm{d}t'\right] = \sum_{n=0}^{\infty}\frac{1}{n!}\begin{pmatrix} 0 & \Theta(t) \\ \frac{1}{u}\Theta(t) & 0 \end{pmatrix}^n \begin{pmatrix} v_r(0) \\ -v_\phi(0) \end{pmatrix}$$
$$= \begin{pmatrix} \cosh\left(\frac{\Theta(t)}{\sqrt{u}}\right) & \sqrt{u}\sinh\left(\frac{\Theta(t)}{\sqrt{u}}\right) \\ \frac{1}{\sqrt{u}}\sinh\left(\frac{\Theta(t)}{\sqrt{u}}\right) & \cosh\left(\frac{\Theta(t)}{\sqrt{u}}\right) \end{pmatrix} \times \begin{pmatrix} v_r(0) \\ -v_\phi(0) \end{pmatrix} \tag{7}$$

The solution of this equation (please see Section 5 of Refs. [6] [7] for details) leads to the usual expression for velocity of vertical free-fall and the implicit velocity as a function of position. For bodies in orbit, it is necessary to add in Equation (3) and in the matrix Equation (4) the effect of the trigonometric angular velocity, due to the orbital motion round the gravitational center.

In Refs. [6] [7] it is shown that the spacedynamic gravity field is spherically symmetric with orbital motions.

The weak ellipticity of the planetary orbits is consequence of the hyperbolic refraction rate (Equation (3) of the very slow r and ϕ velocity components with respect to the QS along the elliptic orbit (see Figure 3).

The Keplerian velocity field of the QS, creating the earth's gravitational field, achieves nearly $8\ \mathrm{km/sec}$ on surface in the sense of the Moon's orbital motion. The earthglobe rotates in the ordinary space only very slowly $(460\ \mathrm{m/sec})$ in the same sense as the QS, so that the earth based laboratories move with respect to the QS oppositely to the earth' rotation at nearly $8\ \mathrm{km/sec}$.

Bodies released within earth based laboratories are observed as vertical free-fall. However, from the viewpoint of the earth's non-rotating reference, they in fact move for a short time interval along highly eccentric elliptic orbits round the center of earth.

3. THE OBSERVED GALACTIC GRAVITATIONAL DYNAMICS WITHOUT THE NEED OF DARK MATTER

From the viewpoint of the present theory of space dynamics, in the galactic gravitational dynamics it is fundamental to take into account the fact that almost the whole matter, creating and ruling the galactic gravitational dynamics, is moving round the galactic nucleus. While in the solar system 99.9% of the mass in concentrated in the sun, in the galaxy more than 99.9% is orbiting round the galactic center. This obviously makes a difference. Each orbiting star carries with it its velocity (gravitational) field given by Equation (1). Likewise in the solar system, these velocity fields are fairly well polarized and rotate all in the same sense. Adequate addition of these velocity fields generates the galactic velocity (gravitational) field that rules the galactic gravitational dynamics. The stable spiraled structure of the galaxies must be seen as a perturbation of this ordered composition.

In order to highlight the relevance of the effect of motion of the individual gravitational sources on the collective velocity field of the Higgs quantum space QS of a system of bodies gravitating in their self-consistent field, let us begin with the simple case of a binary system of two equal stars of mass M moving in the same circular orbit round the center of mass in the equatorial plane of the collective velocity field as shown in Figure 4. The ordinary dynamics of such a system can fairly well be described by Newtonian mechanics. Balance of the mutual Newtonian gravitational forces $\gamma M^2/(2x_o)^2$ and of the centripetal force Mv_o^2/x_o on each star, due to the orbital motion, define the orbital velocity v_o of each star round the center of mass (CM):

$$v_0 = \frac{1}{\sqrt{2}}(\gamma M/2x_0)^{1/2} \quad (8)$$

On the other hand, the velocity of a small test particle, orbiting round an equal isolated (static) mass M at the same distance $2x_o$ in a circular equatorial orbit, is considerably larger:

$$v_0' = (\gamma M/2x_0)^{1/2} = \sqrt{2}v_0 \quad (9)$$

Obviously, from the viewpoint of the current gravitational theories there is nothing wrong with Equation (8) and Equation (9). However, although these equations express exactly what is observed, it must be remembered that the Newtonian gravitational forces are not physically real and moreover that after four decades of efforts, Quantum Gravity has not succeeded in providing physical reality to them. Hence, the explanation in terms of gravitational forces must be abandoned. GR is a much more realistic theory of gravitation, in which no gravitational forces are acting. However, GR too cannot explain the galactic gravitational dynamics. Moreover, now GR is in serious troubles with several recent experimental facts, because it cannot explain the absence of the gravitational time dilation on the GPS clocks, due to the solar gravitational potential as well as other clear-cut experimental observations (please see Section 3 in Refs. [4] [5]). These facts frontally contradict predictions of GR. On the other hand, the theory of space dynamics provides the simple physical origin of the inertial dynamics observed within the gravitational fields. It also accounts in terms of simple and genuine physical effects for all the other effects caused by gravitational fields on the velocity of light and on the rate of clocks, including all the new effects recently discovered with the help of the GPS.

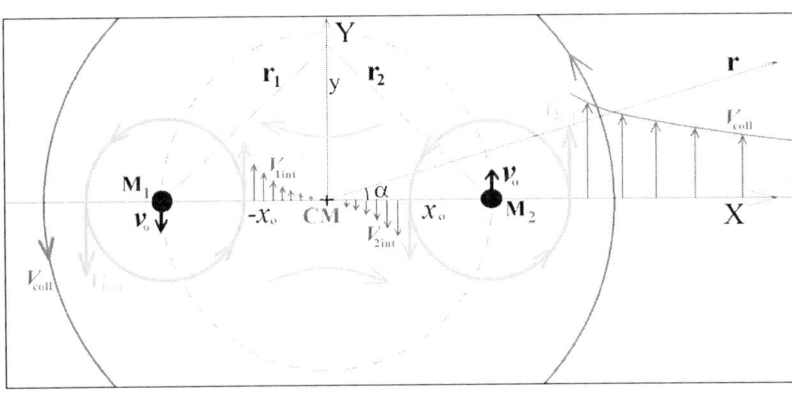

Figure 4. Two equal masses $M_1 = M_2 = M$ moving in the same sense round the center of mass (CM) within the equatorial plane of the combined velocity field. Note that the velocity fields of the QS round each individual mass of the binary are symmetric from the viewpoint of the non-rotating references attached to each moving mass. However, from the viewpoint of the non-rotating [X, Y] coordinate axes with origin fixed to the CM, the velocity of the QS in the velocity fields of M_1 and M_2 are larger outside than inside the binary orbit due to the orbital motion of the sources round the CM.

Here it moreover will be shown that this space dynamic gravitational mechanism predicts the observed gravitational dynamics of galaxies, without the need of dark matter, thereby providing a simple solution to this impasse.

From the viewpoint of space dynamics, Equations (8) and (9) unveil a key feature that discloses the effect of motion of the gravitational sources on their velocity fields of the QS. This is a fundamental ingredient in the gravitational dynamics of a system of bodies gravitating in their self-sustained gravitational field. Equation (9) gives the local velocity of the QS in the Keplerian velocity field round a static mass M. However, this is certainly true also in the case of the binary system. Equation (8) too gives the velocity of the QS, due to each one of the stars at the site of the companion star in the combined velocity field of the binary system. Hence, both Equations (8) and (9) describe the velocity of the QS in the respective velocity fields. However, why are these velocities so different? The only possible reason for the reduced velocity of the QS of M_1 at the position of M_2 and that of M_2 at the position of M_1 in the binary system [Figure 4] is the orbital velocity of respectively M_1 and M_2 round the center of mass (CM). The orbital velocity of the binary system reduces the velocity field $V_{1,int}$ of M_1 at the right-hand side of M_1 and also reduces $V_{2,int}$ of M_2 toward the left-hand side of M_2, as given by Equation (8) and as depicted in Figure 4. In the case of Equation (9) the source and the CM of the system practically remains stationary, and therefore the velocity field of the QS, due to the central source M, is not affected at all and its velocity field is that of a static source, exactly as given by Equation (1). Consider now in addition a small test particle moving in the collective velocity field round the binary system within the orbital plane of the binary, however sufficiently far away in order to minimize the asymmetries of the binary field. The orbital velocity v of such a test particle is of course:

$$v = \left(\gamma 2M / r\right)^{1/2} \tag{10}$$

For large r this expression too represents the velocity of the QS in the collective velocity field round the binary system:

$$V_{coll}\left(r\right) = \left(\gamma 2M / r\right)^{1/2} \tag{11}$$

In order to reconcile the addition of the velocity fields of M_1 and M_2 outside the binary with the collective velocity field, given by Equation (11), the same orbital velocity that reduces the velocity fields toward the inner side must enhance them outward to the binary. It must increase V_1 toward the left-hand side of M_1 and V_2 toward the right-hand side of M_2. At distances r much larger than $2x_o$, addition of the velocity fields of M_1 and M_2 must reproduce the value given by Equation (11). Note that addition of velocity fields must obey the sum rule for velocity fields:

$$V = \left(\left|V_1^2 + V_2^2\right|\right)^{1/2} \tag{12}$$

This rule arises because in the expression for the velocity field Equation (1) only the square root of the source mass appears. Accordingly, addition of the velocity fields as a function of x along the X axis outside of the binary in Figure 4 conforms to the equation:

$$\frac{2\gamma M}{x} = \frac{1}{2}\frac{\gamma M}{x_0 + x} + V_{1,2}^2 \tag{13}$$

Solving for $V_{ext} = V_{1,2}$ along the X axis for large x, which means large r, the velocity fields of M_1 as well as of M_2 outside the binary are given by:

$$V_{1,2} = \sqrt{\frac{3}{2}}\sqrt{\frac{\gamma M}{r}} \tag{14}$$

so that for large x, large r

$$V_{ext} = \sqrt{\frac{3}{2}\frac{\gamma M}{x - x_0} + \frac{1}{2}\frac{\gamma M}{x_0 + x}} : \ (\gamma 2M/r)^{1/2} \tag{15}$$

This shows that effectively the same orbital velocity that reduces the velocity fields of M_1 and M_2 [Equation (8)] by a factor $(1/2)^{1/2}$ toward the inner side in Figure 4, enhances the velocity fields of M_1 and M_2 by a factor $(3/2)^{1/2}$ toward the outer side. This is of course not for nothing. The reason for this clearly is the motion of the sources. This feature is a unique characteristic of the space dynamic gravitational mechanism. None of the current theories of gravitation can give rise to this feature. Of course the situation is very different if the gravitational sources are static (not in orbit). However, in this case the system would be completely unstable and both sources would accelerate (fall) toward the CM because of mutual influence in their velocity fields.

In-between the orbiting bodies of the binary system the velocity fields of M_1 and M_2, are reduced by their orbital velocity. They are opposite to each other and opposite to the external velocity field. Close to M_1 the velocity field of M_1 dominates and close to M_2 the velocity field of M_2 is dominant. The resultant collective velocity field in-between the two masses as a function of x along the axis of X is given by:

$$V_{int}(x) = \sqrt{\left|V_1^2 - V_2^2\right|} = \sqrt{\left|\frac{1}{2}\frac{\gamma M_1}{x_0 + x} - \frac{1}{2}\frac{\gamma M_2}{x_0 - x}\right|} \tag{16}$$

For $x = 0$, $V_{int}(x) = 0$ (if $M_1 = M_2$), while for $x = x_o$, $V_{int}(x)$ points downwards and is large near M_2 (see Figure 4) and for $x = -x_o$, $V_{int}(x)$ points upwards and is large near M_1.

For other directions, adequate vector addition in conformity with Equation (12) must be performed. Along the Y axis the collective velocity field may be expressed by:

$$V(y) = \sqrt{\frac{2\gamma M y^2}{\left(x_0^2 + y^2\right)^{3/2}}} \tag{17}$$

However, this equation is certainly not correct close to the binary because the collective velocity field is considerably affected by the flow continuity and conservation of volume. However, for large y Equation (17) too approximates Equation (10).

Note that within the binary orbit the circulation field of the QS round the CM is opposite to the orbital motion of the binary as well as opposite to the external velocity field as indicated in Figure 4. It is clear that within the orbit of the binary the velocity field is strongly reduced and that a stagnation point exists at the CM where the velocity of the QS falls to zero.

With the increase of the number of masses moving in the same sense in the same orbit, the opposite internal and external velocity fields of the QS become more and more symmetric. However, while the opposite internal velocity field remains reduced, the collective velocity field outside the loop is enhanced by more and more sources. Figure 5 displays a sketch for a system of four masses M orbiting round the center of mass under their collective velocity (gravitational) field.

In order to advance closer toward the realistic situation of a galactic system, consider now multiple concentric circular orbit loops with larger and larger radii and each loop containing a large number of bodies. Each loop acts in the sense of reducing the velocity field toward the inner side while enhancing it outward. The velocity field of the QS increases at each loop from $(-1/2)^{1/2}V$ to $(3/2)^{1/2}V, V=(\gamma\cdot M/r)^{1/2}$ which makes a step by a factor nearly 2V. Note that the squared velocity is equal to the gravitational potential $V^2=\gamma\cdot M/r$. From the viewpoint of the current theories the gravitational potential does not depend on the (small) velocity of the source and therefore the step is only about $2^{1/2}V$. In fact the velocity on from each loop decreases according to $(\gamma M/r)^{1/2}$, where r is the distance on from the loop, the effective increase of the velocity depends on the separation of the loops, which is related with the density of the stellar matter. The positive gradient in the orbital velocity of the stars diminishes with increasing distance (d) between the successive loops and becomes constant for $d:$ 4 times the basic distance $2x_0$ between the inner binary (Figure 4). However, with farther increase of the distance d between the loops, the velocity gradient falls to zero and eventually inverts its direction, where the orbital velocity of the stars becomes constant with the distance to the galactic center and eventually begins to decrease.

The model is of course an idealization. It however shows that, if the density of matter in the galaxy increases with distance, the orbital velocity of the stars may strongly increase. If the matter density becomes constant with distance from the galactic center, the orbital velocity must increase moderately. If the density decreases moderately with distance, the orbital velocity may become constant. However, if the density falls to zero at the galactic border, the velocity field recovers the Keplerian dependence with distance on from this border.

The inner part of the galaxies is a region of hot flowing gases and has low mass density. However, outward the mass density and hence the orbital velocity increases very steeply as observed. Moreover, depending on the rate of the mass density increase, a retrograde rotation with respect to the outer galactic disk may form. In some galaxies such retrograde rotation is effectively observed. NGC 7331 is an example of a galaxy in which the inner bulge rotates in the opposite sense to that of the disk. [13] However, many other known examples exist. Moderate decrease of the mass density with distance, like observed in our Milky Way galaxy is expected to lead to nearly constant velocity as a function of distance from the galactic center.

However, beyond the border of the rotating galaxy the velocity field may retake the decrease according to the Keplerian rotation.

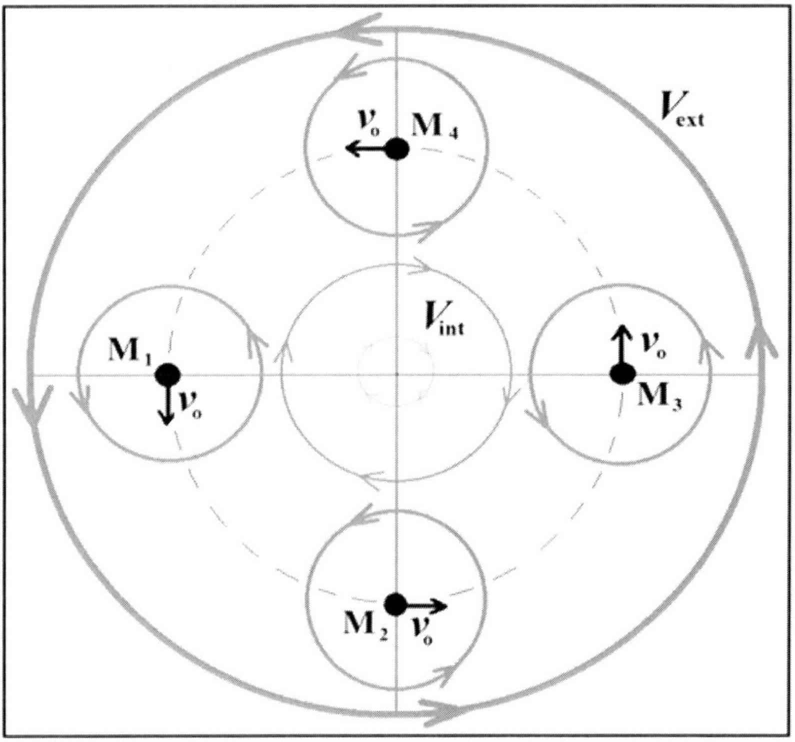

Figure 5. The figure is a qualitative sketch of the collective velocity field generated by four polarized equal masses moving in the same sense in the same circular orbit round the center of mass. Note that with- in the orbit the velocity field of the QS is weak and circulates in the opposite sense with respect to the much stronger external velocity field.

The above analysis evidences that the space dynamic gravitational mechanism provides the base to account for the observed non-Keplerian galactic rotation as a function of the distance from the galactic center and can appropriately settle the galactic gravitational dynamics without the need of dark matter. The velocity field of the QS (gravitational field) that rules the galactic gravitational dynamics of a system of orbiting bodies certainly is much different from the Keplerian velocity field of the QS created by a single, compact, homogeneous, spherically symmetric and nearly static system of matter. The galactic rotation takes place within the self-consistent velocity field of the QS (gravitational field) generated by the whole ensemble of orbiting stars. Hence, from the viewpoint of the above analysis, it is not at all strange that the rotation of galaxies be a non-Keplerian one. Also in the central part of the galaxy the velocity field of the QS may be much reduced, analogously as shown by Equation (16) for the binary and becomes even more evident in the case of loops of four or more bodies [Figure 5]. Even the possibility of retrograde rotation of the inner part of a galaxy, as observed in some galaxies, is predicted. Hence, from the viewpoint of space dynamics, the observed non-Keplerian rotation of the galaxies is in no way a surprise.

It may be interesting to note that within the solar system the planets too are expected to induce a very weak but non-vanishing attenuation in the $(1/r)^{1/2}$ dependence of the solar velocity (gravitational) field. Hence, the radial dependence of the velocity field becomes truly Keplerian only beyond the border of the solar system. This will say that the solar gravitational acceleration increases a little bit beyond the border of the solar system. This may explain the Pioneer anomaly, which is a very small increase in the gravitational attraction of the Pioneer 10 and Pioneer 11 spacecrafts by the Sun, observed beyond the border of the solar system [14] [15] .

Beyond the bulge, up from about 7×10^3 light years, the stars are well known to move round the galactic center within the galactic disk in orbits that are closely circular. From the viewpoint of space dynamics, this is evidence that these stars are commoving with the QS in the galactic velocity field of the QS. Otherwise their orbits could not be so circular. Moreover, the null results of the Michelson experiments searching for light anisotropy due to the orbital and cosmic motion of Earth demonstrate that the velocity of our solar system with respect to the QS, despite its orbital velocity of about 230 km/sec round the galactic center, is closely zero and by consonance the velocity of all the stars of the galaxy within the galactic disk may be closely zero too. This will say that the equator of the galactic velocity field of the QS creating the galactic gravitational field coincides with the galactic disk and that the motion of stars along the circular orbits round the galactic center are in fact commoving with the QS in the galactic velocity field. They simply trace out the motion of the QS in the galactic velocity field. Hence, the function that describes well the observed circular orbital velocities of the stars as a function of distance within the galactic disk round the galactic center also describes well the galactic velocity field of the QS. These stars are resting with respect to the QS and therefore the velocity of light is isotropic with respect to them.

Moreover, clocks locally commoving with these stars, rest with respect to the QS. They run naturally synchronous and show all the same universal proper time. The circular orbital motions of the stars are of course not constrained to these circular orbits by gravitational forces. According to space dynamics, these stars are locally stationary with respect to the moving QS. They are essentially carried around by the moving QS in the galactic velocity field of the QS, analogously as the planets are carried around the sun by the solar Keplerian velocity field of the QS. Therefore, these orbital motions need not to be explained anymore, because it is space itself that so moves.

Within the bulge of the galaxy the orbits of stars are much elliptic and the orbital planes are much inclined with respect to the galactic plane. This will say that the velocity of these stars with respect to the QS is relatively large and this is what gives rise to the bulge.

Another important parameter in the gravitational dynamics is the gravitational potential $[U(r)]$. According to the Virial Theorem, the gravitational potential has a very simple relation with the (circular) orbital velocity $V_{orb}(r)$:

$$U(r) = -V_{orb}^2(r) \tag{18}$$

Hence, knowing the orbital velocity, the gravitational potential can be directly calculated from Equation (18) and vice-versa.

Several authors have computed the gravitational potential $[U(r)]$ as a function of the distance from the galactic nucleus for our Milky Way galaxy as well as for other galaxies with base in the conventional theories of gravitation, taking into account the visible star density as a function of distance (see for instance reference [16]). Systematically these gravitational potentials are all inconsistent with the observed orbital motions of the stars.

As predicted by space dynamics, the gravitational potential determined within the galactic disk, with base in Equation (18) (please see Figure 6), is considerably more leveled than the potentials computed with base in the current gravitational theories. The velocity gradient in the galactic velocity field and hence the gravitational acceleration of a body resting within the galactic disk, predicted by space dynamics, is very low and notwithstanding the stars move along circular orbits round the galactic center (no need of a central force field). They are carried around the galactic nucleus by the moving QS; likewise the planets are carried round the sun by the solar Keplerian velocity field of the QS and need no centripetal force to move so. The impasse of the current theories with these observations arises because, in order to account for the circular motion of the stars round the galactic center they absolutely need a gradient of the gravitational potential (centripetal force), pointing to the gravitational center. The absence of such a gradient gives rise to an unsolvable paradox. Instead of amending the theories, people preferred to believe in the presence of invisible halo of dark matter crating such a gradient.

On from the galactic border, the density of stars falls much more steeply. Beyond the galactic border, the velocity field of the QS may fall according to $(1/r')^{1/2}$, where r' = r ? 40 (in 10^3 light year units) is the distance on from the galactic border. Beyond the galactic border, the gravitational acceleration of a resting body toward the galactic center is very large. The galactic gravitational potential is approximately a disk shaped square potential well.

Although an exact determination of the galactic velocity field with base in the space dynamic gravitational mechanism has not been worked out, the above analysis gives convincing evidence that space dynamics can predict qualitatively, even with details the observed galactic gravitational dynamics without the need of dark matter. Even the possibility of the unbelievable retrograde rotation of the inner part of galaxies, observed in many galaxies, is a natural outcome of space dynamics.

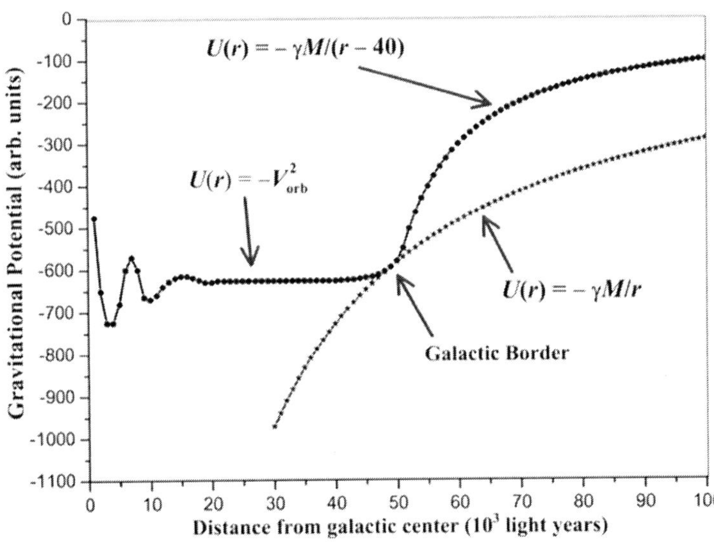

Figure 6. The figure depicts the usual gravitational potential $U(r) = -\gamma M/r$, nor- malized to the value of U at the galactic border, in which r is the distance from the galactic center. It also shows the gravitational potential, calculated using Equation (18) within the galactic disk as well as beyond, assuming that beyond the galactic border U follows the $(1/r)$ dependence, where r' = r ? 40 in light year units.

4. DARK ENERGY A GLOBAL HIGGS QUANTUM SPACE PHENOMENON

The observed accelerated expansion of the universe [17] [18] has no physical explanation to present date.

Usually it is imputed to a hypothetical new kind of positive potential energy, simply dubbed dark energy. The consensus is that dark energy distributes it homogeneously throughout space and is estimated to be 68% of the total mass-energy content of the universe (fourteen times the whole visible matter). The accelerated expansion also has been explained in terms of the cosmological constant in Einstein's equation for the spacetime curvature. This however is not a physical explanation, but only a mathematical account. The present work proposes a suitable physical mechanism, that may be responsible for the accelerated expansion of the universe.

The null results of the Michelson light anisotropy experiments show that earth is resting with respect to Higgs quantum space (QS), ruling the inertial motion of matter and the propagation of light. However, obviously earth cannot be kinematically privileged in detriment to all the other astronomical bodies throughout the universe. From the viewpoint of the present work, the gravitational dynamics is created by motion of the real QS round each astronomical body, according to a Keplerian velocity field, consistent with the local main astronomical motions. Within the galaxy this velocity field is consistent with the galactic gravitational dynamics (see the previous Section 3). However, while this accounts for the null results of the Michelson light anisotropy experiments concerning the motion within the solar system and within the galaxy, it does not explain why the velocity of light is isotropic with respect to earth, viewed the recession between the galaxies. In fact the close isotropy of light with respect to earth and certainly with respect to all the main astronomical bodies, demonstrates that the recession of the galaxies does not make the velocity of light anisotropic with respect to earth. This can be reasonable only if the quantum space (QS), ruling the inertial motion of matter and the propagation of light, is itself expanding consistently with the recession of the galaxies and with the expansion of the matter universe. In this case the red-shift of the radiation, emitted by the distant astronomical objects, is not a usual Doppler shift but is due to stretching of the wavelength, caused by the expansion of the QS, its medium of propagation, as a function of time. The wavelength of radiation from more distant objects stretches during a longer time and so becomes more red-shifted.

Expansion of the QS is a phenomenon that necessarily is related with the physics involved in the creation of the Higgs condensate. In Section 2 it has been asserted that the Higgs condensate is created by Bose-Einstein (BE) condensation of the zero spin Higgs bosons. BE condensation occurs because of the phase correlation between the wavefunctions of the zero spin Higgs bosons, leading to spontaneous breaking of the U(1) gauge symmetry of the system of particles. Phase coherence allows the system of bosons to lower their energy. During BE condensation the bosons undergo a transition to a long-range phase coherent state, the particles all assuming a well defined phase, becoming indistinguishable and tunneling throughout the volume of the condensate. The Higgs condensate (HC) is a quantum fluid analogous to the superconducting condensate. The HC plays in space an analogous role as the superconducting condensate plays is superconductivity. It is described by a complex order parameter $\phi = \phi(r)e^{i\theta}$, where ϕ is the amplitude and θ_0 is the well defined phase. The BE correlation is responsible for creating a negative potential energy (bonding) term proportional to $\phi^*\phi$. In addition to this negative potential energy term there is the positive potential energy (anti-bonding) interaction term, the value of which is proportional to $(\phi^*\phi)^2$ so that the system cannot collapse. The effective potential energy $U(\phi)$ is given by:

$$U(\phi) = -n(\phi^*\phi) + m(\phi^*\phi)^2 \tag{19}$$

However, the negative coefficient of the first (bonding) term is significantly larger than the positive one of the anti-bonding term. Therefore, the effective potential energy forms a negative potential energy well, analogously as the covalent chemical bond, the form of which is a characteristic of the boson system and the deepness depends on the strength of the BE correlation. The minimum of potential energy occurs not for $\phi^*\phi = 0$, as would be usual, but for the finite value $\phi^*\phi = n/m$. As the phase can take values within zero and 2π, the potential energy well has the form of a Mexican sombrero in the complex ϕ_{Re} and ϕ_{Im} plane (see Figure 7). During BE condensation the boson system lowers its total energy by simultaneously creating and going into the potential energy well. However, the energy of the boson system depends on the local volumetric density $\phi^*\phi$ of the condensate. In the case of the Higgs condensate, this density extends homogeneously throughout space. If the volumetric density is lower than n/m (see H_- in Figure 7), the system can reduce its energy by accelerated shrinking of its volume (Higgs mode see red arrow in Figure 7). If the density $\phi^*\phi$ is larger than n/m (see H_+ in Figure 7), the system can reduce its energy by accelerated expansion of its volume. Hence, if initially the density $\phi^*\phi$ is maximum, the boson system may undergo the cycle of successive accelerated expansion, decelerated expansion (after which the density $\phi^*\phi$ is minimum), then accelerated shrinking and finally decelerated shrinking (after which the density becomes maximum). In the language of the QFT the fluctuations of the density $|\phi^*\phi|$ along the red arrows in Figure 7 correspond to the Higgs mode. Fluctuations of θ along the blue line correspond to Goldstone modes.

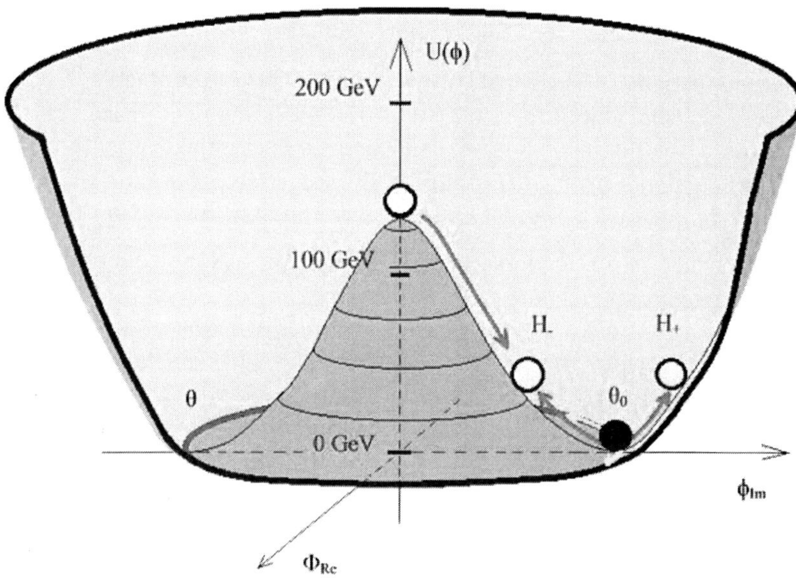

Figure 7. The figure depicts the form of the Mexican sombrero potential, indicating the transition to the lower energy phase coherent Higgs condensate (HC) with the well-defined phase θ_0. The figure also indicates the high volumetric density (H_+) and low volumetric density (H_-) states, driving accelerated expansion or accelerated contraction (red arrows) of the HC and of the universe. This is the Higgs mode. The Goldstone mode also is indicated along the blue bottom circle.

The Higgs condensate is a BE condensate to which such cyclic expansion and contraction may apply. Volumetric expansion of the Higgs condensate (HC) is volumetric expansion of the Higgs quantum space (QS), which entails expansion of the universe. Observations show that actually the universe is in a stage of accelerated expansion. This means that the density $\phi^*\phi$ of the HC is larger than the value n/m and hence that the HC is minimizing its energy undergoing accelerated expansion of its volume. It is very important to note that expansion of the universe must take place without setting the astronomical bodies into motion with respect to the QS, because this would lead to large anisotropy of light with respect to earth and which would have immediately been detected by the Michelson light anisotropy experiments. The fact that the Michelson light anisotropy experiments gave a null result, shows that the universe expands consistently with the Higgs-QS. The Higgs-QS is like a 3D elastic body that is cycling between successive expansions and contractions as a whole, likewise imagined in the theory of the cyclic universe. The question that may be asked is if the universe is a homogeneous breathing universe or if it is a wave-like inhomogeneous universe, in which, while some regions are expanding, others are contracting.

Finally, what is the role of gravity in the life of the universe? Observations show that the direction of the rotation axes of the galaxies is almost random. From the viewpoint of the present work, because of this randomness, the gravitational interaction may be attractive or repulsive. It is attractive between galaxies rotating in the same sense (round rotation axes making an angle lower than 90 degrees), but is repulsive between oppositely rotating galaxies (round axes making angles larger than 90 degrees). In this way attractive gravitation may on the overall be neutralized by anti-gravitation. In order to explain what is here meant by anti-gravitation, consider two well separated astronomical bodies of equally large masses and the Keplerian velocity fields of the QS, creating their gravitational fields (see Section 2), spinning round parallel axes. The important question to be answered is: What will happen if the two bodies approximate each-other? The point here is not giving an exact quantitative answer. A qualitative answer suffices. According to the space dynamic gravitational mechanism, if their velocity fields are spinning in the same sense, they will add up their velocity fields according to Equation (12), thereby increasing the total negative gravitational potential of the system. However, in case the velocity fields are spinning in opposite senses, their velocity fields will gradually add up to zero according to Equation (12) and their total effective negative gravitational potential may gradually cancel. While in the first case approximation of the bodies creates an additional negative (attractive) gravitational potential term, in the second case the approximation adds a positive (repulsive) gravitational potential term that gradually cancels the total negative potentials of the individual bodies. While the behavior in the first case fits into the view of the current gravitational theories, the second case is a novelty predicted by space dynamics. According to conservation of energy, while in the first case kinetic energy can be created by letting the two bodies approximate each-other, because this increases the negative gravitational potential energy of the system, in the second case, kinetic energy can be created by letting the bodies move away from each-other, because this too increases the gravitational potential energy of the system. This behavior evidences that while in the first case the bodies interact attractively, in the second case they repel each other, they antigravitate.

Within the solar system the rotation axes of the velocity fields of the sun, of the planets and of the planetary satellites are fairly well polarized (aligned), their velocity fields spinning in the same sense as can be verified from the motions of the satellites round the planets and of the planets round the Sun. The only anomalous case is with the satellites of Uranus. This however may be due because the satellites of Uranus are captured satellites. An analogous polarization seems to be well present between the stars of each galaxy. Therefore, within the solar system, within the Milky-Way galaxy and within galaxies in general only attractive gravitational interaction is observed. However, astronomical observations show that the direction of the rotation axes of the different galaxies throughout the universe is almost random. Hence, in the large scale dynamics of the universe the gravitational interaction does not play a dominant role. This dominant role belongs to the cyclic expansion-contraction of the Higgs-QS. The gravitational interaction is responsible for the filamentary large scale structure of the universe.

5. CONCLUSION

Jacob Schaf (5,22)">5] -[7] can predict correctly the observed gravitational dynamics of the galaxies, without the need of dark matter. This work also appoints expansion of the Higgs quantum space itself, driven by energy minimization, as the physical origin of the accelerated expansion of the universe. This mechanism has the virtue of being intrinsically homogeneous throughout the universe and does not put the astronomical bodies into motion with respect to the Higgs quantum space, which would run into conflict with the null results of the Michelson light anisotropy experiments.

REFERENCES

1. Rubin, V. and Ford Jr., W.K. (1970) Astrophysical Journal, 159, 379.
2. Rubin, V., Thonnard, N. and Ford Jr., W.K. (1980) Astrophysical Journal, 238, 471.
3. Remmen, G. (2010) Journal of Undergraduate Research in Physics, 1.
4. Schaf, J. (2014) Recent Progress in Space Technology, 4, 44-66.
5. Schaf, J. (2014) Journal of Modern Physics, 5, 407-448.
6. Schaf, J. (2013) Journal of Modern Physics, 4, 676-685.
7. Schaf, J. (2012) Journal of Modern Physics, 3, 714-749.
8. Schaf, J. (2009) Physics Essays, 22, 435.
9. Schaf, J. (2003) Physics Essays, 16, 213.
10. Hatch, R.R. (2007) Physics Essays, 20, 83.
11. Hatch, R.R. (2004) GPS Solutions, 8, 67.

12. Hatch, R.R. (2004) Foundations of Physics, 34, 1725.
13. Prada, F., Gutierrez, C., Peletier, R.F. and McKeith, C.D. (1996) A Counter-Rotating Bulge in the Sb Galaxy NGC 7331.
14. Turyshev, S.G. and Toth, V.T. (2010) The Pioneer Anomaly.
15. Anderson, J.D., Laing, P.A., Lau, E.L., Liu, A.S., Nieto, M.M. and Turyshev, S.G. (2002) Physical Review D, 65.
16. Barbanis, B. and Prendergast, K.H. (1966) The Astronomical Journal, 72, 215.
17. Riess, A., et al. (1998) Astronomical Journal, 116, 1009.
18. Perlmutter, S., et al. (1999) Astrophysical Journal, 517, 565.

CHAPTER 5

Why Energy and Mass Can Be Converted between Each Other? A New Perspective Based on a Matter Wave Model

Donald C. Chang

Macro-Science Group, Division of LIFS, Hong Kong University of Science and Technology, Hong Kong, China

ABSTRACT

It was predicted by Einstein that energy and mass can be converted between each other. But why? Energy and mass are two very different physical concepts. How can they be exchanged with each other? We think the key to answer this question is to recall that a particle can behave like a wave. Particle properties like energy and momentum are known to be related to their corresponding wave properties (frequency and wave vector). Mass is clearly a particle property; is it also related to a wave property? This study suggests that it is. We found that mass and energy appear to share similar physical nature in the wave perspective. Both of them are related to the curvature of bending the vacuum medium during the propagation of the excitation wave. This similarity explains why they are convertible.

Keywords: Energy, Mass, Particle, Wave, Vacuum, Matter Wave, Special Relativity

1. INTRODUCTION

Recent studies of cosmology suggest that our universe is not only composed of visible matter; there are also dark matter and dark energy [1] . The visible matter only accounts for about 5% of the content of the universe, while 25% is composed of dark matter, and about 70% is dark energy [1] . At this point, we have very little knowledge about the physical properties of the dark matter and dark energy. In order to find some hints to hunt for the dark matter, perhaps we should thoroughly examine our understanding of the visible matter first.

At this point, there are still many unanswered questions about the visible matter. For example, we know that energy and mass can be converted between each other. But what is its physical basis? This energy-mass conversion was attributed to the Special Theory of Relativity (STR) [2] . That is, energy and mass are related by the famous relationship $E = mc^2$. But why? Energy and mass are very different physical concepts. How can they be convertible? STR only proposed the energy-mass conversion rule; it did not explain the physical foundation of this rule. Thus, it will be interesting to explore ways to explain this rule on a physical basis.

2. THE WAVE NATURE OF SUB-ATOMIC PARTICLES

We think the key to understand energy-mass conversion is to recognize that matter has wave properties. The concept of matter wave was proposed almost a century ago. Phenomenologically, it was found that a particle can behave like a wave or a corpuscular object depending on the situation [3] . This peculiar phenomenon is called "wave-particle duality". The key question now is about the true physical nature of the particle: Is it a massive point object or a physical wave packet? In the case of photon, there is no question that light is an electro-magnetic wave. But it can behave like a particle (photon) during certain physical interactions, such as in the case of photoelectric effect or Compton scattering [4] . Thus, a packet of electro-magnetic wave can behave like a particle.

Besides light wave, sound waves can also behave like particles. It is well known in condensed matter physics that excitation waves in a solid (or fluid) can be treated like particles (called "phonons"). Phonons can interact with other particles (like electrons) and play a major role in determining the electrical conductivity of many materials [5] .

On the other hand, sub-atomic particles with mass can also behave like waves. It was shown in 1927 that electrons can be diffracted by a nickel crystal [6] . Later, other subatomic particles such as neutrons or helium atoms were also found to behave like wave in diffraction experiments [7] [8] .

The physical nature of matter wave is still not clearly understood today. In most quantum mechanics textbooks, the particle is treated as a pointed object; the wave property is only associated with the probability of finding the particle at a particular space and time. (This is called the "Copenhagen interpretation") [3] . But in view of the findings of the diffraction experiments, it seems more sensible to regard the particle as a physical wave rather than a probability wave. This point can be demonstrated easily using the example of a radiation wave. We know that light is an electro-magnetic wave; its diffraction pattern can be explained very easily based on such wave properties. But if we regard light as a probability wave of a photon, the explanation will be much more difficult.

Furthermore, we know particles can be created or annihilated in the vacuum. If the particle is a real object, how can it appear from nowhere or disappear suddenly with no trace? The only rational explanation is that the particle is an excitation wave, so that it can be excited by an energetic stimulation and it can be transformed from one type of wave into another type of wave. Therefore, we propose that the so-called "particle" is actually an excitation wave of the vacuum [9] . In another word, the particle is a physical wave in the microscopic view; it behaves like a corpuscular object (particle) only in the macroscopic view.

3. PHYSICAL NATURE OF THE VACUUM

If the matter wave is a physical wave, then what is its carrying medium? In the 19th century, the electro-magnetic field was thought to be carried by a medium called "aether" [10] . The aether hypothesis, however, was rejected in the early 20th century due to the following reasons:

1) The mechanical properties of this hypothetical aether were full of contradictions. Aether was supposed to fill all space between matters. Thus, it must be a highly fluidic substance (gas or liquid). Such fluidic medium, however, can only transmit low frequency dilational wave. In order to transmit high frequency transverse wave such as light, the aether must be a rigid solid. These two requirements are contradicting with each other.

2) The aether hypothesis could not explain why large astronomical objects like Earth or Mars can pass through it without experiencing any resistance. This is contradicting to the requirement that aether is a fluid or a solid.

3) The aether hypothesis was not supported by experiments. In late 19th century, many experiments were conducted to detect the motion of aether using optical interferometers [11] . All these experiments failed to detect any movement between aether and the Earth.

4) Later, the aether hypothesis was thought to be unnecessary. Finally in 1905, Einstein proposed the Special Theory of Relativity and showed that one can explain the null results easily without the assumption of aether [2] . With the widely acceptance of STR, the aether hypothesis was abandoned.

Hence, people in the early 20th century believed that the vacuum in our universe is totally empty. But such a view is no longer valid today. With the development of quantum electrodynamics, the vacuum is thought to contain the zero-point energy of all oscillations in the electromagnetic field [10] . In fact, the properties of the vacuum become very complicated in today's quantum field theory. For example, according to the Dirac theory of electron, the vacuum is filled with an ocean of negative energy electrons. The hole created in this ocean will become the anti-particle, positron [12] . The vacuum thus is far from emptiness. Moreover, since there are many types of anti-particles, if one follows the Dirac theory, it would require the vacuum to be filled with multiple oceans of different types of negative energy fermions. (Note: The Dirac theory is even more troublesome for bosons, which do not follow Pauli's exclusion principle. Thus, their negative energy states cannot be fully filled.)

In the more recent quantum field theory, a particle is regarded as the excitation of its field. Different types of particles are associated with different fields. However, these fields do not have the same meaning as classical fields. The physical meaning of quantum field is not clear at present. There is still a debate about what is more fundamental: Is it particle or field? Currently there is no agreement on it [13] . Furthermore, such thinking requires a large number of fields for our universe. It is not so beautiful.

Thus, there is a dilemma now. On the one hand, the old concept of aether is unviable. On the other hand, with the development of quantum field theory, the vacuum can no longer be regarded as "emptiness". What is the way out? We think this dilemma can be solved by assuming that the vacuum is a pre-existing medium in our universe and it fills all space; both matter waves and radiation waves are excitation waves of this medium [14] . In another word, all particles found in nature (with or without mass) are excitation waves of the same vacuum medium.

This new model can overcome all problems of the previous aether hypothesis:

First, we can avoid contradicting mechanical requirements for the vacuum. Since the vacuum is a pre-existing medium occupying the entire universe, not just filling the space between matters, it does not need to be fluidic.

Second, the atoms making up the planet are all waves; there should be no friction between matters and the vacuum. Thus, planets can move within the vacuum medium without dragging.

Third, the null results of the optical interferometer experiments reported by Michelson and Morley (1887) can be easily explained [11] . All one needs is to show that the wave equation describing the excitation waves in the vacuum medium is Lorentz invariant. As we will show in the following, this requirement can indeed be met.

4. THE VACUUM BEHAVES LIKE A DIELECTRIC SYSTEM ACCORDING TO THE MAXWELL THEORY

If particles are excitation waves, what is the physical nature of the vacuum medium? At this point, our knowledge about the vacuum is very limited. However, there is a very good hint from the work of Maxwell, who suggested that the vacuum behaves like a dielectric medium [15] . When Maxwell first proposed his equations in 1862, his description of Ampere's Law was

$$\nabla \times \boldsymbol{H} = \boldsymbol{J} \quad . \tag{1}$$

He later found a problem in it, because this equation would violate the condition of conservation of charge,

$$\nabla \cdot \boldsymbol{J} = -\frac{\partial \rho_e}{\partial t} \quad . \tag{2}$$

In order to fix this problem, Maxwell proposed to add a new term $\partial \boldsymbol{D}/\partial t$ into the right hand-side of Equation (1) [16] . (D is called the "charge displacement".) The justification for this was that, in a dielectric material, there are both positive and negative charges embedded within the medium.

So, when the dielectric medium is exposed to an electric field, it will induce a displacement of the dielectric charges. A time variation of the displaced charge should produce a displacement current (J_d), which can also affect the magnetic field. So that in Equation (1), one should consider not only the externally applied current J, but also the internally induced displacement current J_d. That is

$$\boldsymbol{J}_{\text{total}} = \boldsymbol{J} + \boldsymbol{J}_{\text{d}} = \boldsymbol{J} + \frac{\partial \boldsymbol{D}}{\partial t}.$$

Thus, Maxwell thought that the correct form of Ampere's law should be

$$\nabla \times \boldsymbol{H} = \boldsymbol{J} + \frac{\partial \boldsymbol{D}}{\partial t}. \tag{1A}$$

This becomes a part of the final form of the Maxwell's equations used today.

The addition of a displacement current term was a stroke of genius. This allowed Maxwell to construct his theory of light propagation. When Maxwell investigated the propagation of electro-magnetic waves in the vacuum, he argued that the external current J is zero, but the displacement current term is not zero. This is equivalent to say that he believed the vacuum is a dielectric medium. Thus, Equation (1A) in the vacuum becomes

$$\nabla \times \boldsymbol{H} = \frac{\partial \boldsymbol{D}}{\partial t}. \tag{1B}$$

Using this equation, one can easily derive the wave equation for the electro-magnetic wave,

$$\nabla^2 \boldsymbol{A} - \frac{1}{c^2}\frac{\partial^2 \boldsymbol{A}}{\partial t^2} = 0. \tag{3}$$

where A is the vector potential and $c = 1/\sqrt{\mu_0 \varepsilon_0}$ is equal to the speed of light. (ε_0 is permittivity and μ_0 is permeability in free space). Thus, the key conceptual step in Maxwell's theory of light propagation was to regard the vacuum as a dielectric medium. If one treats the vacuum as pure emptiness and remove the $\partial \boldsymbol{D}/\partial t$ term in Equation (1A), it will not be possible to derive the wave equation of light.

5. WHAT IS THE BASIC WAVE EQUATION FOR AN EXCITATION WAVE?

If both the radiation waves and matter waves are excitation waves of the same vacuum medium, then waves representing different particles (with or without mass) should obey the same wave equation, which is determined by the physical properties of the vacuum medium. One may be surprised that this point is actually supported by the Special Theory of Relativity. STR predicts that no object can travel faster than the speed of light [2] .

This means that the limiting speed of all particles (with or without mass) must be the same. We know that the speed of an excitation wave is determined solely by the physical properties of the transmitting medium [15] . The prediction of STR can be justified only if matter wave and radiation wave are carried by the same vacuum medium.

What is the basic form of this wave equation? At this point, we know the wave equation of at least one particle; that is the photon.

This may provide a very useful hint. We know a photon is an oscillating wave of electro-magnetic fields, which are linear functions of the vector potential $A^{\mu} = (\phi, A)$. If matter waves and radiation waves are both excitation waves of the same vacuum medium, we expect that the wave function of the matter wave ψ may also be a linear function of A^{μ}. This means that ψ should also obey the same wave function as shown in Equation (3), i.e.,

$$\nabla^2\psi - \frac{1}{c^2}\frac{\partial^2\psi}{\partial t^2} = 0 \quad . \tag{4}$$

We propose that this is the basic wave equation for all types of excitation waves of the vacuum medium. ψ can have different solutions, which may represent different types of free particle. (Note: Even though Equation (4) does not look like any wave equations commonly used in quantum mechanics, it is not inconsistent with them. In fact, we can show later in this paper that Equation (4) can be reduced into the Klein-Gordon equation for particles of nonzero mass.

By linearizing the Klein-Gordon equation, one can then derive the Dirac equation for electrons [12] . Under the low energy limit, the Schrödinger equation can be obtained.)

6. GENERAL WAVE FUNCTION OF A FREE PARTICLE

Now let us examine the different solutions of the basic wave equation. The simplest solution of Equation (4) is a plane wave $\psi_k \sim e^{i(\boldsymbol{k}\cdot\boldsymbol{x}-\omega t)}$ where k and ω are the wave vector and frequency, respectively. We propose that this plane wave solution represents a radiation particle (massless), such as a photon. For a particle with non-zero rest mass, the solution is more complicated. Such a particle would behave like a mass point in the classical limit; thus, the probability of finding the particle should be highest at its trajectory. This suggests that the wave function of a massive particle not only depends on the spatial coordinate along its trajectory (i.e., $\hat{\boldsymbol{k}}\cdot\boldsymbol{x}$), it also varies in the transverse plane ($\hat{\boldsymbol{k}}\times\boldsymbol{x}$). Thus, one may assume that the general wave function representing a free particle has the form

$$\psi_{\hat{k}}(\boldsymbol{x},t)=\psi_L\left(\hat{\boldsymbol{k}}\cdot\boldsymbol{x},t\right)\psi_T\left(\hat{\boldsymbol{k}}\times\boldsymbol{x}\right), \tag{5}$$

where ψ_L is the longitudinal component of the wave function which describes the travelling wave along the particle's trajectory, and ψ_T is the transverse component of the wave function. Substituting Equation (5) into Equation (4), one can use the technique of separation of variables to obtain two coupled equations:

$$\begin{cases}\left[\nabla^2-\dfrac{1}{c^2}\dfrac{\partial^2}{\partial t^2}\right]\psi_L\left(\hat{\boldsymbol{k}}\cdot\boldsymbol{x},\,t\right)=\ell^2\psi_L\left(\hat{\boldsymbol{k}}\cdot\boldsymbol{x},\,t\right) & \textbf{(6A)}\\ \nabla^2\psi_T\left(\hat{\boldsymbol{k}}\times\boldsymbol{x}\right)=-\ell^2\psi_T\left(\hat{\boldsymbol{k}}\times\boldsymbol{x}\right) & \textbf{(6B)}\end{cases}$$

where ℓ is a fitting parameter which is subjected to the condition

$$\omega^2=\left(k^2+\ell^2\right)c^2. \tag{7}$$

By solving Equations (6A) and (6B) separately [9] , one can easily show that the general solution of ψ is

$$\psi_{\hat{k}}(x,t) = aJ_n(\ell r)e^{\pm in\theta}e^{i(k\cdot x-\omega t)}, \tag{8}$$

where J_n is Bessel function of the first kind; n is an integer or half integer; r and θ represent the radial distance and the azimuthal angle of the space vector in the transverse plane. (a is a normalizing constant.) As expected, the wave function of a free particle behaves like a travelling wave moving along the direction of its trajectory. But due to the presence of the Bessel function, $\psi_{\hat{k}}$ varies in a diminishing oscillating manner in the directions perpendicular to the particle's trajectory.

7. PHYSICAL MEANINGS OF THE WAVE PARAMETERS

The wave function shown in Equation (8) contains four parameters, ω, k, ℓ and n. What are their physical meanings? Using the correspondence principle $E \to i\hbar\partial/\partial t$ and $p \to -i\hbar\nabla$, one can easily show that ω and k are related to the energy (E) and momentum (p) of the particle, exactly as predicted by the Planck's relation $E = \hbar\omega$ and the de Broglie relation $p = \hbar k$. (Here $\hbar$ is Planck's constant divided by 2π). Now, what is the physical meaning of ℓ? Our recent work suggested that ℓ is related to the mass of the particle [14] . This can be easily shown. Recall that the particle velocity (v) is determined by the group velocity of the wave packet [3] ,

$v = \frac{\partial\omega}{\partial k} = \frac{\partial E}{\partial p}$, and the particle mass M is defined by $p = Mv$ in the classical limit, one can show from Equation (7) that

$$M = \frac{\hbar\ell/c}{\left(1 - v^2/c^2\right)^{1/2}}. \tag{9}$$

At $v \ll c$, M equals the rest mass, m. That means the rest mass is

$$m = \frac{\hbar\ell}{c}. \tag{10}$$

This result makes good sense, since when one substitutes Equation (10) into Equation (7), and recalls that $E = \hbar\omega$ and $p = \hbar k$, one can get

$$E^2 = p^2c^2 + m^2c^4, \tag{11}$$

which is identical to the energy-momentum relationship implied from STR [2] . By combining Equations (9), (10) and (11), one can further obtain

$$E = Mc^2. \tag{12}$$

At $v \ll c$, the above equation becomes $E = mc^2$. Thus, by regarding the particle as an excitation wave, one can naturally obtain the relationship between energy and mass.

8. BOTH ENERGY AND MASS ARE RELATED TO THE BENDING CURVATURE OF THE VACUUM

Our finding that the rest mass m is related to the parameter ℓ in the wave function has important implication. From Equation (8), we see that the transverse component of the free particle wave function is described by a Bessel function, the asymptotic form of which is

$$J_n(\ell r) \to \left(\frac{2}{\pi \ell r}\right)^{1/2} \cos\left(\ell r - \frac{2n+1}{4}\pi\right). \tag{13}$$

Thus, ℓ can be regarded as the "transverse wave number" of the free particle, i.e., it is the inverse of the wavelength in the transverse oscillation, $\ell = 2\pi/\lambda_T$ (see Figure 1). Thus, Equation (10) means that the rest mass of a particle is associated with the oscillation periodicity of the wave function in the transverse plane.

This finding is very interesting; it is closely parallel to the Planck's relation and the de Broglie relation, which show that the energy and momentum are related to the periodicity of oscillation of the vacuum medium. More specifically, E is related to the periodicity of oscillation in the time dimension, and p is related to the periodicity of oscillation in the spatial dimension along the direction of the trajectory. Our finding that the rest mass is associated with the oscillation periodicity in the transverse direction thus appears to make very good sense. In essence, our result suggests that energy, momentum and mass are all related to the curvature of bending the vacuum medium during the propagation of the excitation wave.

This result is not totally surprising, since both energy and mass must be created by “work”. Before the vacuum is excited by a stimulus, it is in the resting state (ground state). In order to trigger an excitation wave, one must create an energetic disturbance at a local region of the vacuum medium. Since it takes work to bend the vacuum medium, the more sharply the bending curvature, more work is required. This is true for bending the medium in all dimensions (spatial and temporal). Thus, a shorter wavelength of a propagating wave should always associate with a higher “energy state”, which may be reflected in an increase in energy, momentum or “mass” of the excitation wave.

This geometrical interpretation of mass and energy suggests that the physical natures of energy and mass are very similar in the wave perspective. This explains why energy and mass can be converted between each other.

9. A FREE PARTICLE WITH NONZERO REST MASS BEHAVES LIKE A VORTEX

Based on the discussions above, we believe that both radiation wave and matter wave are excitation waves of the vacuum medium. The basic wave equation is determined by the physical property of the vacuum medium, which behaves like a dielectric system as suggested by Maxwell.

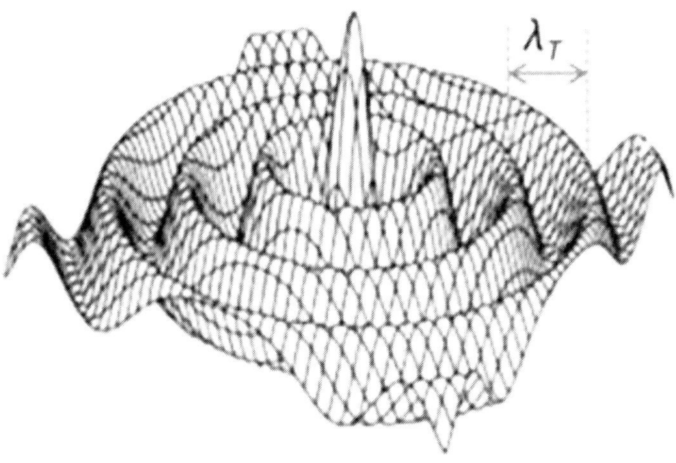

Figure 1. A plot of ψ_T at the transverse plane for n = 0. The wavelength is denoted λ_T.

The solution of this wave equation can represent particles of different types. We think the radiation wave is corresponding to the plane wave solution; while the matter waves are represented by the cylindrical wave solutions,

$$\psi_{\hat{k}}(x,t) = aJ_n(\ell r)e^{\pm in\theta}e^{i(k\cdot x-\omega t)}$$

The various wave parameters in this equation represent different particle properties. The wave vector k is related to the momentum p; the frequency ω is related to the energy E; and the transverse wave number ℓ is related to the rest mass m.

This wave function represents an excitation wave advancing in helical motion. This can be seen by examining the physical meaning of the parameter n. It appears that n is associated with the helicity of the free particle, since it is a quantum number conjugate to the angular coordinate θ. Dimensional analysis thus suggests that n is associated with some sort of angular momentum.

Because of the added phase factor nθ in Equation (8), the wave function representing a free particle actually propagates in a helical fashion (see Figure 2).

We can demonstrate this point more explicitly. Let $\omega = \omega_1 + \omega_2$, substituting it into Equation (8), we have

$$\psi_{\hat{k}}(x,t) = aJ_n(\ell r)e^{i(\pm n\theta-\omega_1 t)}e^{i(k\cdot x-\omega_2 t)} \tag{8A}$$

The first exponential term $e^{i(\pm n\theta-\omega_1 t)}$ represents a rotational motion of the wave packet; the second exponential term $e^{i(k\cdot x-\omega_2 t)}$ represents the translational motion of the wave packet. In this way, when the particle is traveling at different speed, the translational phase velocity ω_2/k can be adjusted at will.

The wave function as a whole behaves like a moving vortex. The phase factor $\pm i\theta$ corresponds to right-handed and left-handed helical motions.

10. OUR PROPOSED WAVE EQUATION IS CONSISTENT WITH OTHER QUANTUM MECHANICAL EQUATIONS

In this work, we propose that all particles (with or without mass) are excitation waves of the vacuum medium. The basic wave equation is Equation (4). For particles without mass (i.e., radiation waves), there should be no problem to accept it, since this equation is identical to the wave equation of light. But for particles with rest mass (i.e., matter waves), can Equation (4) properly describe their excitation waves? In relativistic quantum mechanics, the motion of massive particles is usually described by the Klein-Gordon equation [17] [18] or the Dirac equation. Is Equation (4) compatible with these quantum mechanical equations?

In the following, we will show that this is indeed the case. One should keep in mind that in describing the motion of a particle, there are two categories of wave equations. The first category is the basic wave equation which describes all types of excitation waves in the vacuum medium. Different solutions of this wave equation represent different types of particles. The second category is wave equations that describe the motion of a specific type of particle. We propose that Equation (4) belongs to the first category. Since it is applicable to all types of particles, the particle mass should not appear in the equation. For wave equations of the second category, the particle mass can appear in the wave equation, since it describes the motion of a single type of particle. We believe the Klein-Gordon equation or the Dirac equation belong to this category.

The question now is whether one can derive the Klein-Gordon equation or the Dirac equation from the basic wave equation of the vacuum medium (i.e., Equation (4)). As shown in Equation (5), the wave function representing an excitation wave in the vacuum medium has both a longitudinal component (ψ_L) and a transverse component (ψ_T). When one substitutes Equation (5) into Equation (4) and uses the technique of separation of variables, one can obtain

$$\left[\nabla^2 - \frac{1}{c^2}\frac{\partial^2}{\partial t^2}\right]\psi_L\left(\hat{\boldsymbol{k}}\cdot\boldsymbol{x},t\right) = \ell^2\psi_L\left(\hat{\boldsymbol{k}}\cdot\boldsymbol{x},t\right) \quad . \qquad (6A)$$

Since we know now that ℓ is connected with the rest mass m through Equation (10), the above equation becomes

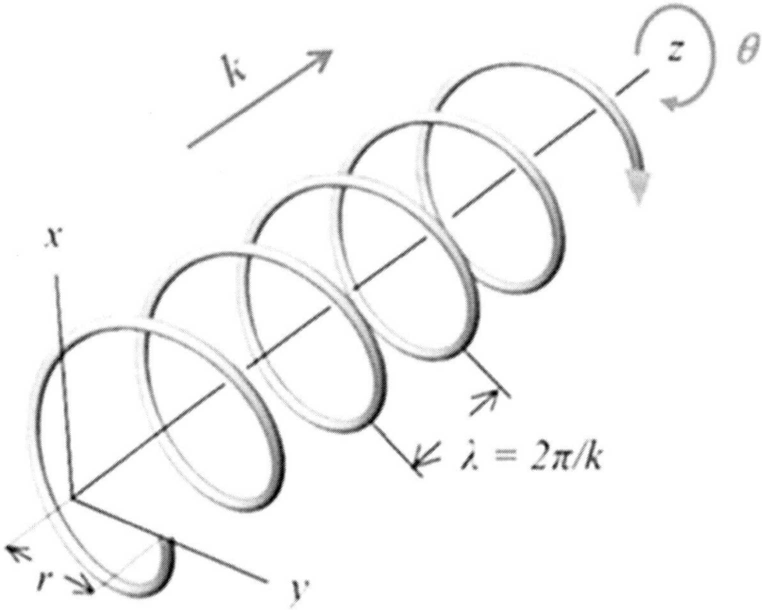

Figure 2. The motion of matter wave representing a massive particle appears as a vortex advancing in the direction k (which is parallel to the z-axis here).

$$\left[\nabla^2 - \frac{1}{c^2}\frac{\partial^2}{\partial t^2}\right]\psi_L - \left(\frac{mc}{\hbar}\right)^2 \psi_L = 0 \quad (14)$$

This is identical to the Klein-Gordon equation if we regard ψ_L as the particle wave function ϕ. This analysis suggests that the wave function described by the Klein-Gordon equation only represents the longitudinal component of the travelling wave of a free particle with a specific mass m. This means that the Klein-Gordon equation can only describe the movement of a particle along its trajectory. Its wave function does not give any information on the oscillation of the vacuum medium along the transverse plane.

Once the Klein-Gordon equation is derived, one can then obtain the Dirac equation to describe the motion of an electron by linearizing Equation (14) [19] . Furthermore, based on the Dirac equation, the Schrödinger equation can be derived under the condition when the Coulomb potential energy and kinetic energy are small in comparison to the rest mass of the particle. This derivation procedure had been shown elsewhere [20] .

Thus, our proposed basic wave equation is consistent with the commonly used quantum mechanical wave equations. The only difference is that the common quantum mechanical equations describe only the motion of a single type of particle. Furthermore, the wave functions in these quantum mechanical equations appear to describe only the motion of the excitation wave along the trajectory of the particle. The basic wave equation (Equation (4)), on the other hand, describes the complete wave function of the excitation wave of a free particle which may have different mass.

We may add that our discussion above only applies to the excitation wave of a free particle, such as a photon or an electron. For composite particles, such as protons or neutrons, the situation is much more complicated. One may need a further examination of the quark model. That will be far beyond the scope of this paper.

11. CONCLUSION

In conclusion, we believe all particles (with or without mass) are excitation waves of the vacuum medium. They behave like corpuscular objects only in the macroscopic view. Each of the particle properties and wave properties appears to have a one-to-one correspondence. Energy, momentum and mass are all related to the curvature of bending the vacuum medium during the propagation of the excitation wave. The similarity between the physical nature of energy and mass explains why the two can be converted between each other.

ACKNOWLEDGEMENTS

I thank the late Profs. John A. Wheeler and H. E. Rorschach for their encouragement when I started this work. I also thank Lan Fu for her assistance. The support of the Hong Kong University of Science and Technology to this work is acknowledged.

REFERENCES

1. Spergel, D.N. (2015) Science, 347, 1100-1102.
2. Einstein, A. (1952) The Principle of Relativity: A Collection of Original Memoirs on the Special and General Theory of Relativity. Dover Publications, New York.

3. Messiah, A. (1965) Quantum Mechanics. Wiley, New York.
4. Shankland, R. (1961) Atomic and Nuclear Physics. MacMillan, New York, 207.
5. Kittel, C. (2005) Introduction to Solid State Physics. 8th Edition, John Wiley & Sons, Hoboken, NJ,.
6. Davisson, C.J. and Germer, L.H. (1927) Nature, 119, 558-560.
7. Halban, H.V.J. and Preiswerk, P. (1936) Comptes Rendus del'Académie des Sciences, 203, 73-75.
8. Estermann, I. and Stern, O. (1930) Zeitschrift Für Physik, 61, 95-125.
9. Chang, D.C. (2004) arXiv:physics/0404044 [physics.gen-ph].
10. Whittaker, E. (1951) A History of the Theories of Aether and Electricity. Thomas Nelson, London.
11. Michelson, A.A. and Morley, E.W. (1887) American Journal of Science, 34, 333-345.
12. Dirac, P.A.M. (1981) Chap. XI. Relativistic Theory of the Electron. In: The Principles of Quantum Mechanics, 4th Edition, Clarendon Press, Oxford, 253-275.
13. Kuhlmann, M. (2013) Scientific American, 309, 40-47.
14. Chang, D.C. (2013) Journal of Modern Physics, 4, 21-30.
15. Chang, D.C. and Lee, Y. (2015) Journal of Modern Physics, 6, 1058-1070.
16. Longair, M.S. (1984) Chap. 3. The origin of Maxwell's Equations and Their Experimental Validation. In: Theoretical Concepts in Physics, Cambridge University Press, Cambridge, New York, 40-49.
17. Klein, O. (1926) Zeitschrift Für Physik, 37, 895-906.
18. Gordon, W. (1926) Zeitschrift Für Physik, 40, 117-133.
19. Sakurai, J.J. (1973) Advanced Quantum Mechanics. Addison-Wesley, Reading, 75-89.
20. Cottingham, W.N. and Greenwood, D.A. (1998) An Introduction to the Standard Model of Particle Physics. Cambridge University Press, Cambridge, 72.

CHAPTER 6

Distribution of Mass and Energy in Closed Model of the Universe

Fadel A. Bukhari

Department of Astronomy, Faculty of Science, King Abdulaziz University, Jeddah, Saudi Arabia

ABSTRACT

The universe's horizon distance and volume are constructed in the closed cosmic model. The universe horizon distance distribution increases constantly for $t < t_{me}$ and decreases for $t > t_{me}$. However, the universe's horizon volume shows a sudden reduction in the range t = 0.5 Gyr – t_{me} due to the change of the universe space from flat to curved then closed in the interval 15.1261 Gyr ≤ t ≤ t_{me}. On the other hand, this distribution exhibits an abrupt rise in the range t = t_{me} – t_* due to the change of the universe space from closed then curved to flat in the interval 39.3822 ≤ t ≤ 40.7521 Gyr. The mass of radiation, matter and dark energy within the horizon volume of the universe are also investigated. These distributions reveal similar noticeable changes as the universe's horizon volume distribution for the same reasons. The mass of radiation dominates up to t = 53221.5 yr, then the mass of matter becomes larger. Afterwards, both distributions of radiation and matter decrease while the distribution of dark energy rises until t = 10.1007 Gyr, where the mass of dark energy prevails up to t = t_{me}. Hence, the distribution of dark energy reduces until t = 40.2892 Gyr, where the mass of matter becomes prominent again. At t = 53.6246 Gyr the masses of both matter and radiation become appreciably high such that the intercluster space will vanish and clusters of galaxies interfere with each other. Furthermore, not only the intergalactic medium will disappear, but also galaxies will collide and merge with each other to form extremely dense and close cosmological bodies. These very dense bodies will undergo further successive collisions and mergers under the action of central gravity, where the interstellar medium will vanish and the universe would develop to big crunch at t_{bc} = 53.6251 Gyr. It is interesting to note that the horizon distance of the universe in the closed model at $t = t_{me}$ is in very good agreement with the maximum horizon distances in the five general cosmic models.

Keywords: Dark Energy, Radiation, Closed Cosmic Model

1. INTRODUCTION

The distribution of density parameters of radiation, matter and dark energy in the closed cosmic model were investigated in a previous study [1] , where we discovered the main epochs of the universe history in this model. It is worthy now to study the distributions of equivalent mass of radiation, mass of matter and equivalent mass of dark energy within the horizon volume of the universe to get deeper sight of the universe evolution in the closed model.

The reason for considering the equivalent mass of radiation in this study is the significant value of the radiation density parameter in the early universe and before the big crunch as we have seen in [1] .

Therefore, it is vital to develop the distributions of the horizon distance and horizon volume of the universe in the closed model at various time ranges depending on the bases presented in [2] . Description of methodology is illustrated in Section 2, while algorithm would be shown in Section 3. Results and discussion are displayed in Section 4. Conclusion is given in Section 5.

2. METHODOLOGY

It is obvious from [2] that the horizon distance and horizon volume of the universe in closed cosmic model at the present time are respectively

$$d_h(t_o) = \frac{c}{H_o}\int_0^1 \frac{1}{a}\left[1 - S^2\Omega''_{\Lambda,t}\left(1-a^2\right) + S^2\Omega''_{m,t}\left(a^2-a^3\right) + S^2\Omega''_{r,t}\left(a^2-a^4\right)\right]^{-1/2} \mathrm{d}a. \tag{1}$$

$$V_h(t_o)=\frac{8\pi}{3}d_h^3(t_o). \tag{2}$$

where $\Omega''_{\Lambda,t}$, $\Omega''_{m,t}$ and $\Omega''_{r,t}$ are given by

$$\Omega''_{\Lambda,t}=\frac{\rho'_{\Lambda,t}}{c^2\rho''_{c,t}}. \tag{3}$$

$$\Omega''_{m,t}=\frac{\rho'_{m,t}}{\rho''_{c,t}}. \tag{4}$$

$$\Omega''_{r,t}=\frac{\rho'_{r,t}}{c^2\rho''_{c,t}}. \tag{5}$$

$$\rho''_{c,t}=\frac{3H'^2(t)}{8\pi G}. \tag{6}$$

$$H'(t)=\frac{H_o}{a}\left[1-s^2\Omega'_{\Lambda,t}\left(1-a^2\right)+s^2\Omega'_{m,t}\left(a^2-a^3\right)+s^2\Omega'_{r,t}\left(a^2-a^4\right)\right]^{\frac{1}{2}}.$$

$$\text{where}\quad s=\frac{H(t)}{H_o},\; S=\frac{H'(t)}{H_o} \tag{7}$$

$$H(t)=\frac{H_o}{a}\left[1-\Omega_{\Lambda,o}\left(1-a^2\right)+\Omega_{m,o}\left(\frac{1}{a}-1\right)+\Omega_{r,o}\left(\frac{1}{a^2}-1\right)\right]^{\frac{1}{2}}. \tag{8}$$

$$\Omega'_{\Lambda,t}=\frac{\rho'_{\Lambda,t}}{c^2\rho_{c,t}}. \tag{9}$$

$$\Omega'_{m,t}=\frac{\rho'_{m,t}}{\rho_{c,t}}. \tag{10}$$

$$\Omega'_{r,t} = \frac{\rho'_{r,t}}{c^2 \rho_{c,t}}. \tag{11}$$

$$\frac{\rho'_{\Lambda,t}}{c^2} = \rho_{c,o}\Omega_{\Lambda,o} - \Delta\left(\frac{\rho_{\Lambda,t}}{c^2}\right). \tag{12}$$

$$\rho'_{m,t} = \rho_{c,o}\frac{\Omega_{m,o}}{a^3} + \frac{1}{2}\Delta\left(\frac{\rho_{\Lambda,t}}{c^2}\right). \tag{13}$$

$$\frac{\rho'_{r,t}}{c^2} = \rho_{c,o}\frac{\Omega_{r,o}}{a^4} + \frac{1}{2}\Delta\left(\frac{\rho_{\Lambda,t}}{c^2}\right). \tag{14}$$

$$\Delta\left(\frac{\rho_{\Lambda,t}}{c^2}\right) = 0.01\frac{\rho_{\Lambda,t}}{c^2}t\ . \tag{15}$$

where t is the cosmic time in Gyr.

$$\rho_{c,t} = \frac{3H^2}{8\pi G}. \tag{16}$$

The horizon distance of the universe in the closed cosmic model at any given time is given by

$$d_h(t) = \frac{c}{H_o}\int_0^a \frac{1}{a}\left[1 - S^2\Omega''_{\Lambda,t}\left(1-a^2\right) + S^2\Omega''_{m,t}\left(a^2-a^3\right) + S^2\Omega''_{r,t}\left(a^2-a^4\right)\right]^{-1/2} \mathrm{d}a. \tag{17-a}$$

Consequently, the change in the horizon distance of the universe in the time interval between two instants of scale factors a_1, a_2 is written as

$$\Delta d_h(t) = \frac{c}{H_o}\int_{a_1}^{a_2} \frac{1}{a}\left[1 - S^2\Omega''_{\Lambda,t}\left(1-a^2\right) + S^2\Omega''_{m,t}\left(a^2-a^3\right) + S^2\Omega''_{r,t}\left(a^2-a^4\right)\right]^{-1/2} \mathrm{d}a. \tag{17-b}$$

The horizon volume of the universe in the closed model at any given time is expressed as

$$V_h(t) = f(d_h(t), k). \tag{18}$$

Equation (18) indicates that the horizon volume of the universe at t is a function of $d_h(t)$ and the curvature of space k at t. Since this curvature could be flat, open and closed from the big bang to big crunch as evident from Table 3 in [1] . Thus, the low of $V_h(t)$ can be determined according to the value of k at t, as explained in the following cases:

(1) Flat space ($k = 0$)

We have seen in [2] that the horizon volume of the universe at time t in this case is given by

$$V_h(t) = \frac{8\pi}{3} d_h^3(t). \tag{19}$$

Therefore, it is obvious from Table 3 in [1] that Equation (19) is used in the time intervals $t \leq 6.5321\,\text{Gyr}$, $10.0751 < t \leq 15.1261\,\text{Gyr}$, $39.3822 < t \leq 40.7521\,\text{Gyr}$, and $53.48 < t \leq t_{bc}$.

(2) Closed space ($k = +1$)

We recall the equation of proper distance of extragalactic object

$$d_p(t) = R(t) f(r_o). \tag{20}$$

where

$$f(r_o) = \int_0^{r_o} \frac{dr}{\sqrt{1 - kr^2}} = \begin{cases} \sin^{-1} r_o & k = +1 \\ r_o & k = 0 \\ \sinh^{-1} r_o & k = -1 \end{cases} \tag{21}$$

And the volume of space within $d_p(t)$ is expressed as

$$V_p(t) = 2R^3(t) \int_0^{2\pi} d\phi \int_0^{\pi} \sin\theta \, d\theta \int_0^{r_o} \frac{r^2}{\sqrt{1 - kr^2}} dr. \tag{22}$$

where $R(t), r_o, r, \theta$ and ϕ are defined as in [3] [4] . For $k=+1$, Equations (21), (20) and (22) yield

$$d_p(t) = R(t)\sin^{-1} r_o \tag{23}$$

Hence Equation (22) gives

$$V_p(t) = 8\pi R^3(t)\int_0^{r_o} \frac{r^2 \mathrm{d}r}{\sqrt{1-r^2}}. \tag{24}$$

Assume

$$r = \sin\alpha,\ \mathrm{d}r = \cos\alpha\mathrm{d}\alpha,\ \cos\alpha = \sqrt{1-r^2} \tag{25}$$

Substituting by (25) in (24) we get

$$V_p(t) = 4\pi R^3(t)\int_0^{\sin^{-1} r_o} (1-\cos 2\alpha)\mathrm{d}\alpha. \tag{26}$$

Let

$$\beta = 2\alpha,\ \mathrm{d}\beta = 2\mathrm{d}\alpha. \tag{27}$$

Substituting by (27) in (26) we have

$$V_p(t) = 4\pi R^3(t)\left(\sin^{-1} r_o\right)^3 \left[\frac{1}{\left(\sin^{-1} r_o\right)^2} - \frac{r_o\sqrt{1-r_o^2}}{\left(\sin^{-1} r_o\right)^3}\right]. \tag{28}$$

Substituting by (23) in (28) yields

$$V_p(t) = 4\pi d_p^3(t)\left[\frac{1}{\left(\sin^{-1} r_o\right)^2} - \frac{r_o\sqrt{1-r_o^2}}{\left(\sin^{-1} r_o\right)^3}\right]. \tag{29}$$

Suppose $r_o = 1$, hence $\alpha_o = \frac{\pi}{2}$ and Equation (29) becomes

$$V_p(t) = \frac{16}{\pi} d_p^3(t). \tag{30}$$

Thus, the horizon volume of the universe in the closed cosmic model at time t in this case is expressed as

$$V_h(t) = \frac{16}{\pi} d_h^3(t). \tag{31}$$

It is evident form Table 3 in [1] that Equation (23) is used in two time intervals extending through $15.1261 < t \le 39.3822$ Gyr.

(3) Open space ($k = -1$)

Equations (21), (20) and (22) give

$$d_p(t) = R(t)\sinh^{-1} r_o \tag{32}$$

$$V_p(t) = 8\pi R^3(t) \int_0^{r_o} \frac{r^2 \mathrm{d}r}{\sqrt{1+r^2}}. \tag{33}$$

Assume

$$r = \sinh\xi,\ \ \mathrm{d}r = \cosh\xi \mathrm{d}\xi,\ \ \cosh\xi = \sqrt{1+r^2} \tag{34}$$

Substituting by (34) in (33) we have

$$V_p(t) = 4\pi R^3(t) \int_0^{\sinh^{-1} r_o} (\cosh 2\xi - 1)\mathrm{d}\xi. \tag{35}$$

Let

$$\eta = 2\xi,\ \ \mathrm{d}\eta = 2\mathrm{d}\xi. \tag{36}$$

Substituting by (36) in (35) we get

$$V_p(t) = 4\pi R^3(t)\left(\sinh^{-1} r_o\right)^3\left[\frac{r_o\cosh\left(\sinh^{-1} r_o\right)}{\left(\sinh^{-1} r_o\right)^3} - \frac{1}{\left(\sinh^{-1} r_o\right)^2}\right]. \quad (37)$$

Substituting by (32) in (37) yields

$$V_p(t) = 4\pi d_p^3(t)\left[\frac{r_o\cosh\left(\sinh^{-1} r_o\right)}{\left(\sinh^{-1} r_o\right)^3} - \frac{1}{\left(\sinh^{-1} r_o\right)^2}\right]. \quad (38)$$

When $r_o = 1$, Equation (38) reduces to

$$V_p(t) = 3.112983\pi d_p^3(t). \quad (39)$$

Therefore, the horizon volume of the universe in closed cosmic model at time t in this case is written as

$$V_h(t) = 3.112983\pi d_h^3(t). \quad (40)$$

It is clear form Table 3 in [1] that Equation (40) is used in the time intervals $6.5321 < t \le 10.0751\ \text{Gyr}$, $40.7521 < t \le 53.48\ \text{Gyr}$.

The total density of the universe in the closed cosmic model at time t is

$$\rho'(t) = \rho'_{m,t} + \frac{\rho'_{r,t}}{c^2} + \frac{\rho'_{\Lambda,t}}{c^2}. \quad (41)$$

Substituting by (3)-(5) in (41) we get

$$\rho'(t) = \rho''_{c,t}\Omega''_{m,t} + \rho''_{c,t}\Omega''_{r,t} + \rho''_{c,t}\Omega''_{\Lambda,t}$$

$$\rho'(t) = \rho''_{c,t}\Omega''(t). \quad (42)$$

where

$$\Omega''(t) = \Omega''_{m,t} + \Omega''_{r,t} + \Omega''_{\Lambda,t}. \quad (43)$$

From Equation (19), (30), (40) and (42) the total mass of the universe within the horizon volume in closed cosmic model at time t is

$$M_h(t) = V_h(t)\rho'(t). \quad (44)$$

The masses of matter, radiation and dark energy within the horizon volume of the universe in closed cosmic model at time t are respectively

$$M_{m,t} = M_h(t)\frac{\Omega''_{m,t}}{\Omega''(t)}. \quad (45)$$

$$M_{r,t} = M_h(t)\frac{\Omega''_{r,t}}{\Omega''(t)}. \quad (46)$$

$$M_{\Lambda,t} = M_h(t)\frac{\Omega''_{\Lambda,t}}{\Omega''(t)}. \quad (47)$$

The time interval between two instants with scale factors a_1, a_2 during the universe expansion is given by Equation (16) in [4] [5] as

$$\Delta t = \frac{1}{H_o}\int_{a_1}^{a_2}\left[1 - \Omega_{\Lambda,o}\left(1 - a^2\right) + \Omega_{m,o}\left(\frac{1}{a} - 1\right) + \Omega_{r,o}\left(\frac{1}{a^2} - 1\right)\right]^{-\frac{1}{2}} da. \quad (48)$$

However, during the universe contraction if $a_2 < a_1$ then modulus of the right hand side of (50) should be taken.

3. ALGORITHM

In determination of the distributions of $d_h(t), V_h(t), M_h(t), M_{m,t}, M_{r,t}$ and $M_{\Lambda,t}$ we use the following steps:

(1) Stage of the universe expansion.

a) Set $t=0, d_h=0$ and insert the value of $a_{max}=0.093188, J=1000$ for t ≤ 0.5 Gyr, $a_{max}=2.3755873$, $J=1600$ for $0.5<t\leq t_{me}$.

b) Calculate $DA=\frac{a_{max}}{DBLE(J)}$.

c) Start general DO loop $I=2$, J which includes the following sub steps:

d) $a_1=DA(I-1), a_2=DA\ I.$

e) Compute new value of cosmic time t numerically using (48), where $t=t+\Delta t$.

f) Obtain new value of the universe horizon distance d_h numerically using (17-b), where $d_h=d_h+\Delta d_h$.

g) Determine the corresponding values of V_h using (19), (31) and (40), in addition to the values of $H'(t), \rho''_{c,t}, \Omega''_{m,t}, \Omega''_{r,t}, \Omega''_{\Lambda,t}, \Omega''(t), \rho'(t), M_h(t), M_{m,t}, M_{r,t}$ and $M_{\Lambda,t}$ using (7), (6), (4), (5), (3), (43), (42), (44), (45), (46) and (47) respectively.

h) Continue the general DO loop.

(2) Stage of the universe contraction

a) Set $t=t_{me}=26.8125327$ Gyr, $d_h=d_h(t_{me})=18.619224$ Gpc, and insert the values of $a_{max}=2.3755873, J1=1600$ for $t_{me}<t\leq t_*, t_*=t_{bc}-0.5$ Gyr, $a_{max}=0.093539, J1=90$ for $t>t_*$.

b) Evaluate $DA=\frac{a_{max}}{DBLE(J1)}$.

c) Start general DO loop $I=1$, $J1$ which includes the following sub steps:

d) Set $J2=J1-I+1, a_1=DA(J2-1), a_2=DA\ J2.$

e) Obtain new value of cosmic time t numerically using (48), where $t=t+\Delta t$.

f) Compute new value of the universe horizon distance d_h numerically using (17-b), where $d_h=d_h-\Delta d_h$.

Sub steps (g) and (h) are similar to (g), (h) mentioned above in the stage of the universe expansion.

4. RESULTS AND DISCUSSION

The distribution of the universe horizon distance in the closed cosmic model until $t = 0.5$ Gyr is shown in Figure 1(a). The distribution increases quite slowly up to $t = 5.8780$ Myr, then the distribution starts raising rapidly. However, the distribution of the universe horizon distance in the range $t = 0.5$ Gyr - t_{me} increases very fast until about $t = 5.7237$ Gyr. Afterwards, it raises gradually as indicated in Figure 1(b). Furthermore, the distribution of the universe horizon distance in the range $t = t_{me} - t_*$, where $t_* = t_{bc} - 0.5$ Gyr, decreases quite slowly up to $t = 46.5790$ Gyr, hence it decreases relatively fast as presented in Figure 1(c). Nevertheless, the distribution of the universe horizon distance in the range $t = t_* - t_n$ decreases slowly until $t = 53.4733$ Gyr, $t_n = t_{bc} - 0.0005$ Gyr, then it starts reduction sharply towards $t = t_{bc}$ as displayed in Figure 1(d).

The distribution of the universe horizon volume in the closed cosmic model up to $t = 0.5$ Gyr is shown in Figure 2(a). The distribution increases very slowly until $t = 18.4785$ Myr. Afterwards, the distribution starts raising appreciably. During this cosmic time range the space of universe is flat. The distribution of the universe horizon volume continues raising up to $t = 14.8$ Gyr, hence the distribution suddenly decreases until $t = 16.6667$ Gyr, then it increases gradually up to $t = t_{me}$ as seen is Figure 2(b). The sharp decrease of the distribution in the range $14.8 < t < 16.6667$ Gyr because the space of the universe changes from flat to curved then closed in the interval $15.1261 \text{ Gyr} \leq t \leq t_{me}$.

The distribution of the universe horizon volume in the range $t = t_{me} - t_*$ is disclosed in Figure 2(c). The distribution decreases until $t = 37.6974$ Gyr, hence it shows abrupt raising up to $t = 40.6579$ Gyr, hence it reduces again until $t = t_*$. The abrupt increase of the distribution in the range $37.6974 < t < 40.6579$ Gyr is due to the fact that the universe space changes from closed then curved to flat in the interval $39.3822 \leq t \leq 40.7521$ Gyr. The distribution of the universe horizon volume in the range $t = t_* - t_n$ is presented in Figure 2(d), which is reducing gradually towards $t = t_{bc}$.

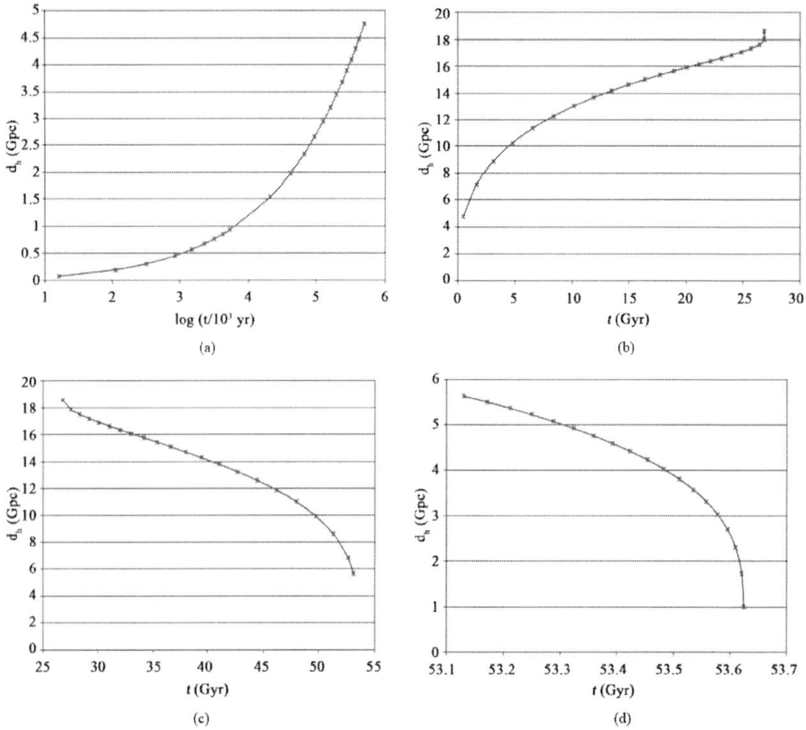

Figure 1. (a) The distribution of the universe horizon distance in the closed cosmic model up to t = 0.5 Gyr; (b) The distribution of the universe horizon distance in the closed cosmic model in the range t = 0.5 Gyr - t_{me}; (c) The distribution of the universe horizon distance in the closed cosmic model in the range t = t_{me} - t_*; (d) The distribution of the universe horizon distance in the closed cosmic model in the range t = t_* - t_n.

The distribution of mass and energy within the horizon volume of the universe in the closed cosmic model until $t = 0.5\text{ Gyr}$ is shown in Figure 3(a). The distributions of radiation and total mass decrease very slowly up to and $t = 162061\text{ yr}$ respectively, then they reduce more rapidly. However, the distribution of matter increases so slowly until $t = 181161\text{ yr}$ and it decreases rapidly afterwards. This distribution intersects with the distribution of radiation at $t = 53221.5$ yr and coincides with the distribution of total mass at $t = 3.2819$ Myr. The distribution of dark energy increases continuously towards $t = 0.5$ Gyr. The distribution of mass and energy within the horizon volume of the universe in the closed model in the range $t = 0.5$ Gyr - t_{me} is illustrated in Figure 3(b). It is noticeable that the distribution of radiation intersects the distribution of dark energy at $t = 0.6618$ Gyr, hence it reduces very fast up to $t = 3.0147$ Gyr.

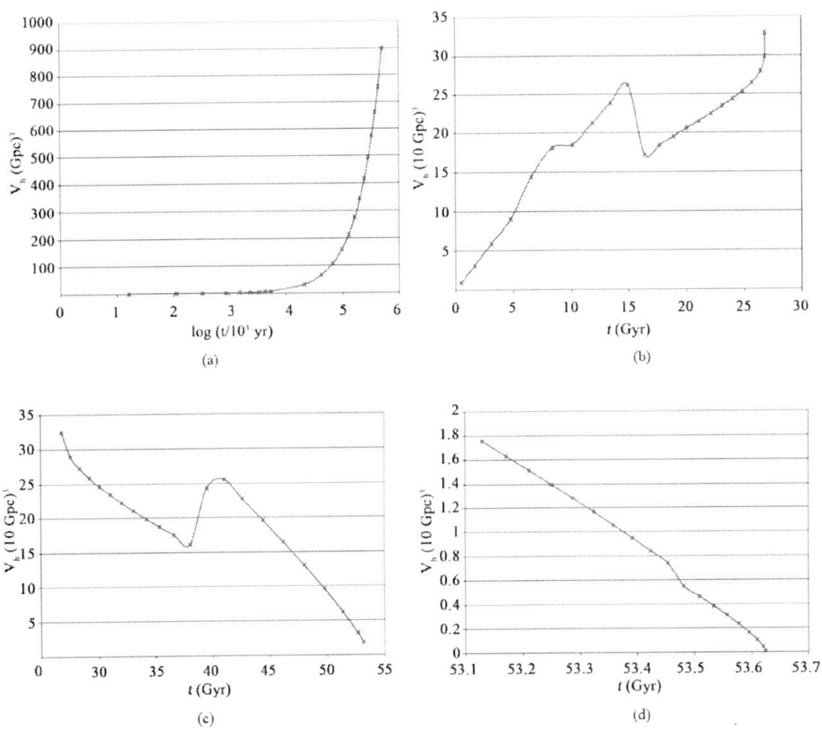

Figure 2. (a) The distribution of the universe horizon volume in the closed cosmic model up to t = 0.5 Gyr; (b) The distribution of the universe horizon volume in the closed cosmic model in the range t = 0.5 Gyr - t_{me}; (c) The distribution of the universe horizon volume in the closed cosmic model in the range t = t_{me} - t_*; (d) The distribution of the universe horizon volume in the closed cosmic model in the range t = t_* - t_n.

Afterwards, it raises gradually until $t = 14.9265$ Gyr, then it decreases abruptly up to $t = 16.3971$ Gyr, where it starts increasing again towards t_{me}. The distribution of matter decreases up to $t = 22.2794$ Gyr, hence it raises very slowly towards t_{me}. This distribution coincides with the distribution of total mass up to $t = 1.9853$ Gyr, and intersects with the dark energy distribution at $t = 10.1007$ Gyr. The matter distribution indicates obvious decrease in the range $t = 14.9265 - 16.3971$ Gyr. The distribution of dark matter increases until $t = 14.9265$ Gyr, hence it also shows marked decrease up to $t = 16.3971$ Gyr. Afterwards, it raises slightly towards t_{me}. The total mass distribution reduces up to $t = 14.9265$ Gyr, then it decreases abruptly until $t = 16.3971$ Gyr. Hence, this distribution increases gradually towards t_{me}. The noticeable decrease of all four distributions in the range about $14.9225 < t < 16.3971$ Gyr is due to the change of the universe space from flat to curved then closed in the range $15.1261 \le t \le t_{me}$.

The distribution of mass and energy within the horizon volume of the universe in the closed model in the range $t = t_{me} - t_*$ is illustrated in Figure 3(c). The distribution of radiation reduces very slowly until $t = 37.9839$ Gyr, then it raises suddenly up to $t = 40.4032$ Gyr. Hence, this distribution decreases again until $t = 52.5807$ Gyr where it intersects with the distribution of dark energy, then it starts raising towards t_*. The matter distribution increases so slowly up to $t = 37.9839$ Gyr, hence it exhibits marked raise until $t = 40.4032$ Gyr and intersects with the dark energy distribution at $t = 40.2892$ Gyr. Afterwards, the distribution of matter shows substantial increase towards t_* and coincides with the total mass distribution at $t = 51.2903$ Gyr hence forth. The dark energy distribution reduces gradually until $t = 37.9839$ Gyr where it displays outstanding raise up to $t = 40.4032$ Gyr, hence it decreases again towards t_*. The total mass distribution reduces also gradually until $t = 37.9839$ Gyr, then it exhibits obvious increase up to $t = 40.4032$ Gyr. Afterwards, it raises very fast as the matter distribution. The prominent increase of the four distributions in the interval about $37.9839 < t < 40.4032$ Gyr is owing to the change of the universe space from closed then curved

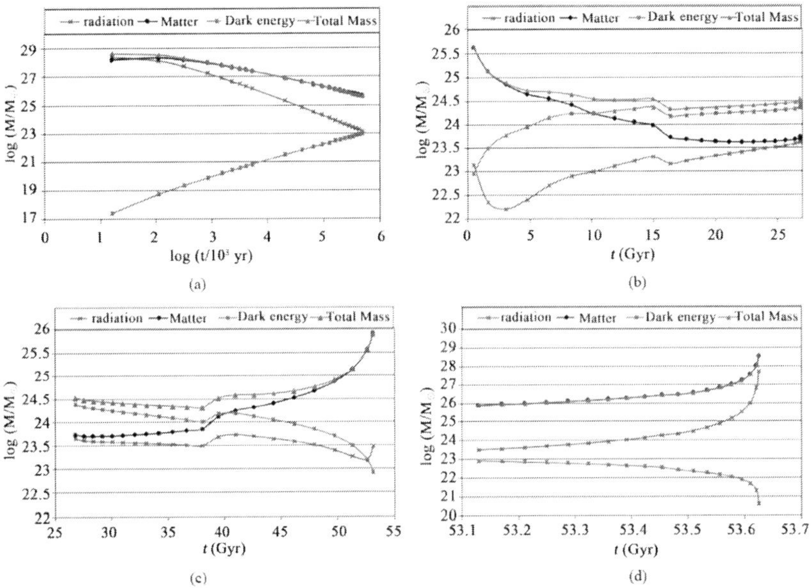

Figure 3. (a) The distribution of mass and energy in the closed cosmic model up to t = 0.5 Gyr; (b) The distribution of mass and energy in the closed cosmic model in the range t = 0.5 Gyr - t_{me}; (c) The distribution of mass and energy in the closed cosmic model in the range t = t_{me} - t_*; (d) The distribution of the mass and energy in the closed cosmic model in the range t = t_* - t_n.

to flat in the range $39.3822 \le t \le 40.7521\,\text{Gyr}$. The distribution of mass and energy within the horizon volume of the universe in the closed model in the range $t = t_* - t_n$ is displayed in Figure 3(d). The radiation distribution increases quite slowly until $t = 53.5032\,\text{Gyr}$, hence it starts raising appreciably fast.

The distribution of both matter and total mass coincide on each other and lie over the radiation distribution. The two distributions increase gradually up to $t = 53.5274\,\text{Gyr}$, then they raise up. However, the dark energy distribution decreases so slowly until $t = 53.5742\,\text{Gyr}$, afterwards it reduces substantially.

Estimations of $d_h(t), V_h(t), M_r(t), M_m(t), M_\wedge(t)$ and the equivalent number of the Coma-like clusters to the mass of matter within the universe horizon volume $N_{COMA}(t)$ in the closed cosmic model at special times are presented in

Table 1. It is interesting to note that at $t = t_n = t_{bc} - 0.0005\ \text{Gyr}$, the horizon volume of the universe $V_h(t_n) = 0.0003(10\ \text{Gpc})^3$.

Since the radius of the Coma cluster is $r_{COMA} = 3.6\ \text{Mpc}$ [5], then $V_h(t_n) = 1.5738\times10^6 V_{COMA}$, where V_{COMA} is the Coma cluster volume. However, the mass of matter within the horizon volume of the universe at $t = t_n$ is $M_m(t_n) = 1.7455\times10^{13} M_{COMA}$. This indicates very clearly that the intercluster medium will disappear at $t = t_n$ and galaxy clusters will interfere with each other. Furthermore, the radius of the Milky Way galaxy is $r_{Mw} = 50\ \text{Kpc}$ [6] . Thus, $V_h(t_n) = 5.8742\times10^{11} V_{MW}$, where V_{MW} is the Milky Way galaxy volume. Nevertheless, it is found that $M_m(t_n) = 4.3638\times10^{16} M_{MW}$. Therefore, not only the intergalactic spaces will vanish at $t = t_n$, but also galaxies will collide and merge with each other to form extremely dense and close cosmological bodies. These very dense bodies will undergo further successive collisions and mergers under the action of central gravity, where the interstellar medium will vanish and the universe would develop to big crunch at $t_{bc} = 53.6251\ \text{Gyr}$.

Table 1. Estimations of the horizon distance, horizon volume, mass of radiation, mass of matter, mass of dark energy and the equivalent number of the Coma-like clusters to the mass of matter within the universe horizon volume in the closed cosmic model at special times.

t	$d_h(t)$ Gpc	$V_h(t)$	$\log(M_r/M_\odot)$	$\log(M_m/M_\odot)$	$\log(M_\Lambda/M_\odot)$	$N_{COMA}(t)$
t_{rm} (53221.5) yr	0.1356	$0.0209u_1$	28.4321	28.4321	18.3292	$1.352u_3$
$t_{m\Lambda 1}$ (10.1007) Gyr	12.9717	$18.2854u_2$	22.9777	24.2252	24.2252	$8.3970u_4$
t_o (13.7) Gyr	14.4420	$25.2348u_2$	23.2298	24.0515	24.3294	$5.6291u_4$
$t_{m\Lambda}$ (26.812.53) Gyr	18.6192	$32.8742u_2$	23.6535	23.7347	24.3906	$2.7141u_4$
$t_{m\Lambda 2}$ (40.2892) Gyr	14.0058	$23.0168u_2$	23.6758	24.1474	24.1474	$7.0198u_4$
t_n (53.6246) Gyr	0.9972	$0.0003u_2$	27.7134	28.5430	20.5951	$1.7455u_3$

where $t_n = t_{bc} - 0\ 0005\ \text{Gyr}$, $u_1 = (\text{Gpc})^3$, $u_2 = (10\ \text{Gpc})^3$, $u_3 = 10^{13}$ and $u_4 = 10^5$.

It is also interesting to note from Table 1 that the horizon distance of the universe at maximum expansion is $d_h(t_{me}) = 18.6192$ Gpc. This horizon distance is in very good agreement with the maximum value of the universe horizon distance in the observed general cosmic model A, $d_h(t'_n) = 19.0103$ Gpc, where $t'_n = 124$ Gyr.

where $t_n = t_{bc} - 0.0005$ Gyr, u$_1$ = (Gpc)^3, u$_2$ = (10 Gpc)^3, $u_3 = 10^{13}$ and $u_4 = 10^8$.

The value of $d_h(t_{me})$ is also in high agreement with the values of $d_h(t'_n)$ in the other four general cosmic models as shown from Table 1 in [7].

5. CONCLUSIONS

In this paper we have investigated the distributions of the universe horizon distance and the universe horizon volume in the closed cosmic model. It is found that the universe horizon distance distribution increases constantly for $t < t_{me}$ and decreases for $t > t_{me}$. However, the universe horizon volume distribution shows sudden reduction in the range $t = 0.5$ Gyr - t_{me} due to the change of the universe space from flat to curved then closed in the interval $15.1261 \text{ Gyr} \le t \le t_{me}$. On the other hand, this distribution exhibits abrupt raise in the range $t = t_{me}$ - t_* because of the change of the universe space from closed then curved to flat in the interval $39.3822 \le t \le 40.7521$ Gyr.

Distributions of mass of radiation, matter and dark energy within the horizon volume of the universe were also investigated in the closed cosmic model. These distributions reveal similar noticeable changes as the universe horizon volume distribution for the same reasons. The mass of radiation dominates up to $t = 53221.5$ yr, then the mass of matter becomes larger. Afterwards, both distributions of radiation and matter decrease while the distribution of dark energy rises until $t = 10.1007$ Gyr, where the mass of dark energy prevails up to $t = t_{me}$. Hence, the distribution of dark energy reduces until $t = 40.2892$ Gyr where the mass of matter becomes prominent again. At $t = 53.6246$ Gyr the masses of both matter and radiation become appreciably high such that the intercluster space will vanish and clusters of galaxies will interfere with each other. Furthermore, not only the intergalactic medium will disappear, but also galaxies will collide and merge with each other to form extremely dense and close cosmological bodies. These very dense bodies will undergo further successive collisions and mergers under the action of central gravity, where the interstellar medium will vanish and the universe would develop to big crunch at $t_{bc} = 53.6251\,\text{Gyr}$.

REFERENCES

1. Bukhari, F.A. (2013) A Closed Model of the Universe. International Journal of Astronomy and Astrophysics, 3, 189-198.
2. Bukhari, F.A. (2013) Cosmological Distances in Closed Model of the Universe. International Journal of Astronomy and Astrophysics, 3, 199-203.
3. Bukhari, F.A. (2013) Cosmological Distances in Five General Cosmic Models. International Journal of Astronomy and Astrophysics, 3, 183-188.
4. Bukhari, F.A. (2013) Five General Cosmic Models. Journal of King Abdulaziz University: Science, 25.
5. Ryden, B. (2003) Introduction to Cosmology. Addison & Wesley, Boston.
6. Schneider, P. (2010) Extragalactic Astronomy and Cosmology. Springer, New York.
7. Bukhari, F.A. (2015) Distribution of Mass and Energy in Five General Cosmic Models. International Journal of Astronomy and Astrophysics, 5, 20-27.

CHAPTER 7

Ground State Nucleon, Dark Energy and Dark Matter

Gintautas P. Kamuntavičius

Vytautas Magnus University, Kaunas, Lithuania

ABSTRACT

The recently introduced model of the nucleon as a system of three point particles predicts all characteristics of the proton and neutron with experimental precision only at condition that nucleon is the loosely bound state in system of deep potential wells. The model's Hamiltonian contains ground state with the same parity and spin as the nucleon but with other different characteristics. Existence of this ground state nucleon means that the visible Universe is composed of excited matter. The direct transition between excited and ground states is strictly forbidden, however, stimulated emission can ignite such process. Most likely, corresponding conditions realize at supernova explosion. It is shown that presence of this matter, composed of ground state nucleons, in Universe gives the chance for consistent explanation of dark matter and dark energy phenomena.

Keywords: Solutions of Wave Equations (Bound States), Potential Models, Proton and Neutron, Dark Matter, Dark Energy

1. INTRODUCTION

The present understanding of fundamental structure of matter is the Standard Model of Elementary Particles (SM), based on assumption that the structure of hadrons is explainable in terms of quark configurations. However, hadrons, excluding only stable proton and neutron, referred as particles, are really extremely short lived resonances, and each of them can be produced over a broad range of excitation energies, and the energies overlap for different resonances. Moreover, the quark model predicts far more resonances than having yet been observed and has a disturbingly large number of parameters whose origins and numerical values are not explained. The SM considers the proton and neutron as the nucleon-the lowest energy state of a family of hadrons that can be constructed from up (u) and down (d) valence quarks. Despite of surprisingly successful SM description of resonances, more than a half of century lasting attempts to apply it for proton and neutron are not so impressive.

On the other side, namely the proton and neutron with slightly modified intrinsic structure compose the atomic nuclei, accounting for nearly all the visible mass in the Universe (not less than 0.999), hence the incompletely understood structure of these particles causes problems with atomic nuclei, but especially with description of nuclear matter of such objects as neutron stars and black holes. The outstanding theoretical challenges in nuclear astrophysics, such as collapse of evolved star to structures, which densities are several times larger than nuclear matter densities, the origin of pulsars beams, structure of black holes, the origin of dark matter and dark energy cannot be explained in terms of recent SM.

Therefore, the possibilities of SM modification for visible matter description are already investigated, and one needs more successful model of the nucleon. The article is namely devoted for brief introduction of recently obtained problem solution and presentation of new point of view to structure and evolution of the Universe, following from this innovation.

The paper is organized as follows. In Section 2, we perform the analysis of origins of SM and suggest the modifications of model, necessary for description of the nucleon. In Section 3, we present the main results of the recently introduced model of nucleon as a system of three point particles, which dynamics is governed by Schrödinger equation able to describe all known characteristics of the proton and neutron with experimental precision. Section 4 is devoted for investigation of ground states of introduced Schrödinger's equations, which excited states describe the proton and neutron. In Section 5, we comment different exotic phenomena, caused by ground state proton and ground state neutron presence in the Universe as particles, composing dark matter and being able to predict dark energy phenomenon, accelerating the expansion of the Universe. Finally, we conclude in Section 6.

2. SM PROBLEMS WITH THE NUCLEON

The earliest electron scattering experiments have shown that electric charge and magnetism distributions correspond the point-like electrically charged constituents present in the proton. The proton behaved as if it was made out of point like objects of spin one-half. The further experiments soon indicated that constituents of the proton had a positive parity and the baryon number equal 1/3. These were the first definitive evidences for the quark substructure of the proton and neutron. Moreover, the point-like behavior of quarks at short distance and the strong confining force at large distance, known as asymptotic freedom [1] was the one of most important implications of quark model.

Nearly all of the known hadrons seem to be explainable in terms of the simplest quark configurations with all quarks in ground states, i.e. states, corresponding angular momentum that equal zero. In this approximation, some of measured properties of hadrons can be reproduced surprisingly well but the challenge to the quark model lay in the spin and statistics puzzle. If the baryon J = 1 octet and J = 3 decuplet are taken to be composites of three quarks, all in relative s-states, then the corresponding wave functions appear to be symmetric in spin-flavor, in conflict with the Pauli exclusion principle. As explicit examples consider the Ω^-, formed of three identical strange (s) quarks, sss, or the Δ^{++}, an isospin 3 state made of three up quarks, uuu.

To reconcile the success of the quark model with the requirement that fermion wave functions are antisymmetric, one of the possibilities is to hypothesize that each quark flavor comes in three distinguishable species, which are labeled by the primary colors red, green, and blue. By definition, all baryons are color singlets, which wave functions may then be antisymmetrized in color. Namely this method of wave function antisymmetrization gives not far of experimental value the relation of magnetic momenta of proton and neutron, equal -3/2 [2] (experimental value is −1.459898). This idea soon was realized in form of Quantum chromodynamics (QCD), called as theory of strong interactions among quarks, derived from the color gauge symmetry $SU(3)_c$ with eight gluons, defined as carriers of this interaction. However, after more than half of a century of intensive investigations QCD cannot even explain the structure of spin of nucleon [2] - [4] , not speaking about the entire set of characteristics of the proton and neutron, including masses, magnetic momenta, charge radii, form factors and so on measured with high precision [5] . The complexity of problem and characteristic results of this kind can be found in recent publications such as [6] [7] and many others.

As it is shown below, the main problem of SM concerning the proton and neutron is namely the definition of nucleon as ground state of hadrons family. The rich set of known and remarkably different characteristics of the proton and neutron requires excited configurations presence in nucleon wave function expansion.

3. THE SCHRÖDINGER'S NUCLEON

Let us consider nucleon as an usual quantum system, i.e. system, composed of point-like particles (PP), corresponding recommendations for quark structure of neutron and proton, which dynamics is governed by Schrödinger equation. These particles cannot be called as quarks because widely accepted definition of quarks in SM and QCD is based on assumption that they are particles with intrinsic characteristics such as flavor and color. For the successful description of nucleon only two of them: u and d, considered as different structureless (point-like) particles are necessary. The sum of masses of constituents of such system has to be larger than the mass of the entire system. However, for nucleon the difference of these masses, defined usually as binding energy, is senseless due to baryon number conservation, forbidding the split down of the nucleon into composing particles. The Galilei invariance of formalism is crucial; hence elimination of center of mass kinetic energy operator from Hamiltonian is necessary. After this operation the Hamiltonian contains only two Jacobian variables, hence the same dependence has to be in wave functions. All operators of observables have to be rewritten in these variables [8] [9] . Obviously, the potential of point particles interactions contains Coulomb part and some simple potential, keeping the nucleon as bound system, are also necessary. The nucleon does not have bound excited states; hence the potentials of PP interactions have to be of finite depth. The best form of these potentials for translation-invariant Hamiltonian is three-dimensional harmonic-oscillator, together with kinetic energy operator invariant in respect of SU(3) transformations in corresponding space.

Three different pairs of PP are possible (uu, ud and dd), hence one needs at least six parameters (depth and range for every one) for the definition of these potentials. Together with two undefined masses of PP for fitting eight free parameters of nucleon's Hamiltonian appear. However, only six characteristics of the proton and neutron (masses, magnetic momenta and charge radii) are known with high enough precision. Thus, two parameters have to be eliminated from the list. In [9] this problem is solved by defining equal ranges of all three potentials.

More precisely, the potentials have the following form:

$$V_{\alpha\beta}\left(\left|\mathbf{r}_j-\mathbf{r}_k\right|\right)=-V_{\alpha\beta}+\frac{1}{2}k_{\alpha\beta}\left(\mathbf{r}_j-\mathbf{r}_k\right)^2, \tag{3.1}$$

where αβ = uu, ud and dd, and $V_{\alpha\beta}>0$ is depth of a well, while $k_{\alpha\beta}>0$ is the spring constant. The condition for these potentials is that they cannot gain positive values. When $\left(\mathbf{r}_j-\mathbf{r}_k\right)^2=2V_{\alpha\beta}/k_{\alpha\beta}$, the potential equals zero and remains equal zero at larger distance. As mentioned, this condition is necessary to ensure the absence of bound excited states of nucleon. Namely this relation defines the range of potential, taking the same value for all pairs of PP. Therefore, only depths of these potentials are different.

After this modification the basic Hamiltonian (without Coulomb potential, considered as perturbation) of model has the form, analogous for both, proton and neutron (σ = 1 and 2 respectively):

$$H_{\sigma}^{(0)}\left(\xi_1,\xi_2\right)=H_{\sigma 1}\left(\xi_1\right)+H_{\sigma 2}\left(\xi_2\right). \tag{3.2}$$

Here

$$H_{\sigma\rho}\left(\xi_\rho\right)=\begin{cases}-\dfrac{(\hbar c)^2}{2v_{\sigma\rho}c^2}\Delta_{(\rho)}+\dfrac{1}{2}\varkappa_{\sigma\rho}\left(\xi_\rho^2-\xi_{\sigma\rho}^2\right), & \text{if } \xi_\rho<\xi_{\sigma\rho}\\ -\dfrac{(\hbar c)^2}{2v_{\sigma\rho}c^2}\Delta_{(\rho)}, & \text{if } \xi_\rho\geq\xi_{\sigma\rho}\end{cases} \tag{3.3}$$

Reduced masses $v_{\sigma\rho}$ and $\varkappa_{\sigma\rho}$, $\xi_{\sigma\rho}$, all expressible as functions of six introduced parameters-two PP masses, three depths and one range of potentials of PP interactions, are different for proton and neutron.

The eigenfunction of Hamiltonian (3.3) equals the product of spherical harmonic and radial function. Radial function, corresponding the eigenvalue $E_{n\lambda}^{(\sigma\rho)}$ (here λ is angular momentum quantum number, while n = 0, 1, 2 ... equals the number of eigenfunction nodes) is proportional the degenerate hypergeometric function in the area where a potential does not equal zero. In the area where a potential equals zero, radial function is spherical Hankel function of imaginary argument [9] .

The eigenfunction of intrinsic Hamiltonian (3.2) equals the product of eigenfunctions of basic Hamiltonians, corresponding to different Jacobi coordinates, with bound orbital angular and spin momenta. Angular momentum of nucleon is obtained by binding Jacobi angular momenta and spin momenta of PPs, thus spatial and spin part of the wave function has the expression:

$$\left|\left(\left(\lambda_1,s_1\right)j_1,\left(\lambda_2,\left(s_2,s_3\right)s_{23}\right)j_{23}\right)j=1/2,m\right\rangle. \tag{3.4}$$

Here the parentheses indicate the operation of momenta binding; λ_1 is angular momentum of the first Jacobi coordinate ξ_1; s_1 = 1/2 is spin momentum of the first PP; λ_2 indicates the angular momentum of these cond Jacobi coordinate ξ_2, for which, according to Jacobi tree structure [8] , the second and third PPs spinmomenta $s_2 = 1/2$ and $s_3 = 1/2$ need to be set. The j_1 , j_{23} are total momenta of respective Jacobian subsystems. Their sum j = 1/2 equals the nucleon momentum, i.e. its spin.

Obvious restrictions for quantum numbers, caused by nucleon characteristics, appear. Firstly, the total momentum of nucleon equals 1/2, this has already been expressed in the wave-function. Secondly, orbital parity of a nucleon is positive, therefore $\lambda_1+\lambda_2$ has to be an even number. Finally, the second and third particles are identical; therefore, the corresponding part of the wave-function needs to be antisymmetric with regard to spin and orbital degrees of freedom permutations. So, λ_2+s_{23} also has to be an even number.

These wave functions were applied for different characteristics of nucleon calculation and expression in terms of parameters of Hamiltonian. The fitting has shown that for the neutron and proton description applying the eigenfunctions of corresponding Hamiltonians necessary is superposition of two excited configurations with Jacobi angular momenta $\lambda_1=\lambda_2=1$ and nodeless radial functions:

$$\frac{1}{\sqrt{2}}\left[\left|\left(\left(1,\frac{1}{2}\right)\frac{1}{2},\left(1,\left(\frac{1}{2},\frac{1}{2}\right)1\right)1\right)\frac{1}{2},m\right\rangle+\left|\left(\left(1,\frac{1}{2}\right)\frac{3}{2},\left(1,\left(\frac{1}{2},\frac{1}{2}\right)1\right)2\right)\frac{1}{2},m\right\rangle\right]. \tag{3.5}$$

The set of parameters of model, predicting the known data of nucleon with experimental precision, is published in [9] . The other characteristics of neutron and proton, such as elastic form factors, different rms radii and Zemach momenta, are present in [10] and [11] . All obtained results do not contradict the known information about the structure of the proton and neutron.

4. GROUND STATE NUCLEON

Therefore, successful model of the proton (p) and the neutron (n) predicts they are loosely bound states in system of deep potential wells. As mentioned, the intrinsic Hamiltonians of model are free of bound excited states of p as well as n; however, they have ground states of this kind. As it will be shown later, namely presence of these states, ground state proton $\overline{p}$ and ground state neutron $\overline{n}$ can explain obvious but not completely clear connection between nuclear properties and the nature of the Universe and give interesting information about new sort of matter and some phenomena, which origin is not yet understood applying traditional approaches. For the sake of simplicity let us call these particles as dark proton and dark neutron.

The angular and spin part of the wave function of dark nucleon has the same structure as the wave function of nucleon, present in Equation (3.4) with both angular momenta $\lambda_1 = \lambda_2 = 0$, i.e.:

$$\left|\left(\left(0,\frac{1}{2}\right)\frac{1}{2},\left(0,\left(\frac{1}{2},\frac{1}{2}\right)0\right)0\right)j=\frac{1}{2},m\right\rangle. \tag{4.1}$$

The corresponding radial wave functions again are nodeless, like wave functions of ordinary nucleon.

The energies and other main characteristics of these particles are present in Table 1.

Masses of $\overline{p}$ and $\overline{n}$ are the most interesting result. They are negative due to character of potential wells, necessary to describe the proton and neutron. Moreover, absolute values of these masses coincide with good precision with masses of proton and neutron. However, these particles cannot be identified as Dirack's antiparticles because they correspond the solutions of the same Schrödinger equation with absent mechanism of charge inversion.

As it will be shown later, the best consideration of obtained results is as follows. Let us define the potentials of PP interaction, following from nucleon parameters fitting, as a short range gravitation.

Really, the putative graviton has not yet been found, therefore this is only choice for point-like particles interaction, not stranger in comparison with long range gravitation, tractable as space curvature. If so, the next conclusion is obvious: one has to define the mass, corresponding negative energy of dark particles, as gravitational mass according to definition

$$m_g = E/c^2 . \tag{4.2}$$

Let the interaction of masses is defined by well-known formula with gravitational constant, modified only by signs of interacting masses, resulting in repulsion of masses, composed of ordinary and dark nucleons.

Table 1. Main characteristics of dark proton and dark neutron.

	$\overline{p}$	$\overline{n}$
Mass (MeV)	−938.2720454	−939.5653813
Magnetic momentum(μ_N)	−0.9248409550	1.946549588
Charge radius (fm)	0.4036284062	
Mean-square charge radius (fm^2)		−0.02474114545
Mass radius (fm)	0.3584056883	0.3570377735

For ordinary nucleons the gravitational mass is positive and with high precision equals the inertial mass. Therefore, the inertial mass, by definition taking only positive values, has to be defined as

$$m_i = |m_g| . \tag{4.3}$$

Let us study now, what interesting conclusions follow from this definition.

5. EVOLUTION OF A STAR REMNANT

Obviously, both new particles are very compact and $\overline{n}$ is the most stable system, composed of three PPs. Once created, the dark proton after a weak process

$$\overline{p} = \overline{n} + e^+ + \nu \tag{5.1}$$

appears as dark neutron $\overline{n}$.

Therefore, dark neutron is the best candidate for particles, composing the extremely dense matter, invisible in usual regions of electromagnetic radiation. Due to compactness of these particles (mass radius of neutron equals 0.768 fm [10] , while mass radius of $\overline{n}$ is more than two times smaller and equals 0.357 fm) such matter can reach densities, by order larger than nuclear matter density.

The interaction of these particles is very weak, governed only by gravitation. The electromagnetic interaction is absent due to neutrality of $\overline{n}$. The strong interaction, which carriers are mainly mesons, it is quark and antiquark pairs, cannot play significant role due to extreme binding of PPs, composing the dark nucleon. For ordinary nucleons this interaction is really strong due to loosely bound PPs; therefore, our model predicts that dark neutrons are extremely stable and weakly interacting particles.

The main question is this: How do these particles appear in the Universe?

The reactions, responsible for creation of dark nucleons, are

$$\begin{cases} n \to \overline{n} + \gamma(1.879 \text{ GeV}) \\ p \to \overline{p} + \gamma(1.877 \text{ GeV}) \end{cases} \tag{5.2}$$

The following weak process (5.1) converts dark protons to dark neutrons. The proton and the corresponding dark proton, exactly as neutron and dark neutron, are eigenstates of the same Hamiltonian, thus the interaction with electromagnetic field can cause transitions between these states. The problem is that any transitions of this kind are strictly forbidden. Electric dipole transition operator expectation value equals zero due to positive parity of both wave functions, (3.5) and (4.1). Magnetic dipole transition operator, written in intrinsic variables [9] , expectation value vanishes due to orthogonality of both Jacobi subsystems of mentioned wave functions. The operators of higher multiplicity expectation values equal zero due to Wigner-Eckart theorem because total momenta of both wave functions are equal 1/2.

Only possibility for this transition to happen is finite lifetime of nucleon in respect of such transformations and stimulated emission, developing at some conditions. These conditions realize, when massive star-supernova progenitor-exhausts the nuclear fuel and gravitation starts iron core conversion to easier structures, such as alpha particles and neutrons. Density of core grows and onion-skin structure of progressively lighter elements from the inside out can serve as a mirror for stimulated emission of gamma quanta and produce in center of remnant the clamp of negative gravitational mass. The energy, generated during this process, is enough for observable extremely fast motion outside the outer shells of progenitor, composed of visible matter and at the same time the collapse of core due to gravitational attraction of dark nucleons. This scenario is best possible for the core-col- lapse supernova mechanism explanation because "...the current status of the theory for the mechanism and the systematics of core-collapse explosions is ambiguous, if not confusing" [12] .

Further development of remnant is dependent mainly on progenitor mass. If this mass is large enough for all core conversion to system, composed of negative gravitational mass, and happens the event, called as black hole creation after hypernova explosion. Therefore, the massive progenitors generate massive cores, composed of dark neutrons. At smaller mass of progenitor core, created after supernova explosion, together with negative gravitational mass contains also some amount of ordinary mass. Due to angular momentum conservation such remnant, called as neutron star, fastens rotation and takes prolate form like disc or more complex, not axially symmetric, shape. Namely along these largest dimensions the best conditions for stimulated emission appear. The crystalline crust of neutron star can serve as semitransparent mirror and the process like the one in laser starts. This scenario gives a perfect explanation of observable orientation of rotation and emission axes of pulsars (for axially symmetric pulsar the emission axis is perpendicular to the rotation axis) and coherence of corresponding radiation. Young pulsars with significant amounts of ordinary matter produce powerful beam of γ-quanta with defined above energies. Later on, when the intensity of beam reduces, the crystalline crust of pulsar can redshift the penetrating radiation, converting it for old pulsars to radio waves.

The further evolution of pulsar can develop in a following way. The born dark neutrons accumulate in central region of ordinary matter disc. This region is stable in respect of negative mass outflow to space until the amount of negative mass is small enough. Growing amounts of negative gravitational mass in these regions can cause different phenomena, happening during star remnant evolution. These are the powerful shock waves that travel outward through the mantle of the star remnant and ejections of visible matter to space after negative mass amount riches critical value.

Therefore, after supernova explosion, excluding complete disassembling of remnant at minimal possible mass of progenitor, sooner or later the core appears as structure, composed of dark neutrons. The mass of these structures can grow due to attraction and collapse of remnants with negative gravitational mass. This scenario can explain masses of merging black holes, producing recently observed gravitational waves [13] , about twice as massive as they should be according to the current theories of how black hole form a star. For the first time in history, scientists have physical proof that pairs of black holes will sometimes collide and mush together to form a single, bigger black hole. Moreover, all of the observed black hole systems are found in low stellar density galactic fields and dynamically formed merging binary black holes are ejected from the host cluster before merger. These observations are in full accordance with our explanation, stating that no ordinary, i.e. positive gravitational masses can be present in closest vicinity of these black holes due to gravitational repulsion.

6. DARK ENERGY AND DARK MATTER

Finally, when star remnant produces big enough amount of dark neutrons, the observable matter and matter with negative gravitational mass will separate. The ordinary matter spreads out in galaxy forming clouds of gas and dust for following generation of stars. The clumps of negative mass due to gravitational repulsion move out of galactic discs to extragalactic space. Namely this scenario can explain the stability of galactic discs and not growing large galaxy mass due to the absorption of dwarf galaxies. From the other side, the appearance of these clumps before escape from disc interacts with visible matter, present in the disc; hence regions of galaxy with older stars are spread in thicker layer and have larger velocity spreads than younger stars do. This distribution of stars (thick disk) cannot be explained as result of ordinary gravitational attraction, see [14] . Obviously, these clumps move to extragalactic space, compacting nearby dwarf galaxies and preventing them from tidal destruction due to big galaxy gravitation. The attraction of these masses creates bigger and bigger structures of this kind and presence of these objects in space causes expansion of the Universe. Due to the absence of mechanism for disappearance of this matter, the growing amount of negative gravitational mass in the Universe has caused the originally slowing expansion during the first half of the Universe life to reaccelerate [15] . The accelerated expansion of the Universe has been the biggest recent challenge, not consistent with dark matter phenomenon and not explainable yet in another way than given above. The best characteristic of existing situation is given in [16] : "the label "dark energy" encapsulates our helplessness".

Concerning the development of Galactic Center supermassive black hole; according to our suggestion, it is composed of dark neutrons and looks like the black hole, defined according to a standard consideration. This object of negative gravitational mass, situated in the center of galactic bulge, absorbs clumps of negative mass matter, produced by pulsars. This structure grows until mass reaches value, destroying the stability of Galactic center. The final stage of such scenario is escape of black hole from bulge, creating ultrafast outflows of remnants of visible matter, observable as active galactic nuclei and bright regions around its central black hole identified as quasars.

Moreover, our model can explain dark matter consistence with dark energy. According to the traditional approach, for the explanation of measured galactic rotation curves, significantly bigger mass of galaxy is necessary. For our model the key to explain this phenomenon is the evaporation of matter from galaxy disc to halo and later on to extragalactic space. Strange character of visible matter movement in halo can be caused by repulsion of visible matter and present in halo clumps of dark matter, which gravitational mass is negative. On the other side, the infrared observations showed that stellar mass assembly favored more massive systems at earlier epochs [17] . Bright galaxies of the early universe, built from clouds, composed mainly of protons and alphas, were big structures with many gigantic stars, significantly more massive as now. Evolution of these stars resulted in big amounts of negative mass appearance, hence evaporation and lowering the mass and moment of inertia of galaxy. This effect due to angular momentum of galaxy conservation causes observable movement of halo stars faster than following from Kepler's laws. Other effect, explanation of which needs bigger than visible masses is stability of galactic clusters. Again, this can be explained as result of gravitational repulsion of negative masses, moving in outer regions of cluster. Finally, one of first observations, which explanation required that the total mass of the Milky Way and Andromeda galaxies had to be much larger than their stellar mass, is character of movement of these galaxies, currently approaching each other. From our point of view, this movement can be caused by nearly parallel discs of both galaxies, hence the streams of clumps of negative gravitational mass move in the same directions, perpendicular to the galaxies discs and cannot resist this process. By a way, some merging big galaxies, present in Hubble gallery, have parallel discs.

7. CONCLUSIONS

Therefore, a successful model of nucleon has shown that proton and neutron are the systems of point particles, occupying excited states of three-particle Schrodinger equation. The basic of our approach is taken into account Galilei invariance of nucleon Hamiltonian and wave functions. This invariance, fundamental for relativity theory, until the last time due to problems with identical particles wave functions antisymmetrisation and problems with different operators of observables expansion in Jacobi variables, was ignored in quantum mechanics. The second basic feature of our approach is introduction of potentials of PP interaction, equal three dimensional harmonic oscillator ones, so that Hamiltonians dependent on different Jacobi variables, are symmetrical in respect of SU (3) transformations, characteristic for Standard Model dynamics. This symmetry is significant at conditions, characteristic for resonances because at high excitations, i.e. high kinetic energies, the fine distinctions of particle masses and interaction potentials do not play a significant role. The character of applied potentials shows some links between theory of resonances—Standard Model and present model of bound states of hadrons. The third basic feature of our approach is finite depth of all potential wells, necessary to ensure absence of excited bound states of our Hamiltonian because the only stable hadrons are the proton and neutron. As a result, this approach eliminates necessity to mark quarks as colored objects because if excited orbital states are allowed for nucleon, they can also be present in expansions of the wave functions of mentioned resonances, composed of identical quarks, such as Δ^{++} and Ω^{-}.

The excited bound states of our Hamiltonians for the proton and neutron are absent but ground states with quantum numbers (parity and spin), coinciding with ones for proton and neutron are present. Jacobian orbital angular momenta equals zeroes, hence these states are the best for nucleon of SM identification. However, other characteristics of these states do not correspond to any known particle. Moreover, due to deep potential wells, the energies of these states are negative and absolute values of ones are equal the masses of ordinary neutron and proton. The best consideration of this result is to define this energy as a corresponding the negative gravitational mass of these particles. The inertial mass, in complete correspondence with known relation of these masses for ordinary particles, has to be defined as equal the absolute value of gravitational mass.

To some extent, the reactions (5.2) look like annihilation of particle and antiparticle. The nucleon disappears, and the born dark nucleon moves out due to repulsion and the only observable result is gamma quantum of defined energy. The difference of these acts is as follows: at annihilation of particle and antiparticle due to momentum conservation, necessary are two quanta, every having energy, equal to the mass of annihilating particles. The best known process of this kind is annihilation of electron and positron, producing quanta with energy, equal to the mass of electron. Our process creates only one gamma quantum of double energy. Several groups have found strong evidence for extended emission of the gamma rays from the Galactic Center using data from the Large Area Telescope aboard the Fermi Gamma-ray Space Telescope and have found an excess in γ rays over known astrophysical backgrounds which peak at energy, equal approximately 2 GeV [18] . One possible source of this radiation is a previously unidentified population of millisecond pulsars in the Galactic Center bulge [19] [20] .

Therefore, from our consideration it follows that exotic objects, such as big stars remnants and dark matter, are composed mainly of dark neutrons and have negative gravitational mass. Gravitation is the main force acting between these masses. They are weakly interacting elements of matter, able to compose larger structures of significantly larger than nuclear matter densities. It looks like they finally accumulate in voids-big cavities of the Universe, having no visible matter. Most probably, the gravitational interaction of these absolutely stable particles converts these clumps into structures such as ordinary protostars, i.e. with permanently growing kinetic energy and temperature. In ordinary star, the thermonuclear synthesis starts with deuteron production due to gain in energy converting one proton to neutron because the difference of masses of proton and neutron, equal 1.3 MeV, is less than binding energy of deuteron, equal 2.2 MeV. No reactions like this can start in clump of negative gravitational mass due to mentioned above extreme binding of PPs, composing $\overline{n}$, hence the process of negative gravitational mass deuteron, i.e. the system, composed of $\overline{n}$ and $\overline{p}$, and more complex systems creation cannot stop this process. After some time, when temperature reaches necessary value, the energy of this gigantic clump of negative mass can be enough to excite and convert this system to ordinary neutrons and protons and start process like Big Bang without any problems with antimatter annihilation.

One more conclusion, following from present consideration, is that reaction (5.1) can explain the excess of positrons in the Universe, necessary for intensive annihilation with electrons, producing the photons with characteristic well defined energy, equal 511 keV. Almost 40 years after discovery, the origin of this phenomenon remains unknown and makes this problem one of the most intriguing ones in high energy astrophysics [21] . This emission is strongly concentrated toward the Galactic bulge, and hence follows directly from our model of negative gravitational mass appearance. Moreover, the distribution of emitted positrons kinetic energies present in [21] , do not exceed 1.3 MeV in complete correspondence with our prediction. By the way, this reaction also explains the origin of intensive jets of neutrinos after the supernova explosion.

Finally, our model explains numerous unsuccessful efforts to observe dark particles in laboratories on Earth [22] . These particles are absent here due to gravitational repulsion with ordinary matter, composing all observable world. Moreover, for obvious reasons, they cannot be produced in colliders. On the other side, one has indirect evidences for these particles existence-accelerating expansion of the Universe and reduction of star formation rate in our Galaxy [23] , due to mass evaporation to extragalactic space.

Concluding, present results are the best illustration of unity of the Universe. The successful model of the smallest quantum system-nucleon, taking into account cosmological principle, i.e. invariance of the Universe in respect of spatial translations and rotations, creates unique possibility to understand the evolution of the biggest structures-stars, galaxies and finally-evolution of entire Universe.

REFERENCES

1. Gross, D.J. (2005) Nobellecture: The Discovery of as Symptotic Freedom and the Emergence of QCD. Reviews of Modern Physics, 77, 837-849.
2. Thomas, A.W. and Weise, W. (2001) The Structure of the Nucleon. Wiley-VCH, Berlin.
3. Bernard, V., Epelbaum, E., Krebs, H. and Meiβner, U.-G. (2013) New Insights into the Spin Structure of the Nucleon. Physical Review D, 87, Article ID: 054032.
4. Aidala, C.A., Bass, S.D., Hasch, D. and Mallot, G.K. (2013) The Spin Structure of the Nucleon. Reviews of Modern Physics, 85, 655-691.
5. Olive, K.A., et al. (Particle Data Group) (2014) Review of Particle Physics. Chinese Physics C, 38, Article ID: 090001.
6. Capitani, S., Morte, M.D., Djukanovic, D., von Hippel, G., Hua, J., J&aumlger, B., Knippschild, B., Meyer, H.B., Rae, T.D. and Wittig, H. (2015) Nucleon Electromagnetic Form Factors in Two-Flaw or QCD. Physical Review D, 92, Article ID: 054511.
7. Green, J., Meinel, S., Engelhardt, M., Krieg, S., Laeuchli, J., Negele, J., Orginos, K., Pochinsky, A. and Syritsyn, S. (2015) High-Precision Calculation of the Strangenucleonelectro Magnetic Form factors. Physical Review D, 92, Article ID: 031501.
8. Kamuntavicius, G.P. (2014) Galilei Invariant Technique for Quantum System Description. Journal of Mathematical Physics, 55, Article ID: 042103.
9. Kamuntavicius, G.P. (2014) Nucleon as a Nonrelativistic Three Point Particles System. SOP Transactions on Theoretical Physics, 1, 44-56.
10. Kamuntavicius, G.P. (2015) Structure of Schr&oumldinger's Nucleon: Elastic Form-Factor Sand Radii. Journal of Applied Mathematics and Physics, 3, 1352-1360.
11. Kamuntavicius, G.P. (2015) Structure of Schr&oumldinger's Nucleon: Density Distributions and Zemach Momenta. Journal of Applied Mathematics and Physics, 3, 1412-1421.
12. Burrows, A. (2013) Colloquium: Perspectives on Core-Collapse Supernova Theory. Reviews of Modern Physics, 85, 245-261.
13. Abbott, B.P., et al. (2016) Astrophysical Implications of the Binary Black Hole Merger GW150914. The Astrophysical Journal Letters, 818, L22.
14. Sellwood, J.A. (2014) Secular Evolution in Disc Galaxies. Reviews of Modern Physics, 86, 1-46.
15. Perlmutter, S. (2012) Nobel Lecture: Measuring the Acceleration of the Cosmic Expansion Using Supernovae. Reviews of Modern Physics, 84, 1127-1149.
16. Frenk, C.S. and White, S.D.M. (2012) Dark Matter and Cosmic Structure. Annalen der Physik, 524, 507-534.
17. Silk, J. and Mamon, G.A. (2012) The Current Status of Galaxy Formation. Research in Astronomy and Astrophysics, 12, 917-946.
18. Calore, F., Cholis, I., McCabe, Ch. and Weniger, Ch. (2015) A Tale of Tails: Dark Matter Interpretations of the Fermi GeV Excess in Light of Background Model Systematics. Physical Review D, 91, Article ID: 063003.
19. Bartels, R., Krishnamurthy, S. and Weniger, Ch. (2016) Strong Support for the Millisecond Pulsar Origin of the Galactic Center GeV Excess. Physical Review Letters, 116, Article ID: 051102.
20. Lee, S.K., Lisanti, M., Safdi, B.R., Slatyer, T.R. and Xue, W. (2016) Evidence for Unresolved γ-Ray Point Sources in the Inner Galaxy. Physical Review Letters, 116, Article ID: 051103.
21. Prantzos, N., et al. (2011) The 511 keV Emission from Positron Annihilation in the Galaxy. Reviews of Modern Physics, 83, 1001-1056.
22. Undagoitia, T.M. and Rauch, L. (2016) Dark Matter Direct-Detection Experiments. Journal of Physics G: Nuclear and Particle Physics, 43, Article ID: 013001.
23. Haywood, M., et al. (2016) When the Milky Way Turned-Off the Lights: APOGEE Provides Evidence of Star Formation "Quenching" in Our Galaxy. Astronomy and Astrophysics. arXiv: 0601.03042 [astro-ph.GA].

CHAPTER 8

Cosmogonic Speculations: Particle Creation from Energy Conservation in the Universe Evolution

Paolo Christillin

Dipartimento di Fisica, Università di Pisa and I.N.F.N., Sezione di Pisa, Pisa, Italy

ABSTRACT

Information about the Universe from Hubble's law is consistent with that coming from the evaluation of inertial forces, supporting the picture of a steady state expansion of a black hole Universe. Backed up also by the consideration of the black body self energy, the post big bang temperature decrease is consistent with particle creation if energy conservation applies at every scale. This process is shown to provide a gravitational repulsive force which can counterbalance gravitational attraction thus allowing the possibility of a steady non-inflationary expansion. That seems to provide an alternative coherent scheme for our picture of the Universe evolution, disposing of the cosmological term, of dark energy and of the bulk of dark matter.

Keywords: Cosmogony, Black Holes, Particle Creation, Self Energy, Inertial Forces

1. INTRODUCTION

The present scenario for our Universe evolution is based first of all on two sound experimental facts: Hubble's law [1] and the CMB [2] (although the latter was predicted semiquantitatively by Alpher, Herrman and Gamow [3]). In more recent times finer details have come from BOOMERAanG [4] , Cobe [5] , Planck [6] and the Supernovae data [7] [8] which hint at an accelerated expansion.

The prevailing fit is based on the GR equations where the theoretical interpretation of the notorious cosmological term seems to represent a formidable problem. Quoting e.g. Susskind [9] "the best efforts of the best physicists, using our best theories predict the cosmological constant... equivalent to the vacuum energy... incorrectly by 120 orders of magnitude".

That its theoretical treatment leaves many people unsatisfied is reflected e.g. in Perlmutter's words [10] : "Dark energy has now been added to the already perplexing question of dark matter [11] . One is tempted to speculate that these ingredients are add-ons, like the Ptolemaic epicycles, to preserve an incomplete theory".

In that respect, let also quote Feynman's [12] words "Although in ordinary circumstances a physicist would rebel against a theory which so cavalierly ignores our cherished laws about the conservation of matter and energy, one must remember that here we are not dealing with an ordinary problem but with a cosmological problem". Although these words were referring to the steady state theory of Hoyle [13] , the warning applies in general.

That a revision is needed had also already been pointed out in [14] in connection with the treatment of one of our theoretical milestones i.e. black body radiation. Indeed at cosmological scales the gravitational interaction of photons cannot be neglected and this results in a substantial role of the self energy, especially in the radiation dominated era.

If we enforce energy conservation at any scale, one is thus forced to admit photon (and particle) non-con- servation, which explains the non-gravitational collapse of the Universe. This provides an alternative scheme where our causal Universe has evolved from a total zero energy quantum fluctuation at the Planck scale to the present dimensions, again with total zero energy, without any need to invoke inflation.

The crucial points will be :

1) energy vs. particle number conservation,
2) the presence of a gravitational repulsive force due to particle creation,
3) the black body handling,

4) the consideration of inertial forces.

2. PRESENT PICTURE: THE UNITY OF THE UNIVERSE

a) inertial forces

We summarize the results of the calculations backing up the conjecture put forward mainly by Sciama [15] and further developed by Berry [16] concerning the constraints provided by the inertial forces on the mass of the Universe.

Consider first how the Coriolis force is generated in the rotating frame by the counter rotating Universe. Since a rotating matter distribution produces a gravitomagnetic field $\boldsymbol{h}$ (according to the parameter free effective vector formulation of gravitation [17]) proportional to the angular velocity of rotation $\boldsymbol{\omega}$

$$\boldsymbol{F}_{GM} = m\boldsymbol{v}\times\left(\frac{2GM}{c^2R}\boldsymbol{\omega}\right) = 2m\boldsymbol{v}\times\left(\frac{GM}{c^2R}\right)\boldsymbol{\omega} \tag{1}$$

the Coriolis force on a mass m

$$\boldsymbol{F}_{Cor} = 2m\boldsymbol{v}\times\boldsymbol{\omega} \tag{2}$$

is recovered

$$\boldsymbol{F}_{Cor} = \boldsymbol{F}_{GM} \tag{3}$$

if for the Universe,

$$\frac{GM_U}{c^2R_U} = 1 \quad . \tag{4}$$

The relevant point in this argument is that in the relative rotation, the magnetic field generated by distant layers of matter goes as $1/R$ i.e. the same behavior of radiation, rather than the usual $1/R$ of Newtonian forces. Therefore a relative more important role even of distant stars is a matter of fact.

As regards linear inertial forces, the same effective vector gravitation accounts for them via a different mechanism i.e. dipole radiation [18] .

It is well known that in the non inertial frame S' of a particle m_1 attracted by another M, where the said particle is at rest and manifestly $a_1' = 0$, non inertial forces have to be introduced. They come from all other particles which acquire the opposite acceleration $-a_1$ generating an additional force given by

$$F'_{1,I} = m_1 g_{rad} = -m_1 \frac{\mathrm{d}b}{\mathrm{d}t} \simeq m_1 \left(\frac{G}{c^2} \int \frac{2\rho}{|r'|} \mathrm{d}V' \right) a_1 \simeq m_1 \left(\frac{2GM_U}{c^2 R_U} \right) a_1 \tag{5}$$

where I stands for inertial and b for the gravitomagnetic vector potential. Therefore if

$$\frac{2GM_U}{c^2 R_U} = 1 \tag{6}$$

the inertial force generated in the non inertial reference frame S' of particle 1 by the accelerated masses of the rest of the Universe exactly compensates the "true" Newtonian (N) force $F_{1,N} = m_1 a_1 = -\frac{GMm_1 r}{r^3}$

$$F_{1,N} + F'_{1,I} = 0 \tag{7}$$

Thus the two previous conditions on the energy content of the Universe, which taken at their face value seem to be contradictory, confirm instead at a semiquantitative level the fact that we are indeed living inside a black hole. All the more so since they come from two different mechanisms.

Let us now come to an apparently disconnected issue i.e. the cosmological expansion.

b) the gross features of the expansion

Hubble's law [1] of the recession velocity of galaxies reads

$$v = H_0 R \tag{8}$$

where $H_0 \simeq 2.3\times10^{-18}\ \sec^{-1}$ stands for the present value of the Hubble's constant, which defines the radius of our present causal Universe

$$R_U = c/H_o \simeq 10^{26}\ \mathrm{m} \tag{9}$$

Whether our Universe would eventually contract or keep expanding should depend on a critical density ρ_c. In classical terms, following e.g. [19] [20] , consider a sphere of radius R of constant mass density in flat space. The kinetic energy of a particle m at its surface is $K = 1/2\,m(\mathrm{d}R/\mathrm{d}t)^2$ and is accompanied by the attraction of the particles inside the sphere with potential energy $U = -G\dfrac{4\pi R^3\rho m}{3R}$.

The steady state thus takes place when these two energies balance each other or

$$E = K + U = 0 = 1/2\,m(\mathrm{d}R/\mathrm{d}t)^2 - G\frac{4\pi R^3\rho_c m}{3R} \tag{10}$$

In this case

$$(1/R\times \mathrm{d}R/\mathrm{d}t)^2 = 8\pi/3G\rho_c \tag{11}$$

and

$$\rho_c \simeq 3H_o^2/(8\pi G) \simeq 1.1\times10^{-26}\ \mathrm{kg/m^3} \simeq 6m_p/\mathrm{m^3} \tag{12}$$

m_p standing for the nucleon mass.

It is evident that these considerations represent a rough non relativistic schematization of the process. As a matter of fact in the energy density no trace of the kinetic energy of the expanding particles within the sphere appears nor their contribution to the potential energy because of the mass energy equivalence; the mass M is just an "effective" static source which is thought to represent reality and hence to determine the motion of m.

Nevertheless data seem to confirm [4] to the percent the critical value for the density.

Let us then check the consistency of Equation (11) with the results from inertial forces by using Hubble's relation predictively at R_U so that the Universe density will enter the r.h.s.

$$\left(\frac{1}{R}\frac{\mathrm{d}R}{\mathrm{d}t}\right)^2\Bigg|_{R_U} = \frac{c^2}{R_U^2} = \frac{8\pi G\rho_U}{3} = \frac{2GM_U}{R_U^3} \tag{13}$$

whence

$$\frac{2GM_U}{c^2 R_U} = 1;\; M_U c^2 - \frac{2GM_U^2}{R_U} = 0 \tag{14}$$

which implies that the total mass is balanced by its gravitational self energy (up to a factor of 2) to make the total energy of the Universe zero.

We thus see that the extrapolation of Hubble's law to the whole of our causal Universe, circumventing critical estimates of the contribution of distant regions to the critical density, yields the same black hole condition entering the evaluation of inertial forces and validates Feynman's conjecture that the critical density is just about the best density to use in cosmological problems because it allows for free particle creation [12] , the mass being swallowed by its gravitational self energy. Thus Hubble's law is not only fundamental in its vector form in proving that there is no center in the Universe but its extrapolation is paramount in determining its fate.

Let us then see how mass creation embodied in the previous equation enters consistently the equation of motion for the acceleration. We differentiate with respect to t to get from

the velocity

$$(\mathrm{d}R/\mathrm{d}t)^2 = 8\pi/3 G\rho R^2 \tag{15}$$

the acceleration

$$\left(\mathrm{d}^2 R/\mathrm{d}t^2\right) = -GM/R^2 + G/R\left(\mathrm{d}M/\mathrm{d}R\right). \tag{16}$$

This last equation differs fundamentally from the standard treatment rewriting completely our picture of the Universe. Indeed whereas the first term in the r.h.s. yields the Newtonian (and GR) attractive acceleration due to the mass M enclosed within the radius R, the second term gives a repulsive effect due to matter creation. Therefore even without a cosmological term we have a built-in repulsive gravitational term which may make the Universe not to collapse, without dark energy i.e.

$$\Omega_\Lambda = 0 \quad (17)$$

The preceding energy conserving cosmogonic equations (ECCE) represent the minimal parameter-free self- consistent model of the gross features of the Universe in which extra parameters may be introduced only if necessary for fitting finer details of its evolution. In the r.h.s. of Equation (16) the two terms can indeed compensate thus yielding a steady state evolution if

$$\mathrm{d}M/M = \mathrm{d}R/R \quad (18)$$

equivalent to

$$\mathrm{d}\rho/\rho = -2\,\mathrm{d}R/R \quad (19)$$

The scale factor which links mass and radius is manifestly G/c^2 which explains why the self energy can- cellation holds true only for the Universe: in other words only for very small and dense objects (Planck) or for dilute but very big regions i.e. the Universe itself during its evolution. Thus we have a double constraint on the velocity and on the acceleration: if in a black hole condition the Universe expands, mass and radius are linearly linked.

Notice that although the equation of energy conservation looks like the traditional one, the consistency requirement coming from the acceleration, implies a totally different scenario. In other words whereas in a traditional picture (without the cosmological term) the fate of the Universe resembles the description of the motion of a body thrown up in the air in a gravitational field, so that according to how its velocity compares to the escape velocity determined by the density it will collapse or not, here the expansion velocity is

$$(\mathrm{d}R/\mathrm{d}t) \simeq \mathrm{const} \quad (20)$$

It follows that

$$R(t) \simeq t \quad (21)$$

implying at the same time for the density evolution

$$\rho \simeq 1/R^2 \quad \text{or} \quad \frac{\mathrm{d}}{\mathrm{d}t}\left(\rho R^2\right) = 0 \tag{22}$$

which is still another way to express the linear relation between mass and radius of a black hole, confirming the flatness of space. The previous relation can be also directly got from Equation (14) by imposing total energy conservation for the interacting Universe at all times

$$M(t)\left(c^2 - 2GM(t)/R(t)\right) = 0 \tag{23}$$

This has to be contrasted with the requirement one gets from thermodynamic considerations as well as from GR

$$\frac{\mathrm{d}\rho}{\mathrm{d}t}\bigg/\left(\rho + p/c^2\right) = -3\frac{\mathrm{d}R}{\mathrm{d}t}\bigg/R \tag{24}$$

which expresses matter conservation (i.e. essentially particle number) as it can be simply obtained by imposing

$$\frac{\mathrm{d}}{\mathrm{d}t}\left(\rho R^3\right) = 0 \tag{25}$$

and by taking into account the energy content of pressure (which plays a role essentially only for radiation) with the well known behaviors for matter (m) and radiation (rad)

$$\rho_m \simeq 1/R^3, \quad \rho_{rad} \simeq 1/R^4 \tag{26}$$

The ratio of pressure to density $w = p/\rho$, parameter which enters the standard phenomenology for the various forms of energy has an irrelevant role with respect to the pressure by particle creation as it will be further clarified in the following.

In conclusion the standard scenario has to cope with the problem of energy conservation, the proposed one clashes with our belief of particle number conservation. However this picture does not at all correspond to Hoyle's [13] . There, particle creation is invoked to obtain a steady constant density state and to disprove the big bang. The present proposal rather somewhat resembles, apart from the fundamental difference due to particle creation, to the primeval egg by Lemaître [21] .

Just to get an order of magnitude, the estimated number of particles would have resulted from

$$\frac{\Delta N}{\Delta t \Delta V} \simeq \frac{10}{1\,\text{year} \times 1\,\text{km}^3} \tag{27}$$

(which we get allowing for creation of the present number minus the negligible Planck one over the age of the Universe) i.e. the creation of only ten hydrogen atom per cubic kilometer per year (since it has been referred to the present dimensions the creation rate would have been higher in the past). Therefore in that respect the present proposal does not go against any experimental evidence.

3. SELF ENERGIES: THE PLANCK SCALE

The fundamental Planck quantities are

$$R_P = \left(\hbar G/c^3\right)^{1/2} \simeq 10^{-33}\ \text{cm} \tag{28}$$

$$t_P = R_P/c \simeq 10^{-44}\ \text{sec} \tag{29}$$

$$E_P = M_P c^2 = \frac{\hbar}{t_P} \simeq 10^{19}\ \text{GeV} \tag{30}$$

As well known one way to derive them is to ask in which units our “fundamental” coupling constants $c, \hbar, G$ are equal and of the order of 1. In particular G enters through Newton’s law which is therefore assumed to be valid at any scale.

In another derivation they are determined by requiring the Compton wavelength of a particle to coincide with its Schwarzschild radius. Thus the Planck energy corresponds to the energy contained in the minimum (because it cannot be compressed farther without violating the uncertainty principle) quantum Schwarzschild radius i.e. the smallest quantum black hole.

This gravitational quantum "fluctuation" should, according to the prevailing picture,

$$\Delta E \Delta t \geq \hbar \tag{31}$$

last for times of the order of Planck's time.

But since

$$M_P c^2 - GM_P^2/R_P = 0 \tag{32}$$

the Planck fluctuation is energetically stable in the sense that it does not have to decay in Planck times (up to numerical factors of order 1).

The equations written for the Universe at the present time can therefore be rewritten for Planck quantities

$$\left(\mathrm{d}R/\mathrm{d}t\right)^2\Big|_P = 8\pi/3G\rho R^2\Big|_P \tag{33}$$

etc. with the same considerations.

Thus

$$G\rho_P \simeq \frac{GM_P}{R_P^3} = \frac{c^2}{R_P^2} = \frac{1}{t_P^2} \simeq H_P^2 \tag{34}$$

When compared to the present quantity

$$G\rho_U \simeq \frac{GM_U}{R_U^3} \simeq \frac{c^2}{R_U^2} = \frac{1}{t_U^2} \simeq H_0^2 \tag{35}$$

one has

$$\frac{H_0^2}{H_P^2} = \frac{t_P^2}{t_U^2} \simeq 10^{-122} \simeq \left(10^{-61}\right)^2 \simeq \left(\frac{R_P}{R_U}\right)^2 \tag{36}$$

From that we can immediately obtain an expression for Hubble's parameter at a generic time t and Universe radius R in terms of the present quantities H_0, t_U, R_U. We thus see that e.g. at half the time and radius of the Universe, the corresponding Hubble's parameter would be twice H_0.

In general

$$H(t) = \frac{c}{R_{U(t)}} \simeq \sqrt{G\rho(t)} \quad . \tag{37}$$

Thus Hubble's parameter would have decreased in the course of the expansion by the above $\simeq 60$ orders of magnitude due to the density dilution.

A hypothetical observer in the past, say 10 billion years ago, would have of course "measured" a younger Universe by roughly a factor ten, which would have reflected in a corresponding bigger Hubble's parameter(t). These formulas confirm the argument [22] that an observer in the past ($t < t_U$) would have seen a different history, although that has been stated for the cosmological term which we have not needed to invoke. It goes without saying that, in case, given the different role attributed to the vacuum which is the expanding Universe itself, this unnecessary extra parameter Λ would not be constant but time dependent and should be also of order c^2/R^2 .

That would answer Perlmutter's ([10]) worry that in the standard picture it appears "a remarkable and implausible coincidence that the mass density, just in the present epoch, is within a factor of 2 of the vacuum energy density because the mass density is supposed to become ever more dilute in the course of the expansion, whereas the vacuum energy density Λ, a property of empty space itself, stays constant". It would dispose as well of a similar statement [23] "to make things worse, the energy density $(10^{-3}\text{ eV})^4$ happens to be of the same order of magnitude of the matter density of the Universe ρ_m $\left(\text{indeed}\left(10^{-3}\text{ eV}\right)^4 \simeq m_p c^2/m^3\right)$.

4. SELF ENERGIES: EVOLUTION OF RADIATION AND MATTER

Let us inquire how much this parameter free scenario can account for the present 6-parameter cosmological phenomenology in the reconstruction of the past . As regards matter, its total energy in the Universe coming from Hubble's law is

$$E_m \simeq 10^{80}\ \text{GeV} \simeq 10^{80} m_p . \tag{38}$$

The present energy density of radiation, coming from the CMB, is

$$\epsilon_\gamma \simeq \left(kT_\gamma\right)^4 \simeq 10^5\ \text{eV/m}^3 \tag{39}$$

which yields a total energy for the Universe at present

$$E_\gamma \simeq \epsilon_\gamma \left(R_U\right)^3 \simeq 10^{74}\ \text{GeV} \simeq 10^{74} m_p \tag{40}$$

so that

$$E_\gamma / E_m \simeq 10^{-6} . \tag{41}$$

Thus "matter" (by which we mean loosely speaking non radiation) dominates over radiation by such a typical order of magnitude.

The total number of photons is given by

$$N_\gamma = n_\gamma \times R^3 \simeq \left(kT_\gamma R\right)^3 \tag{42}$$

and is of the order of 10^{87}.

Self energy would correct these energies in accord with the black body condition. The relevant point is that the self energy dependence on R is not the same as that of the energy so that this correction does not factorize. Thus self-energy can act as a gauge in the reconstruction of the history of the Universe, undisputedly proving that a creation mechanism must have been constantly at work.

As regards nucleons traditionally from energy conservation of non interacting particles (which amounts to particle number conservation)

$$E_m \simeq \epsilon_m R^3 \simeq \left(N_m \left(mc^2 + 3/2kT\right)\right) \simeq n_m R^3 \left(mc^2 + 3/2kT\right) \tag{43}$$

one predicts

$$\epsilon_m \simeq n_m \simeq \frac{1}{R^3} \qquad (44)$$

(this conclusion is somewhat altered when taking into account the temperature and the velocity dependence of baryons. It is however easy to see that the former is totally irrelevant up to a temperature of $10^{10}\ \mathrm{K} = 1\ \mathrm{MeV}$ and that the latter even for $v = c$ would just give a factor of 2).

On the contrary the requirement of positivity for the energy coming from the self energy implies that the present black hole condition should hold, if we enforce energy conservation, also going backward in time. Thus as a function of R

$$1 - \frac{GM_m}{c^2 R} \simeq 0 \Rightarrow \rho \simeq \frac{1}{R^2} \qquad (45)$$

($\rho = \epsilon / c^2$) where the irrelevant factor of 2 has already been commented upon. This is the same behavior derived from the equations of motion in par. 2).

Again energy conservation of gravitationally interacting particles and particle number conservation are mutually contradictory. Indeed the requirement Equation (44) would make the self energy correction of Equation (45) to turn negative. In other words the fact that less particles were present in the past in the energy conserving case with respect to their constancy of the traditional picture means once more that particles have been created in the expansion.

As regards photons, because of the energy mass equivalence, they are also obviously source of a sizable gravitational field when sufficiently energetic and/or numerous. Their black body treatment at ordinary conditions is absolutely unquestionable. Indeed if we consider e.g. a black body of one cubic meter at room temperature $kT \simeq 3 \times 10^{-2}\ \mathrm{eV}$, its total energy is $(kT)^4\, m^3 \simeq 10^8\ \mathrm{MeV}$. Its mass M is hence $10^{-22}\ \mathrm{kg}$ and its gravitational self energy, assuming a uniform distribution, $GM^2/R \simeq 10^{-10} \times 10^{-44}/1 \simeq 10^{-54}\ \mathrm{J}$, completely negligible with respect to its "bare" energy $Mc^2 \simeq 10^{-5}\ \mathrm{J}$!

For the Universe at present

$$\frac{GM_\gamma}{c^2 R_U} \simeq O\left(10^{-6}\right) \tag{46}$$

which is also negligible. And this must have been so for a long time since the energy density of the Planck spectrum which comes to us from the CMB

$$u(\omega)\mathrm{d}\omega \simeq \frac{\hbar\omega^3}{\mathrm{e}^{\hbar\omega/kT}-1}\mathrm{d}\omega \tag{47}$$

is remarkably accurate.

So the spectrum has kept its black body shape (which is sort of astonishing in view of the absence of significant later scattering processes) in agreement with scaling properties of the former angular frequency ω'

$$\omega = \omega'/(1+z) \tag{48}$$

(where z is the cosmological redshift), provided also the photon temperature T_γ scales by the same amount

$$T_\gamma \simeq 1/R\,. \tag{49}$$

This is consistent with photon number $N_\gamma \simeq (kT_\gamma R)^3 \simeq n_\gamma (R)^3$ conservation

$$n_\gamma \simeq \frac{1}{R^3} \tag{50}$$

from which the preceding result for the energy density ϵ_γ (and for $u(\omega)$)

$$\epsilon_\gamma \simeq \frac{1}{R^4} \tag{51}$$

follows. Thus

$$\frac{\epsilon_\gamma}{\epsilon_m} \simeq \frac{1}{R^2} \tag{52}$$

and from the previous estimates, matter and photon energy density would equalize at recombination (rec)

$$R_{rec} \simeq 10^{23} \text{ m} \tag{53}$$

which corresponds to the standard estimate for $T_\gamma \simeq 3000 \text{ K}$, although through a completely different mechanism. The total mass and radius at that time would be one thousandth of the present one. At the same time, given the constancy of the photon number in this lapse of time, it is immediate to see that gravitational self-energy completely swallows the radiation energy

$$1 - \frac{GM_\gamma}{c^2 R} \simeq 0 \tag{54}$$

at

$$R' = \frac{GE'_\gamma}{c^4} \simeq 10^{23} \text{ m} \tag{55}$$

Thus at

$$R' = R_{rec} \tag{56}$$

we have, at the same time, equality of matter and photon energy densities and zero total energy for both. This is gratifying since it proves that the self energy mechanism offers an alternative doubly constrained prediction. In former times the self energy mechanism would work both for radiation and for matter thus keeping their ratio constant at the value of recombination.

The different behavior of radiation and matter is then apparent. In the first case, because of the null mass, the wavelength is closely related to the Universe dimensions. This explains why we can reach $z \simeq 1000$. As regards nucleons the dominance of the mass in the energy makes them rather insensitive to the expansion (the typical z is much smaller; in essence their number varies but not their energy.) As mentioned, at R' only $O(10^{-3})$ of the present estimated number should have been present.

To summarize, back to R_{rec} we have conservation of N_γ but decrease of N_m till the different energy densities equalize. In still former times both energy densities increase in the same way because of the self energy constraint so that there is no radiation dominated era! But their number density varies as

$$n_\gamma / n_m \simeq m_p c^2 / kT_\gamma \,. \tag{57}$$

Thus typically at a temperature $T \simeq 10^9$ K of primordial nucleosynthesis, about 10^4 photons per nucleon exist. Whether this poses a problem for nuclear abundances will be examined elsewhere.

Further backward in the reconstruction of the history of the "dark ages", from the self energy condition for photons

$$(kT)^4 R^3 / c^2 R \simeq 1 \tag{58}$$

one gets

$$T^2 \simeq 1/R(t) \simeq 1/t \tag{59}$$

where the second equality comes from the results of the equations of motion.

The conclusions of this paragraph are therefore a linear connection between matter creation and the dimensions of the Universe throughout, and a linear dependence between the temperature of radiation and radius R back to recombination and quadratic before.

5. STRUCTURES

a) CMB and the event horizon

As well known [24] [25] CMB temperature fluctuations $\frac{\delta T}{T} \simeq 10^{-5}$ reflect in the angular size $\theta \simeq 0.01$ (which

corresponds to the multipole moment $l \simeq 200$ of the first peak of the observed angular power spectrum) of cold and hot spots of the remarkably homogeneous map of the CMB temperature. They are usually accounted for in terms of the angular diameter distance d_A and of their dimensions of wavelength $\lambda \simeq ct_{rec}$

$$\theta = \frac{\lambda}{d_A} = \frac{2}{3\sqrt{3}\left(\left(1+z_{rec}\right)^{1/2}-1\right)} \simeq 1/\sqrt{z} \tag{60}$$

where the Universe expansion in the matter dominated era goes, according to the standard treatment, as $R \simeq t^{2/3}$ with the ensuing $z \simeq 1100$.

Acoustic oscillations are clearly due to inhomogeneities in the baryon-photon fluid before recombination resulting in "standing waves" connected to the Universe dimensions and hence for us to the event horizon. In the present scenario photons emerging from the last scattering surface at R_{rec} in order to subtend today the said angle of $\theta \simeq 0.01$ of temperature fluctuations in the CMB, must have originated at $R_{ac} \simeq 10^{21}$ m (ac = acoustic) since indeed the subtended angle would be

$$\theta \simeq \frac{R_{ac}}{R_U} \simeq \frac{10^{21}}{10^{26}}. \tag{61}$$

Correspondingly that would imply a much bigger z than at recombination.

Notice that in terms of the corresponding quantities at recombination one has

$$\frac{R_{ac}}{R_{rec}} \simeq \frac{10^{21}}{10^{23}} \simeq \frac{t_{ac}}{t_{rec}} \simeq \frac{T_{rec}^2}{T_{ac}^2}. \tag{62}$$

So the acoustic peaks are directly connected to density fluctuations in a much smaller event horizon than that at recombination with corresponding shorter wavelengths. With CMB anisotropies we can "penetrate" the last scattering barrier.

The above prediction would entail a temperature an order of magnitude bigger of the usual 3000 K (i.e. kT of the order of 4 eV which is reasonable given the value of $\simeq 10$ eV for hydrogen ionization) and thermali- zation would have begun much earlier and have lasted from 0.1 million yr to 10 million yr.

These spots which originated well before, when interpreted in terms of the Universe mass at recombination, seem to imply a contribution of the baryonic mass to the total energy

$$\Omega_B \simeq 10^{-2} \tag{63}$$

which is one of the parameters entering the angular distribution fits. This is simply due to a theoretical "snapshot" of the Universe disregarding its time dependence and therefore the mass creation mechanism at work. In other words we have invisible matter in a time independent description which disappears when considering its time dependence.

b) supernovae : an accelerated expansion?

In the standard approach far away objects are "measured" through the luminosity distance

$$d_L = R_U r(1+z) \simeq \frac{z}{H_0}(1+..) \tag{64}$$

which comes from an expansion of the relevant quantities from their present value and where the dots stand for terms involving derivatives of the Hubble parameter. Thus measurements of luminosity distances and redshift parameters z of a number of galaxies have led to the discovery of the accelerated expansion of the Universe. This was once more unexpected since it clashed with our traditional picture of an attractive gravity.

In the present scenario on the other hand no expansion is necessary, since a linear dependence of the Hubble parameter on time has been predicted.

The farther the objects, the older but also within a Universe with a smaller radius $R_U(t') = R'$. Thus the extension of Hubble's law to any epoch means that their velocity is c times the ratio of their distance to the causal Universe radius R' at the given r (the one in the distance ladder determined by the luminosity, with the additional assumption that standard candles be really standard in the course of time) with respect to the present R_U.

$$v = \frac{c}{R'}r = H'r \tag{65}$$

Thus the velocity would be greater than that predicted by H_0 in a time independent description,

$$\Delta v = (H' - H_0)r = \left(\frac{c}{R'} - \frac{c}{R_U}\right)r \qquad (66)$$

the distance would consequently appear to be greater for very distant objects, supernovae would look fainter and this might be ascribed, following Riess and Turner [26] , to an acceleration in the past.

c) galaxy clusters

Galaxy clusters involve dimensions of Mpc i.e. factors 10^{-3} with respect to R_U.

Their distance from us are e.g. $O(16\ \text{Mpc})$ for the Virgo cluster and $O(100\ \text{Mpc})$ for the Coma cluster, with a mass -to-light ratio of the order of 1000.

The gravitational fields are weak; indeed from

$$\beta = v^2/c^2 \simeq GM/c^2R = \epsilon \qquad (67)$$

$\beta \simeq 10^{-5}$ ($v \simeq 1000$ km/s) making GR corrections inessential.

The finding that the dynamic mass entering the kinetic energy be much higher than the one determined by luminosity measurements or

$$K > |U| \quad i.e \quad E > 0 \qquad (68)$$

has led to speculate on the existence of missing mass or dark matter. These speculations are based on the application of the well known virial theorem which states that

$$\langle U \rangle = -2\langle K \rangle \qquad (69)$$

This relation for average values applies for systems remaining within a finite volume, by averaging over "sufficiently" long times. However with the given values, the orbital period of individual galaxies around the common center of mass would be of the order of some billion years. Thus the application of the virial theorem in the preceding form seems at least questionable. More than that, if energy conservation holds true and E is therefore time independent

$$E=\langle E\rangle=\langle K\rangle+\langle U\rangle=-\langle K\rangle \tag{70}$$

contradicting the the initial inequality for the energy, which is just a learned way of expressing the simple fact that a system of total positive energy cannot be bound.

Thus it seems plausible that in this domain the formation of gravitationally bound objects out of negative capacity gravitational fluctuations be overcome by the effect of the Universe expansion. As a matter of fact the virial theorem reads in reality

$$K_{tot}-1/2\frac{\mathrm{d}}{\mathrm{d}t}\left(\Sigma M_i \boldsymbol{r}_i\cdot\boldsymbol{v}_i\right)=-1/2U_{tot}\ . \tag{71}$$

An estimate of the second term, which cannot be averaged over many cycles in this case, yields

$$1/2\frac{r^2}{H^2} \tag{72}$$

i.e. a velocity of the indivual masses M_i

$$v=cr/R_U\simeq c\times10^{-3} \tag{73}$$

for the present value of the H, which would be increased by an order of magnitude when correctly considering its value some ten times bigger at the formation time thus substantially decreasing the value of the kinetic energy term. This brings velocity and luminosity measurements towards agreement challenging the necessity of invoking a sizable amount of dark matter.

In conclusion the virial theorem in an expanding Universe gets substantial contributions from the expansion questioning the bound state picture [27] [28] .

d) the rotation curves of galaxies

As repeatedly underlined the present scenario concerns the bulk features of the Universe evolution. It is however natural to wonder whether it can account also for the missing mass which seems necessary for the description of galaxies where Hubble's expansion plays practically no role.

Application of Newton's law to the rotation curves of masses m (hydrogen atoms emitting the 21 cm line) orbiting a galaxy of total bulge mass M within a radius R reproduces reasonably well experimental data in the inner part i.e. for $r \leq R$

$$v^2/r = GM/R^3 r \tag{74}$$

whereas in the exterior predicts a decrease of v

$$v^2/r = GM/r^2 \tag{75}$$

contradicted by experimental data which show a roughly constant behavior [29].

It has been noted that the presence of extra matter (missing mass) in the halo (h) with a density a

$$\rho \simeq 1/r^2 \tag{76}$$

which implies

$$M_h(r) \simeq r \tag{77}$$

would provide the extra attraction for the rotation curves and an extra deflection for lensing ("This distribution is spacially more extended than that of the visible baryons, demonstrating clearly the existence of dark matter" [30]), although alternative explanations have been proposed as well as for the bullet cluster [31].

Indeed that would result in an extra $1/r$ term in the previous equation (accounting for a constant v in the halo) equivalent to a modification of the acceleration in Newton's law [32] . Such an additional term to Newton's force appears quite naturally for rotating objects from gravitomagnetism (of which Coriolis is a relativistic example). However, given the value of ϵ, it seems to provide too low an acceleration. The problem is under investigation.

The overall situation of the proposed history of the Universe is summarized in Table 1 and Figure 1.

The three most relevant situations i.e. Planck, total zero energy of radiation and present Universe are represented for a CMB observer, not in scale.

- The Planck "primeval egg" evolves in time, causally connected (c.c.), at the speed of light. No inflation has to be invoked.
- The evolution is accompanied by matter and photons creation which decreases the temperature.
- In all three cases the total energy is zero: we are inside an expanding black hole. This is backed up by the account of inertial forces.
- The evolution of the Universe is determined by the time dependent Hubble's parameter which measures its age and, within each bubble, the expansion velocity of its constituents.
- The Hubble radius and the event horizon coincide.

Table 1. Parameters of the Universe evolution. The relevant part of our gravitational history has been spaced.

Age	Total mass	Temperature (K)	R_U (*m*)	Process
10^{-44} s	10^{19} GeV	10^{32}	10^{-35}	Planck origin
1 s	10^{63} GeV	10^{10}	10^{9}	Neutron freezout
1 min	10^{64-65} GeV	10^{9}	10^{11}	Primordial nucleosynthesis
1 yr	10^{70} GeV	10^{7}	10^{12}	Atomic processes
0.1 million yr	10^{75} GeV	30.000	10^{21}	BAO
10 million yr	10^{77} GeV	3.000	10^{23}	Decoupling
......				
1 billion yr	10^{79} GeV	20	10^{25}	Gravitation
14 billion yr	10^{80} GeV	2.7 (CMB)	10^{26}	Now

6. CONCLUSIONS

In the present paper a cosmogonic scenario has been proposed. At variance with respect to the standard matter conservation treatment, energy conservation, with the essential role played by self energy, has been enforced and seen to account for the puzzling non-gravitational collapse due to matter creation balancing gravitational attraction.

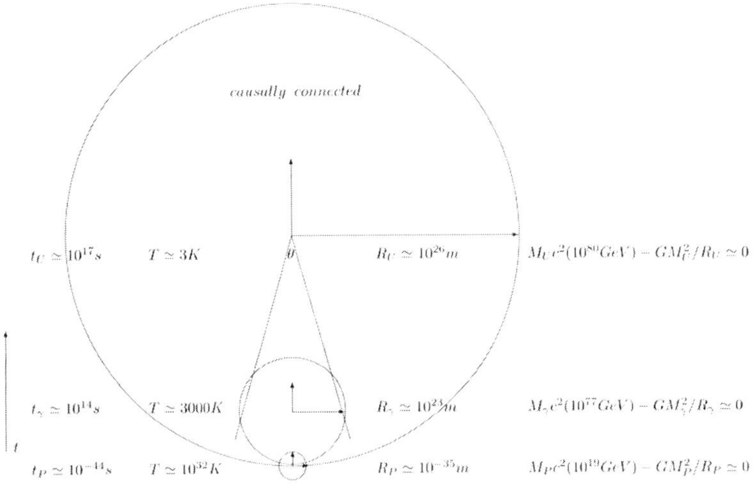

Figure 1. The Universe evolution.

In this parameter free phenomenological treatment, the gross features of our Universe evolution from the stable quantum fluctuation at the Planck epoch have been accounted for without any need for the cosmological constant Λ, entailing a different chronology.

The stringent connection among a zero total energy Universe, (an extended use of) Hubble's law, and matter creation in the Universe expansion consistent with the equations of motion for the velocity and the acceleration, has been underlined and emphasized.

We can hence summarize our results as follows:

1) the consideration of self gravitational effects is essential,
2) they play a crucial role both at a classical and at a quantum level and both for matter and for radiation. Indeed the gravitational interaction of photons at cosmic scales necessarily alters the standard black body treatment,
3) the main features of the evolution of our Universe from the Planck to the present scale reasonably follow from energy conservation and (no need for an unbelievably) fine tuning of the cosmological parameters,
4) the evidence for the black hole nature of our Universe coming from the reproduction of inertial forces supports Hubble's formulation of its expansion. More than that the black hole condition $GM/c^2R \simeq 1$ implies the constancy in time of these laws,

5) finally the constancy of the power necessary to create matter in the course of the Universe evolution

$$W_U = E_U/t_U = M_U c^2/t_U = c^5/G = W_P = 10^{52}\ \text{W} \tag{78}$$

where the usual black hole relation has been used (seems to provide a further), piece of support for the consistency of the present approach.

In conclusion, the cosmological term, dark energy and (the bulk of) dark matter seem to be reasonably due only to our use of inadequate equations to account for the gross features of the actual evolution of the Universe.

ACKNOWLEDGEMENTS

It is a pleasure to thank G. Cicogna for a critical reading of the manuscript and C. Bonati, G. Morchio, P.G. Prada Moroni, S. Degl'Innocenti for useful discussions.

REFERENCES

1. Hubble, E. (1929) Proceedings of the National Academy of Sciences of the United States of America, 15, 168-173.
2. Penzias, A.A. and Wilson, R.W. (1965) Astrophysical Journal, 142, 419-421.
3. Kragh, H. (1999) Cosmology and Controversy: The Historical Development of Two Theories of the Universe. Princeton University Press, Princeton.
4. de Bernardis, P., et al. (2000) Nature, 404, 955-959. arXiv:astro-ph/0004404.
5. COBE Mission Design, Spacecraft and Orbit. Goddard Space Flight Center, 18 April 2008.
6. Planck Collaboration. ar-Xiv:1502.01582.
7. Riess, A.G., et al. (1998) The Astronomical Journal, 116, 1009.
8. Perlmutter, S., et al. (1999) The Astrophysical Journal, 517, 565.
9. Susskind, L. (2006) The Cosmic Landscape. Hachette Book Group, New York.
10. Perlmutter, S. (2003) Physics Today, 56, 53.
11. Rubin, V. (1998) Dark Matter in the Universe. Scientific American, Inc., New York, 106.
12. Feynman, R.P. (1999) Lectures on Gravitation. Penguin Books, London.
13. Hoyle, F. (1948) Monthly Notices of the Royal Astronomical Society, 108, 372-382.
14. Christillin, P. (2011) The European Physical Journal Plus, 126, 88.
15. Sciama, D.W. (1959) The Unity of the Universe. Doubleday, Garden City, NY.
16. Berry, M. (1976) Principles of Cosmology and Gravitation. Cambridge University Press, Cambridge.
17. Christillin, P. and Barattini, L. (2013) ArXiv:1205.3514v3.
18. Christillin, P. (2014) The European Physical Journal Plus, 129, 175.
19. Okun, L.B. (1982) Leptons and Quarks. North Holland, Amsterdam.
20. Maoz, D. (2006) Astrophysics in a Nutshell. Princeton University Press, Princeton.
21. Lema&icirctre, G. (1927) Annales de la Société Scientifique de Bruxelles, A47, 49-59.
22. Barrow, J.D. and Shaw, D.J. (2011) General Relativity and Gravitation, 43, 2555-2560. ArXiv:1105.3105.

23. Zee, A. (2003) Physics Letters B, 594, 8-12. ArXiv:hep-th/0309032v1.
24. De Bernardis, P. and Bersanelli, M. (2013) Il Nuovo Saggiatore, 29, 25.
25. Mennella, A. and Natoli, P. (2015) Il Nuovo Saggiatore, 31, 63.
26. Riess, A.G. and Turner, M.S. (2004) Italian edition of Scientific American, 427, 56.
27. Ambartsumian, V.A. (1961) Astronomical Journal, 66, 536-540.
28. Van den Bergh, S. (2000) ArXiv:astro-ph/0005314v1.
29. Roos, M. (2010) ArXiv:1001.0316v1.
30. Ellis, R.S. (2010) Philosophical Transactions of the Royal Society A, 368, 967-987.
31. Markevitch, M., et al. (2002) The Astrophysical Journal, 567, L27-L31.
32. Milgrom, M. (1983) Astro-physical Journal, 270, 365-370.

CHAPTER 9

Kähler Dark Matter, Dark Energy Cosmic Density and Their Coupling

Mohamed S. El Naschie

Department of Physics, Faculty of Science, Alexandria University, Alexandria, Egypt

ABSTRACT

We utilize homology and co-homology of a K3-Kähler manifold as a model for spacetime to derive the cosmic energy density of our universe and subdivide it into its three fundamental constituents, namely: 1) ordinary energy; 2) pure dark energy and 3) dark matter. In addition, the fundamental coupling of dark matter to pure dark energy is analyzed in detail for the first time. Finally, the so-obtained results are shown to be in astounding agreement with all previous theoretical analysis as well as with actual accurate cosmic measurements.

Keywords: Kähler Topology, Dark Matter, E-Infinity, Super Strings, Golden Mean Computer, Kerr Black Hole Geometry, Accelerated Cosmic Expansion, Fractal Cantorian Spacetime

1. INTRODUCTION AND BACKGROUND INFORMATION

In many previous publications [1] - [80] , we considered the major problems of dark energy as well as the related quest to elucidate the missing mass of the universe which was dubbed "Dark Matter" [6] [38] [39] . In short it was found that a highly accurate estimate of these densities may be found by equating the Lorentzian Gamma factor γ of Einstein's equation of maximal energy (E = mc^2 as given for $\gamma = 1$ where m is the mass and c is the velocity of light) to the ratio of the Betti number b_2 of Einstein's 4D smooth manifold (b_2 = 1) and that of a K3-Kähler manifold b_2 = 22 [81] - [108] . Thus inserting in E leads to an ordinary energy density [2] [44]

$$E(O) = \frac{b_2(\text{Einstein})}{b_2(\text{k3} - \text{Kähler})} mc^2$$
$$= \frac{1}{22}(mc^2) \tag{1}$$

This is almost 4.5% of Einstein's classical energy density E = mc^2 from which one is naturally led to conclude that the "hidden" or missing energy density of the universe must be equal to 1 – (1/22) = 21/22 i.e. 95.5% of that of Einstein's maximal energy in an astounding agreement with the highly accurate measurements of COBE, WMAP, Type L_2 Supernova and Planck [1] - [10] . Furthermore, comparing the result with other theoretical derivations, for instance the exact transfinite one obtained from the multiplicative five dimensional value of a zero set quantum particle ϕ^5 and additive five dimensional value of an empty set quantum wave 5ϕ [24] [27] namely

$$E(O) = (\phi^5/2)mc^2 = mc^2/(22+k)$$
$$E(D) = (5\phi^2/2)mc^2 = mc^2(21+k/22+k)$$
$$= 1 - E(O), \tag{2}$$

we find that setting the transfinite irrational tail $k = \phi^3(1-\phi^3) = 0.18033989$ equal zero leads to the very same result obtained using the Betti numbers b_2 of the said K3 Kähler manifold [2] [44] . From the above it was concluded sometime ago that b_2 measures the amount of "space" voids in the manifold [10] [24] [28] [50] [52] . Consequently since the classical Einstein spacetime manifold is a smooth voidless manifold for which $b_2 = 1$ and K3-Kähler manifold is a highly non-smooth four dimensional structure with $b_2 = 22$, we see that the topological index b_2 is a quite accurate measure for the fractal ruggedness involved for any manifold [81] - [86] . Therefore the used ratio $E(\text{Einstein})/b_2(\text{Kähler})$ will account for the reduction of $E = mc^2$ to $E(O) = mc^2/22$ and its dissection into two components as expressed in the by now reasonably well known equation

$$\begin{aligned} E &= E(O) + E(D) \\ &= (mc^2/22) + mc^2(21/22) \\ &= mc^2 \\ &= E(\text{Einstein}) \end{aligned} \tag{3}$$

In the preceding equation E(D) makes no distinction between the dark matter and the pure dark energy components of E(D). By contrast in the present derivation we show how the topology of K3-Kähler and its various Betti numbers can discriminate between not only ordinary and dark energy but also between dark matter and pure dark energy. This remarkable result will be achieved here by utilizing the signature of K3- Kähler τ and the entire set of all non-zero Betti numbers [24] [81] - [86] . In addition we will be able to reveal the subtle coupling between dark matter and pure dark energy [2] - [35] .

2. ANALYSIS

The present analysis rests almost entirely on the intuitively understandable fact that similar to the distinction between the zero set and the various empty set with increasing degree of emptiness used in E-Infinity theory [83] - [103] the various topological Betti numbers measure subtly different degrees of ruggedness caused by the "voids" in the manifold [96] [101] [116] . In the spirit of the above, the heuristic dissection of E (Einstein)

$$\begin{aligned} E(\text{Einstein}) &= (mc^2/22) + mc^2(21/22) \\ &= E(O) + E(D) \\ &= mc^2 \end{aligned} \tag{4}$$

may be rewritten to become

$$E(\text{Einstein}) = E(O) + E(DM) + E(PDE) \\ = (mc^2/22) + mc^2(5/22) + mc^2(16/22) \quad (5)$$

In turn this may be interpreted in terms of the topological invariant of our K3- Kähler, namely [81] - [91]

$$\begin{aligned} &b_0 = b_4 = 1, \quad b_1 = b_3 = 0, \\ &b_2^- = 19, \quad b_2^+ = 3, \\ &b_2 = b_2^+ + b_2^- = 3 + 19 = 22 \text{ and} \\ &\tau = b_2^+ - b_2^- = 3 - 19 = -16 \end{aligned} \quad (6)$$

to mean that

$$\begin{aligned} E(\text{Einstein}) &= \left(\frac{1}{b_2}\right)mc^2 + \left(\frac{b_0 + b_3^+ + b_4}{b_2}\right)mc^2 + \frac{|\tau|}{b_2}mc^2 \\ &= \frac{mc^2}{22} + \left(\frac{5}{22}\right)mc^2 + \left(\frac{|-16|}{22}\right)mc^2 \\ &= mc^2\left(\frac{1+5+16}{22}\right) \\ &= mc^2 \end{aligned} \quad (7)$$

This is a new profound confirmation of our previous result and reinforces the confidence in the K-(or K3) Kähler gained from its use in superstring theory and E-Infinity theory alike [2] - [108] . In particular one should note that the negative sign of the signature $\tau = -16$ can be interpreted as a clear hint that pure dark energy works in the opposite direction to dark matter and ordinary energy [1] - [80] . It is also note worthy to observe that the above result agrees almost completely with that obtained from the geometry of Kerr black holes [8] .

3. THE COUPLING OF DARK MATTER TO PURE DARK ENERGY

There is a subtle point in the previous integer approximation of the triple partitions of E-Einstein. The point is that the ordinary energy part E(O) = $mc^2/22$ and the total dark energy part E(D) = $mc^2/22$ may be found directly from the exact transfinite value $E(O) = mc^2/(22+k)$ and $E(D) = mc^2(21+k/22+k)$ by setting k = 0 and revealing E(O) to be $1/22 \simeq 4.5\%$ and E(D) to be $21/22 \simeq 95.5\%$ of the total energy [2] - [10] . However doing the same for the triple dissection, we do not find E(DM) to be equal to 22% nor E(PDF) to be $\simeq 73.5\%$ as should be because $5/22 \simeq 22.7\%$ and $16/22 \simeq 72.7\%$ [1] - [15] . Thus while our "integer" solution is quite close to the exact value they are not exact nor can they be made exact by setting small irrational tails such as k equal zero similar to the analysis involving two components dissection [1] - [80] .

The deeper reason for this mathematical "anomaly" is that the "mathematics" is trying to tell us something of physical value and meaning. This "something" is actually that dark matter and pure dark energy are subtly coupled as the following computation will reveal. To do that we introduce the irrational transfinite correction coupling constant $\Delta = (8+k^2)/100$.

Rewriting our previous triadic E using Δ we find:

$$E = mc^2\left[\frac{1}{22+k} + \frac{5-\Delta}{22+k} + \frac{16+k+\Delta}{22+k}\right] \tag{8}$$

The remarkable fact is that in the above formalism everything falls in place and we find our exact solution that was found long ago using elaborate and occasionally terse reasoning namely that [8] [24]

$$\begin{aligned} E(O) &= mc^2/22+k \simeq mc^2/22 = 4.5\% \\ E(DM) &= mc^2\left(\frac{5-\Delta}{22+k}\right) = (22+k)\% \simeq 22\% \\ E(PDE) &= mc^2\left(\frac{16+k+\Delta}{22+k}\right) \simeq 73.5\% \end{aligned} \tag{9}$$

exactly as obtained in numerous previous publications [1] - [80] and in full agreement with actual cosmic measurements [1] - [80] . In view of the above one could understand the huge difficulties of detecting dark matter to the extent of doubting its very real existence [132] . The present Author is however of the opinion that there is sufficient observation and theoretical underpinning to convince us that dark energy and dark matter are real [24].

4. CONCLUSION

By design or providence, superstring theory struck a lucky accord with nature by hitting on the marvelous K3-Kähler manifold. However, the inherent inconsistency of using a classical continuum in the form of vibrating and rotating strings coupled to a fundamental discrete ten dimensional network spacetime might be the reason of the failure of string theory in utilizing this K3-Kähler to its utmost capacity. By contrast when introducing a fuzzy-fractal version of K3-Kähler, we are able to use it beyond the compactification procedure of superstring theory to model spacetime itself. Proceeding in this way, a theory was developed which is essentially an E-Infinity-like theory with a twist. In this theory, the fundamental invariants of K3-Kähler, namely the Betti numbers, the signature and the Euler invariants, play a pivotal role not only in deriving the actually measured cosmic ordinary and dark energy density i.e. E(O) and E(D) but also of discriminating between pure dark energy E(PDE) and dark matter E(DM) as well as revealing its coupled nature.

REFERENCES

1. El Naschie, M.S. (2013) Journal of Quantum Information Science, 3, 23-26.
2. El Naschie, M.S. (2013) International Journal of Modern Nonlinear Theory & Application, 2, 43-54.
3. El Naschie, M.S. (2013) Journal of Modern Physics, 4, 354-356.
4. El Naschie, M.S. (2014) Journal of Quantum Information Science, 4, 83-91.
5. El Naschie, M.S. (2013) Open Journal of Microphysics, 3, Article ID: 35744.
6. Lahanas, A.B., Mavromatos, N.E. and Nanopoulosd, D.V. (2003) International Journal of Modern Physics D, 12, 1529.
7. El Naschie, M.S. and Helal, A. (2013) International Journal of Astronomy and Astrophysics, 3, 318-343.
8. El Naschie, M.S. (2015) Natural Science, 7, 210-225.
9. Marek-Crnjac, L., El Naschie, M.S. and He, J.-H. (2013) International Journal of Modern Nonlinear Theory and Application, 2, 78-88.
10. El Naschie, M.S. (2013) International Journal of Modern Nonlinear Theory and Application, 2, 107-121.
11. Lahanas, A.B., Mavromatos, N.E. and Nanopoulosd, D.V. (2007) Physics Letters B, 649, 83-90.
12. El Naschie, M.S. (2014) American Journal of Astronomy & Astrophysics, 2, 72-77.
13. El Naschie, M.S. (2013) International Journal of Astronomy and Astrophysics, 3, Article ID: 40590.
14. El Naschie, M.S. (2013) Journal of Modern Physics and Applications, 2, No. 1.
15. El Naschie, M.S. (2014) Open Journal of Fluid Dynamics, 4, 15-17.
16. El Naschie, M.S. (2014) International Journal of Modern Nonlinear Theory and Application, 3, 1-5.
17. El Naschie, M.S., Marek-Crnjac, L., Helal, M.A. and He, J.-H. (2014) Applied Mathematics, 5, 1780-1790.
18. El Naschie, M.S. (2015) The Open Astronomy Journal, 8, 1-17.
19. El Naschie, M.S. (2014) Journal of Modern Physics Applications, 6, 1-10.

20. El Naschie, M.S. (2014) Journal of Quantum Information Science, 4, 284-291.
21. El Naschie, M.S. (2014) Journal of Applied Mathematics and Physics, 2, 803-806.
22. Helal, M., Marek-Crnjac, L. and He, J.-H. (2013) Open Journal of Microphysics, 3, 141-145.
23. El Naschie, M.S. (2014) International Journal of Astronomy and Astrophysics, 4, 80-90.
24. Marek-Crnjac, L. and He, J.-H. (2013) International Journal of Astronomy and Astrophysics, 3, 464-471.
25. El Naschie, M.S. (2014) Problems of Nonlinear Analysis in Engineering Systems, 20, 79-98.
26. El Naschie, M.S. (2013) American Journal of Modern Physics, 2, 357-361
27. El Naschie, M.S. (2013) Journal of Modern Physics, 4, 591-596.
28. El Naschie, M.S. (2014) World Journal of Mechanics, 4, 153-156.
29. El Naschie, M.S. (2015) Open Journal of Applied Sciences, 5, 313-324.
30. El Naschie, M.S. (2014) Journal of Modern Physics, 5, 743-750.
31. El Naschie, M.S. (2013) Gravitation and Cosmology, 19, 151-155.
32. El Naschie, M.S. (2014) Journal of Applied Mathematics and Physics, 2, 634-638.
33. El Naschie, M.S. (2016) Quantum Matter, 5, 1-4.
34. Marek-Crnjac, L. and El Naschie, M.S. (2013) Applied Mathematics, 4, 22-29.
35. El Naschie, M.S. (2014) World Journal of Condensed Matter Physics, 4, 74-77.
36. El Naschie, M.S. (2015) Natural Science, 7, 287-298.
37. He, J.H. and Marek-Crnjac, L. (2013) International Journal of Modern Nonlinear Theory and Application, 2, 55-59.
38. Sidharth, B.G. (2003) Chaos, Solitons & Fractals, 18, 197-201.
39. Sidharth, B. and Altaisky, M.V. (2012) Frontiers of Fundamental Physics 4. Springer, Berlin.
40. El Naschie, M.S. (2015) World Journal of Condensed Matter Physics, 5, 249-260.
41. El Naschie, M.S. (2015) International Journal of Astronomy & Astrophysics, 5, 243-247.
42. El Naschie, M.S. (2015) Natural Science, 7, 483-487.
43. El Naschie, M.S. (2015) Journal of Modern Physics, 6, 1321-1333.
44. He, J.H. and Marek-Crnjac, L. (2013) Fractal Spacetime and Noncommutative Geometry in Quantum & High Energy Physics, 3, 130-137.
45. El Naschie, M.S. (2014) International Journal of Astronomy and Astrophysics, 4, 332-339.
46. El Naschie, M.S. (2013) Journal of Quantum Information Science, 3, 55-57.
47. El Naschie, M.S. (2013) International Journal of Astronomy and Astrophysics, 3, 205-211.
48. El Naschie, M.S. (2015) Open Journal of Microphysics, 5, 11-15.
49. El Naschie, M.S. (2015) World Journal of Nano Science and Engineering, 5, 57-67.
50. El Naschie, M.S. (2015) Advances in Pure Mathematics, 5, 560-570.
51. Marek-Crnjac, L. (2015) Natural Science, 7, 581-598.
52. El Naschie, M.S. (2015) Open Journal of Philosophy, 5, 123-130.
53. El Naschie, M.S. (2015) Open Journal of Philosophy, 5, 319-326.
54. El Naschie, M.S. (2013) Journal of Modern Physics, 4, 1417-1428.
55. El Naschie, M.S. (2014) World Journal of Nuclear Science and Technology, 4, 216-221.
56. El Naschie, M.S. (2014) Journal of Electromagnetic Analysis and Applications, 6, 233-237.
57. Capozziello, S. (2002) International Journal of Modern Physics D, 11, 483-491.
58. El Naschie, M.S. (2013) Journal of Modern Physics, 2, 18-23.
59. llis, J.R., Mavromatos, N.E., Mitsou, V.A. and Nanopoulos, D.V. (2007) Astroparticle Physics, 27, 185-198.
60. El Naschie, M.S. (2013) Journal of Modern Physics, 4, 31-38.
61. Iovane, G., Laserra, E. and Giordano, P. (2004) Chaos, Solitons & Fractals, 22, 521-528.
62. Tang, W., Li, Y., Kong, H.Y. and El Naschie, M.S. (2014) Bubbfil Nanotechnology, 1, 3-12.
63. Diamandis, G.A., Georgalas, B.C., Mavromatos, N.E. and Papantonopoulos, E. (2002) International Journal of Modern Physics A, 17, 4567-4589.
64. El Naschie, M.S. (2014) Advances in Pure Mathematics, 4, 641-648.
65. El Naschie, M.S. (2015) American Journal of Nano Research and Application, 3, 33-40.
66. El Naschie, M.S. (2015) World Journal of Nano Science & Engineering, 5, 26-33.
67. Marek-Crnjac, L. (2013) American Journal of Modern Physics, 2, 255-263.
68. Iovane, G. (2004) Chaos, Solitons & Fractals, 20, 657-667.

69. El Naschie, M.S. (2005) Chaos, Solitons & Fractals, 25, 969-977.
70. El Naschie, M.S. (2016) Journal of Quantum Information Science, 6, 1-9.
71. Fred, Y.Y. (2009) Chaos, Solitons & Fractals, 42, 89-93.
72. Ellis, J., Mavromatos, N.E. and Nanopoulos, D.V. (2005) Physics Letters B, 619, 17-25.
73. El Naschie, M.S. (2016) World Journal of Condensed Matter Physics, 6, 63-67.
74. Iovane, G., Laserra, E. and Tortoriello, F.S. (2004) Chaos, Solitons & Fractals, 20, 415-426.
75. El Naschie, M.S. (2015) American Journal of Nano Research & Application, 3, 66-70.
76. El Naschie, M.S. (2015) Journal of Modern Physics, 6, 348-395.
77. Greljo, A., Julio, J., Kamenik, J.F., Smith, C. and Zupan, J. (2013) Journal of High Energy Physics, 2013, 190.
78. Diamandis, G.A., Georgalas, B.C., Lahanas, A.B., Mavromatos, N.E. and Nanopoulos, D.V. (2006) Physics Letters B, 642, 179-186.
79. El Naschie, M.S. (2006) World Journal of Nano Science and Engineering, 5, 49-56.
80. Sidharth, B.G. (2001) Chaos, Solitons & Fractals, 12, 1101-1109.
81. El Naschie, M.S. (2005) International Journal of Nonlinear Sciences and Numerical Simulation, 6, 95-98.
82. Gibbons, G., Shellard, E. and Rankin, S. (2003) The Future of Theoretical Physics and Cosmology. Cambridge University Press, Cambridge.
83. El Naschie, M.S. (2006) Chaos, Solitons & Fractals, 30, 579-605.
84. El Naschie, M.S. (2006) Chaos, Solitons & Fractals, 29, 876-881.
85. El Naschie, M.S. (2007) International Journal of Nonlinear Science & Numerical Simulation, 8, 11-20.
86. El Naschie, M.S. (2005) Chaos, Solitons & Fractals, 26, 477-481.
87. El Naschie, M.S. (2006) Chaos, Solitons & Fractals, 27, 9-13.
88. El Naschie, M.S. (2006) Chaos, Solitons & Fractals, 27, 39-42.
89. El Naschie, M.S. (2006) Chaos, Solitons & Fractals, 27, 843-849.
90. El Naschie, M.S. (2005) Chaos, Solitons & Fractals, 26, 1-6.
91. El Naschie, M.S. (2006) Chaos, Solitons & Fractals, 30, 636-641.
92. El Naschie, M.S. (2005) Chaos, Solitons & Fractals, 25, 509-514.
93. El Naschie, M.S. (2006) Chaos, Solitons & Fractals, 29, 816-822.
94. El Naschie, M.S. (2007) Chaos, Solitons & Fractals, 31, 537-547.
95. El Naschie, M.S. (2007) International Journal of Nonlinear Sciences and Numerical Simulation, 8, 445-450.
96. El Naschie, M.S. (2005 Chaos, Solitons & Fractals, 26, 665-670.
97. El Naschie, M.S. (2007) Chaos, Solitons & Fractals, 32, 427-430.
98. El Naschie, M.S. (2006) Chaos, Solitons & Fractals, 30, 1025-1033.
99. El Naschie, M.S. (2006) Chaos, Solitons & Fractals, 29, 845-853.
100. El Naschie, M.S. (2005) International Journal of Nonlinear Sciences and Numerical Simulation, 6, 331-333.
101. El Naschie, M.S. (2005) Chaos, Solitons & Fractals, 26, 303-311.
102. El Naschie, M.S. (2005) Chaos, Solitons & Fractals, 26, 291-294.
103. El Naschie, M.S. (2007) Chaos, Solitons & Fractals, 32, 370-374.
104. El Naschie, M.S. (2006) Chaos, Solitons & Fractals, 30, 525-531.
105. El Naschie, M.S. (2006) Chaos, Solitons & Fractals, 29, 803-807.
106. El Naschie, M.S. (2005) Chaos, Solitons & Fractals, 25, 521-524.
107. El Naschie, M.S. (2008) Chaos, Solitons & Fractals, 35, 7-12.
108. El Naschie, M.S. (2008) Chaos, Solitons & Fractals, 35, 67-70.
109. Krauss, L.M. (2001) Quintessence. Published by Vintage.
110. El Naschie, M.S. (2015) International Journal of High Energy Physics, 2, 13-21.
111. Mills, R. (1994) Space, Time and Quantra. W. H. Freeman, New York.
112. El Naschie, M.S. (2015) International Journal of Innovation in Science and Mathematics, 3, 254-265.
113. El Naschie, M.S. (2016) Journal of Quantum Information Science, 6, 57-61.
114. Kragh, H. (1996) Cosmology and Controversy. Princeton University Press, Princeton, NJ.
115. Mumford, D., Series, C. and Wright, D. (2002) Indra's Pearls. Cambridge University Press, Cambridge.

116. Rogers, B., Pennathur, S. and Adams, J. (2008) Nanotechnology. CRC Press, Boca Raton.
117. El Naschie, M.S. (2015) Problems of Nonlinear Analysis in Engineering Systems, 2, 1-16.
118. Hsu, J.-P. and Hsu, L. (2006) A Broad View of Relativity. World Scientific, Singapore.
119. El Naschie, M.S. (2015) American Journal of Nano Research and Applications, 3, 1-5.
120. El Naschie, M.S. (2016) Natural Science, 8, 152-159.
121. El Naschie, M.S. (2016) Journal of Modern Physics, 7, 729-736.
122. El Naschie, M.S. (2016) International Journal of Astronomy and Astrophysics, 6, 135-144.
123. El Naschie, M.S. (2016) American Journal of Computational Mathematics, 6, 185-199.
124. El Naschie, M.S. (2015) American Journal of astronomy and Astrophysics, 3, 77-86.
125. Babchin, A.J. and El Naschie, M.S. (2015) World Journal of Condensed Matter Physics, 6, 1-6.
126. El Naschie, M.S. (2016) Advances in Pure Mathematics, 6, 446-454.
127. Ho, M.E.N. and Giuseppe Vitiello, M.W. (2015) Global Journal of Science Frontier Research, 15, 61-80.
128. El Naschie, M.S. (2016) International Journal of Astronomy and Astrophysics, 6, 56-81.
129. Auffray, J.P. and El Naschie, M.S. (2016) Journal of Modern Physics, 7, 156-161.
130. El Naschie, M.S. (2016) Journal of Modern Physics, 7, 1420-1428.
131. El Naschie, M.S. (2016) Journal of Applied Mathematics and Physics, 4, 979-987.
132. Brownstein, J.R. and Moffat, J.W. (2007) Monthly Notices of the Royal Astronomical Society, 382, 29-47.

CHAPTER 10

Ultra Low Energy Results and Their Impact to Dark Matter and Low Energy Neutrino Physics

E. Bougamont[1], P. Colas[1], J. Derre[1], I. Giomataris[1], G. Gerbier[1], M. Gros[1], P. Magnier[1], X. F. Navick[1], P. Salin[2], I. Savvidis[3], G. Tsiledakis[1], J. D. Vergados[4]

[1]*IRFU, Centre d'Etudes de Saclay, Gif sur Yvette Cedex, France*

[2]*APC, Université Paris Denis Diderot, Paris, France*

[3]*Aristotle University of Thessaloniki, Thessaloniki, Greece*

[4]*University of Ioannina, Ioannina, Greece*

ABSTRACT

We present ultra low energy results taken with the novel Spherical Proportional Counter. The energy threshold has been pushed down to about 25 eV and single electrons are clearly collected and detected. To reach such a performance two low energy calibration systems have been successfully developed: a pulsed UV lamp extracting photoelectrons from the inner surface of the detector and various radioactive sources allowing low energy peaks through fluorescence processes. The bench mark result is the observation of a well resolved peak at 270 eV due to carbon fluorescence, which is a unique performance for such large massive detector. It opens up a new window in dark matter and low energy neutrino searches and it may allow the detection of neutrinos from a nuclear reactor or from supernova via neutrino-nucleus elastic scattering.

Keywords: Low Energy Neutrino; Dark Matter; Spherical Detector

1. INTRODUCTION

The development of massive low-background, low-energy threshold detectors is a challenge in contemporary physics.

The search of dark matter in the form of hypothetical Weakly Interacting Massive Particles (WIMP's) is under intense development and relies on the detection of low energy (keV scale) recoils produced by the elastic interaction of WIMP's with the nuclei of the detector [1-4]. The need to go to very low energies may become even more crucial, if the WIMP's turn out to be very light [5], since then, the energy transfer to the nucleus is expected to be smaller.

New generation of directional detection of dark matter is under development to perhaps allow for an unambiguous observation even in the presence of backgrounds by observing directional anisotropy of the recoils [6-9].

The scattering off nuclei of low-energy neutrinos (few MeV) via the neutral current interaction remains undetected thirty five years after its first description [10,11], a great challenge in neutrino physics. The resulting cross section is expected to be very large compared to other lowenergy neutrino interaction channels. The reason is that all neutrons in the nucleus contribute coherently. Again the detection consists in observing the recoiling nucleus. Because the neutrino is light, the nuclear recoil energy is extremely small and the challenge is to achieve a very low energy threshold (typically below 100 eV).

2. BRIEF DESCRIPTION OF SOME LOW ENERGY PROCESS REQUIRING A LOW ENERGY THRESHOLD

We will now consider some cases with desired low threshold, which are of special interest:

1) Supernova Neutrinos. Development of large mass detectors for low-energy neutrinos and dark matter may allow supernova detection via

neutrino-nucleus elastic scattering [12]. In this case the emitted neutrinos have a spectrum, which for each flavor is characterized by a FermiDirac distribution with essentially two parameters a temperature and a chemical potential [13]. The peak energies are approximately 15, 25 and 35 MeV for electron neutrinos, electron antineutrinos and all other flavors respectively. The distribution becomes negligible for energies above 70 MeV. For a detector of radius 4 m with a gas under 10 atm and a typical supernova in our galaxy, i.e. 10 kpc away, one finds 1, 30, 150, 600 and 1900 events for He, Ne, Ar, Kr and Xe respectively [13]. The threshold effect is shown in **Figure 1**.

2) The Oak Ridge neutron spallation source [14]. This has the great advantage that the neutrino spectrum is well understood. In this instance one has a) an electron neutrino spectrum with a peak energy at 35 MeV and a maximum at 50 MeV; b) a muon antineutrino spectrum that extends from zero to 50 MeV, which then falls abruptly to zero; and c) a discreet muon neutrino with an energy of 30 MeV. Thus this source may be the best place to test the coherent neutrino nucleus cross section mediated by the neutral current [15]. The total rate for a spherical detector filled with Xe and placed 50 m away from the source, assuming zero threshold is 18 events per ton per year. The effect of threshold is shown in **Figure 2** in the case of the electron neutrino.

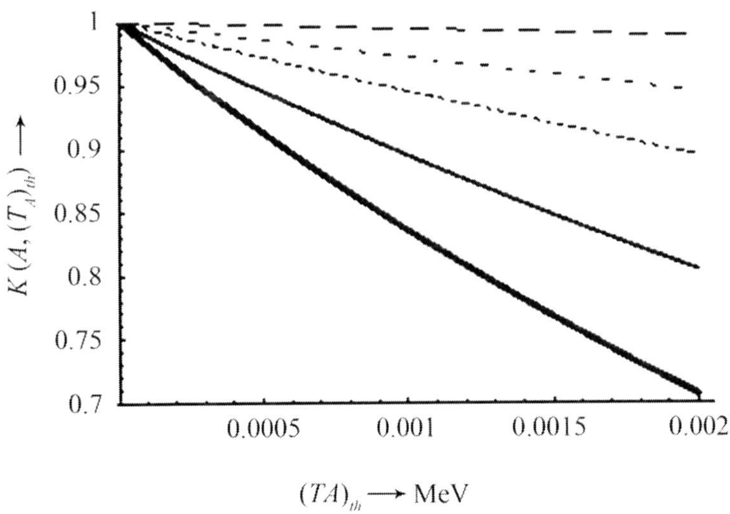

Figure 1. For neutrinos from a Supernova, the ratio of the total rate with a given energy threshold divided by that with zero threshold as a function of the threshold energy. The quenching factor has not been included. From top to bottom for He, Ne, Ar, Kr and Xe respectively.

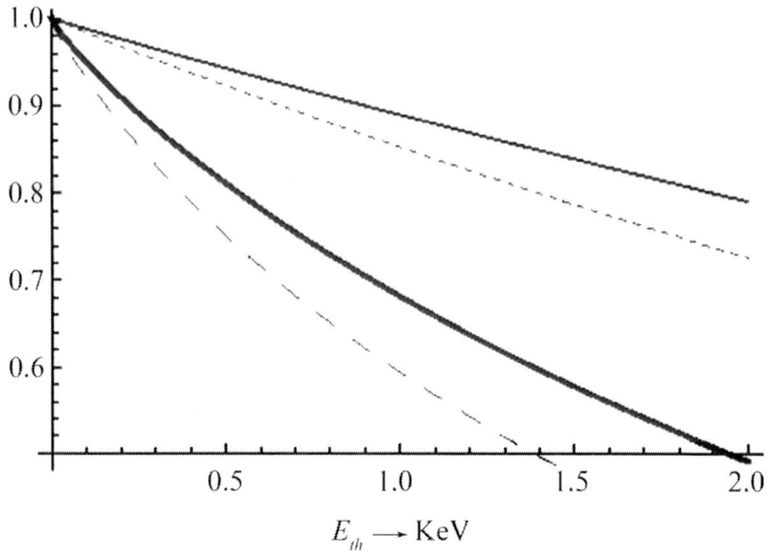

Figure 2. For electron-neutrinos from neutron spallation source, the ratio of the total rate with a given energy threshold divided by that with zero threshold as a function of the threshold energy for a Xe target. The solid (dashed) curves correspond to no nuclear form factor (nuclear form factor) respectively. The two lower curves take into account the effect of quenching.

3) Reactor Neutrinos. The spectrum of these neutrinos depends on the reactor and its mode of operation. This is known quite well for some reactors, e.g. that at Bugey. A reasonable approximation to a typical reactor spectrum can be found in the literature [16]. The normalized spectrum is given in **Figure 3**. We see that the peak energy (~1 MeV) and the average energy (~2.1 MeV) are quite low. So such neutrinos can be used to check the coherent neutrino nucleus scattering, if a low threshold (less than 0.1 keV) and high resolution are achieved.

4) Geoneutrinos. Neutrinos offer a good prospect for an accurate and detailed study of the Earth's interior [17]. Neutrinos of this origin have an energy distribution with a maximum of around 3 MeV and an average of 3.3 MeV. The shape [18,19], can be approximated by an analytic function shown on **Figure 4**. Up to now the detectors utilize the interaction of electronic antineutrinos with a proton to produce via charged current exchange a positron and a neutron. The kinematical neutrino threshold energy is 1.4 MeV. But

experimental thresholds are higher (typical 1.8 MeV) such as LENA's [20]. A neutral current experiment can improve such threshold at lower value as long as the nuclear target is relatively light and the threshold of the recoil nucleus very low (less than 0.1 keV). Admittedly the charged reaction has the advantage that by observing both the positron and the neutron one may have directionality information. It is however possible that this can also be achieved by neutral current experiments in which the direction of recoiling nucleus is also observed.

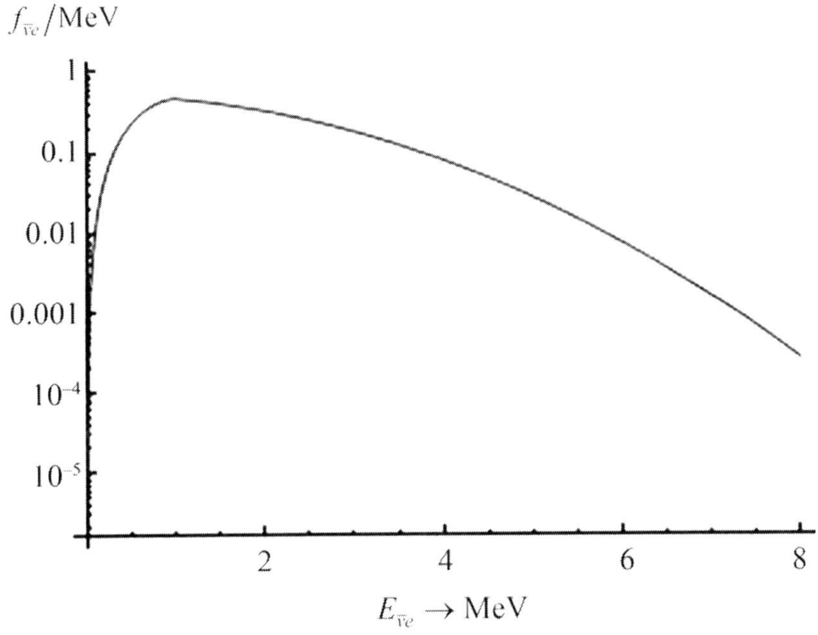

Figure 3. A typical reactor electron antineutrino spectrum.

Ultra low-noise germanium detectors with sub-keV energy capability have been recently developed and are operating at underground laboratories [21,22]. The small excess observed around 1 keV needs to be clarified and verified by detectors having a lower energy threshold.

Now we will report results at low energy obtained by using the novel Spherical Proportional Counter (SPC) which has been recently developed.

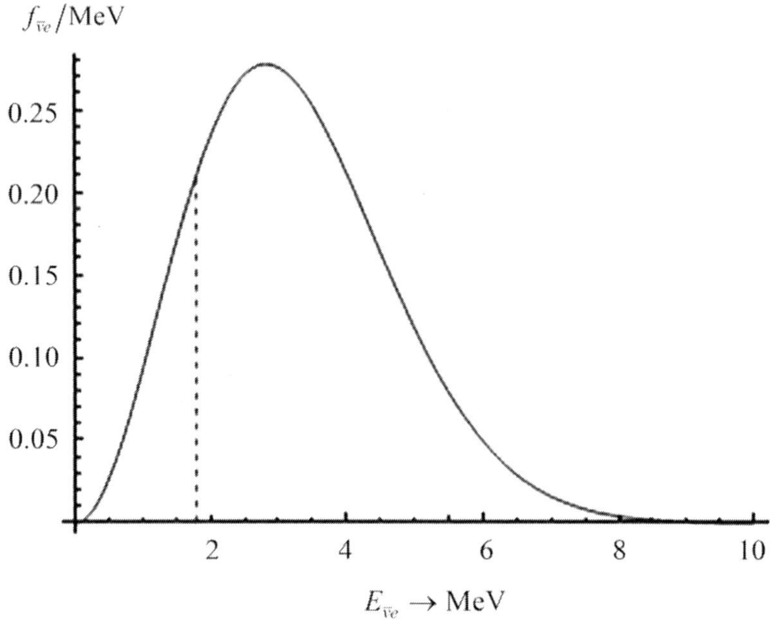

Figure 4. The expected normalized geoneutrino spectrum. The dashed line indicates the typical threshold (1.8 MeV) in charged current experiments.

3. DETECTOR DESCRIPTION

The detector consists of a large spherical copper vessel 1.3 m in diameter and a small metallic ball 16 mm in diameter located at the center of the drift vessel, which is the proportional counter. The ball is maintained in the center of the sphere by a stainless steel rod and is set at high voltage. A second electrode (umbrella-shaped) that is placed 24 mm away from the ball, is powered with an independent but lower high voltage, serving as electric field corrector. The detector operates in a seal mode: the spherical vessel is first pumped out and then filled with an appropriate gas at a pressure from few tens of mbar up to 5 bar. Electrons originating from ionization of the gas in the volume drift to the region of the central ball where the intense electric field allows gas amplification to occur. The produced signal is amplified through a charge amplifier and a shaper and is read-out by a 14 bit ADC. Detailed description of the detector, its electronics, its operation and its performance could be found in references [23-25].

In a previous paper [25] we reported results obtained at higher energy and we pointed out the excellent energy resolution obtained with radon nuclide and its daughters. In this work we will focus our studies to detect very-low energy gamma or X-rays emitted by radioactive sources or fluorescence process.

4. LOW ENERGY CALIBRATION AND RESULTS

4.1. Results with Radioactive Sources and Fluorescence X-Rays

For this study we are using a gas filling of Argon with 2% admixture of CH_4 at various pressures. The pressure (P in mb), the high voltage of the ball (HV1 in V), the high voltage of the umbrella field corrector (HV2 in V) and the gain (G) of the amplifier-shaper in all measurements in the section, are written in the top-right box on each of the following figures.

Calibration of the counter was initially performed using a ^{109}Cd source by irradiating the gaseous volume through a thin 200 mm aluminum window.

Figure 5 shows the energy spectrum with the 22 keV line from the ^{109}Cd source and the 8 keV line, which is an induced fluorescence at the copper vessel. The energy resolution is quite satisfactory 6% and 9% (FWHM) respectively.

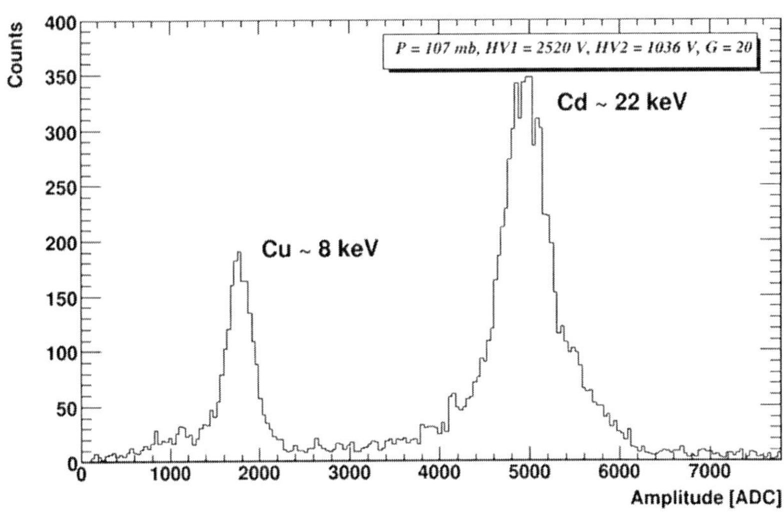

Figure 5. The spectrum of the ^{109}Cd source with the energy lines of 8 and 22 keV respectively.

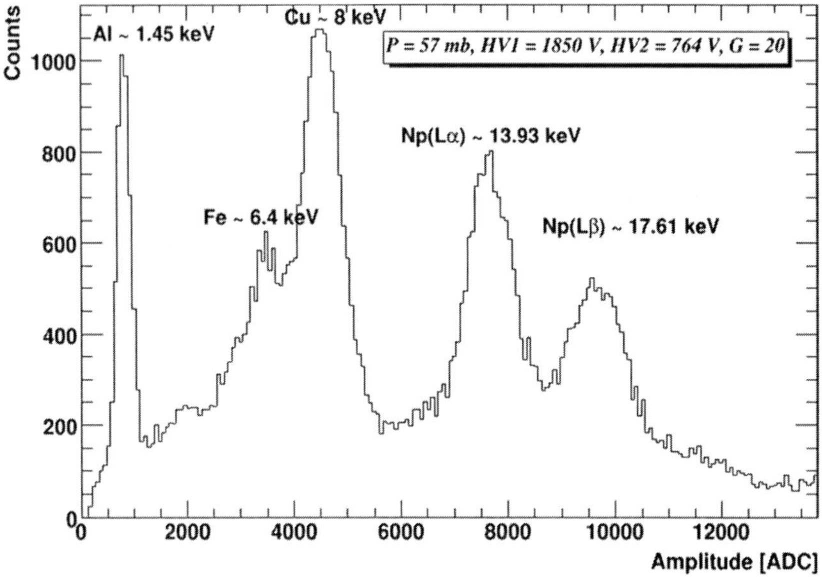

Figure 6. Peaks observed from the ^{241}Am radioactive source. From left to right we observe the Aluminium, Iron and Copper peaks followed by the Neptunium peaks.

Gamma fluorescence is adequate for producing fluorescence lines in the range above several keV. However, it is very difficult to produce low energy calibration lines below a few keV. In order to create lower energy X-rays we have used an ^{241}Am source, which decays by the following process: ^{241}Am $\rightarrow(^{237}\text{Np})^{*}$ + ^{4}He + 5.6 MeV. The ^{237}Np nucleus then decays into a lower energy state by emitting a 59.537 keV gamma ray and other L rays, which are used to fluoresce the elements. The source was evaporated to a stainless steel holder and attached to the sensor rod at middle distance; the source then is covered by thin foils with adequate thickness to totally absorb the 5.6 MeV alpha emitted, leaving only gamma rays and fluorescence induced X-rays to pass into the gas volume.

Covering the source with a 20 mm thick aluminum foil we were able to fluoresce the K X-rays ranging from Aluminum whose K electron has a binding energy of 1.56 keV, to Cu whose K electron has a binding energy of 8.98 keV.

Figure 6 shows the obtained low energy spectrum. From left to right we observe the aluminum line (1.45 keV), the iron line (6.4 keV), the copper line (8.0 keV) and two lines emitted by $^{237}N_p$nucleus at 13.93 keV (X-ray La) and 17.61 keV (X-ray Lb).

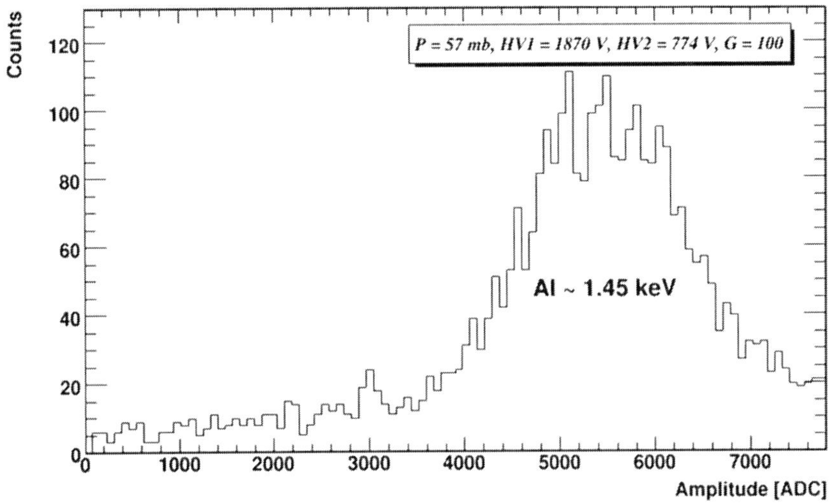

Figure 7. An X-ray spectrum showing its characteristic Aluminum K X-ray energy peak.

By increasing the gain of the amplifier 5 times and adjusting its settings, we were able to push the higher part of the spectrum outside the ADC range and keep in the ADC acceptance only the aluminum 1.45 keV peak as shown in **Figure 7**. The energy threshold is clearly below 100 eV and the low energy background level is flat.

In order to obtain even lower energy calibration lines we replaced the aluminum foil by a thinner 10 mm one and we attached a 20 mm polypropylene foil. Therefore, the alpha particle that is crossing the aluminum foil, is fully absorbed by the polypropylene foil and induces both aluminum and carbon fluorescence as shown in **Figure 8**.

By increasing the gain of the amplifier by a factor of 5, we are able to push the higher part of the spectrum outside the ADC range and keep in the ADC acceptance only the Carbon 270 eV peak as well as the Aluminum (1.45 keV) as shown in **Figure 9**.

High gain combined with low electronic noise can provide energy thresholds clearly below 100 eV. **Figure 10** shows the energy spectra at different gains of the amplifier when no source is used. By applying a cut at the rise time of the signal (which actually provides the depth of the ionized electrons produced in the gas) we can exclude the signal induced by cosmic rays and measure only the Copper energy line of 8 keV (plotted with dashed line in **Figure 10**). Then, by increasing the gain of the amplifier 20 times (from Gain = 10 to Gain = 200), we keep in the ADC acceptance only the ultra low energy region (plotted with full line in Figuew 10).

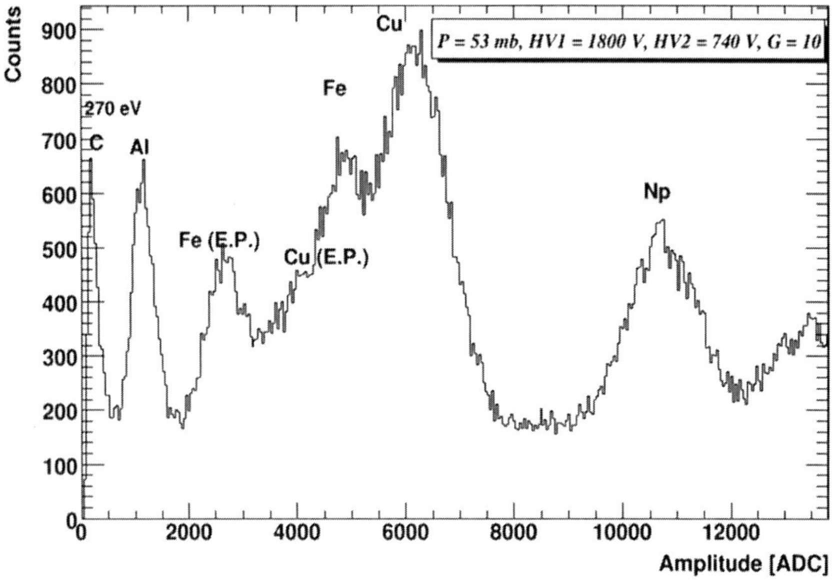

Figure 8. Peaks observed from the ^{241}Am radioactive source through aluminium and polypropylene foil. On the left the Carbon (270 eV) peak is shown, followed by the Aluminium peak (1.45 keV), the escape peak (E.P.) of Iron in Argon (3.3 keV), the escape peak of Copper in Argon (5 keV), the Iron peak (6.4 keV), the Copper peak (8 keV) and the Neptunium peak (13.93 keV).

The value of 1000 ADC corresponds to ~150 eV and the peak at ~50 eV is compatible with single electrons (see Section 2). Thus, the detection threshold of the Spherical Proportional Counter is as low as 25 eV. From ~750 eV to 1200 eV the energy spectrum looks flat (**Figure 10**—full line). At energies lower than ~750 eV, an increase of the slope of the energy spectrum is observed and is followed by a steep rise at very low energies. The exponential shape of the spectrum has to be confirmed by underground measurements in the LSM laboratory, which are on-going.

4.2. Results Using a UV Flash Lamp

A UV window made by MgF_2 crystal was installed in one of the sphere openings.

A hydrogen relaxation flash lamp has been used to produce UV photons in the far ultra violet range. The discharge is also producing a fast signal which is observed through a capacitor and after adequate attenuation can serve as a fast trigger.

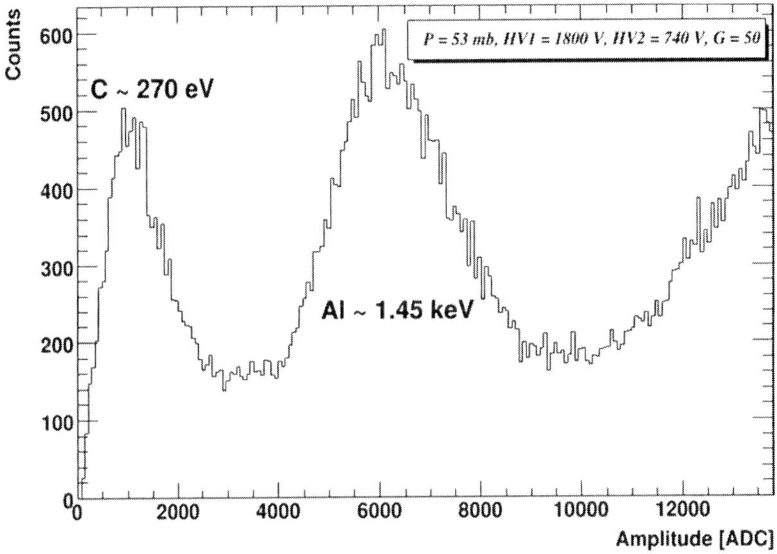

Figure 9. From left the Carbon (270 eV) peak followed by the Aluminum peak (1.45 keV).

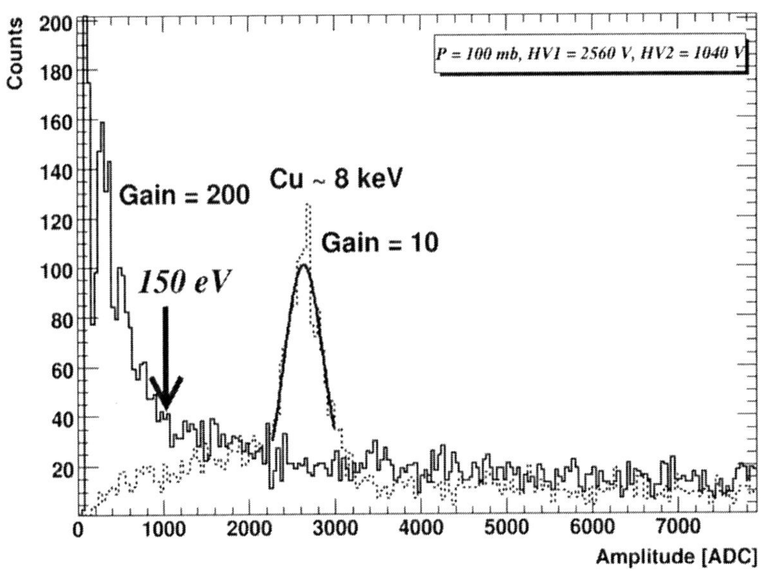

Figure 10. The energy spectrum at Gain = 10 with the Copper peak of 8 keV (dashed line) and at Gain = 200 where the single electron peak at ~50 eV is clearly visible (full line).

For the present study, all results have been obtained without trigger. Photons are crossing the UV entrance window and hitting copper of the internal vessel producing photoelectrons by photoelectric effect. Electrons are being extracted and drifting through the radial field to the central ball where they are amplified and collected. The energy calibration has been done using the 22 keV photon X decay of ^{109}Cd source and the 8 keV Cu fluorescence.

The UV light flux of the lamp was adjusted by introducing adequate light attenuators down to the single photo-electron extraction on the Copper. Each data sample was acquired without trigger during the same running time. The subtraction of the amplitude spectrum of the sample lamp off from that of lamp on suppresses the huge background due to cosmic rays. The gas used was a mixture of Ne with 7% CH_4 at a pressure of 100 mbar.

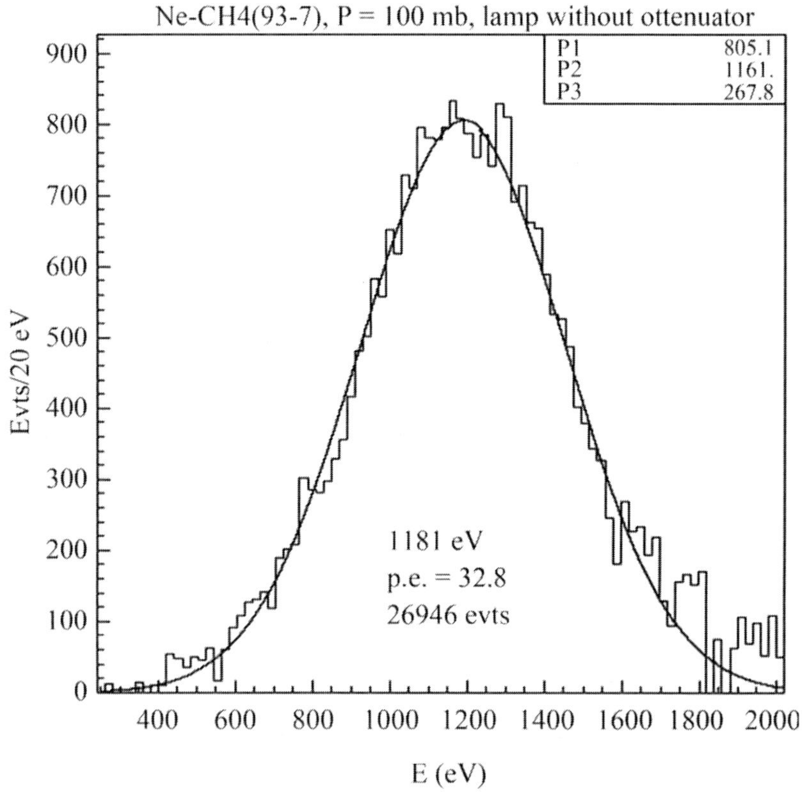

Figure 11. Energy spectrum due to photelectron extraction from the copper by the UV lamp without attenuation.

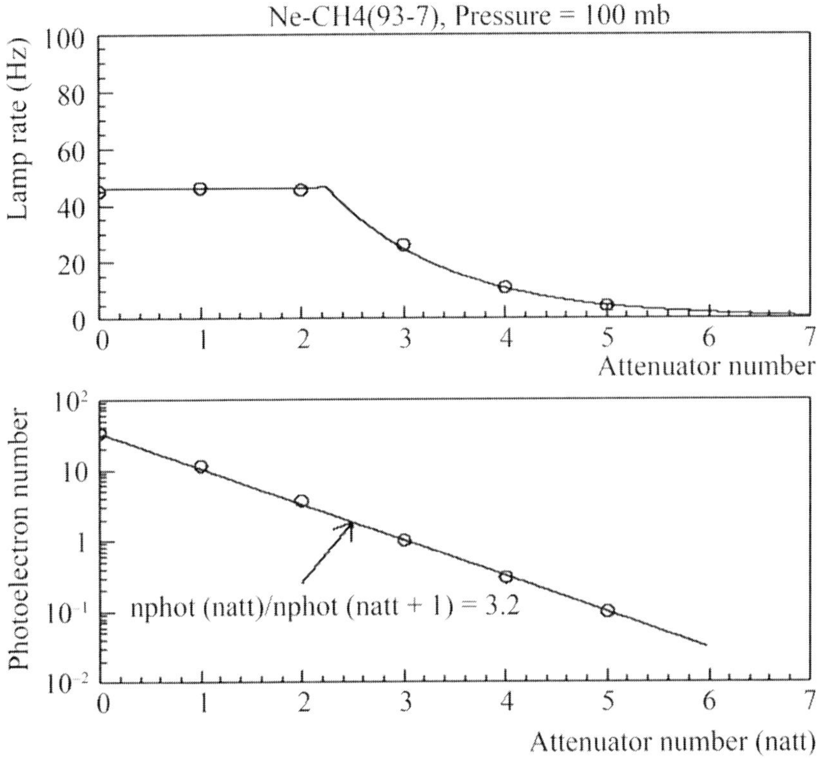

Figure 12. The frequency (top) of detected lamp flashes and the mean number of photoelectrons (bottom) as a function of the number of attenuators.

Results are shown on **Figure 11** without light attenuator. The amplitude spectrum of the lamp signal is fitted with a Gaussian function. The mean energy value divided by 36 eV, the photon energy for ionization of gas, gives a mean value of 32.8 photoelectrons extracted by each flash of the lamp.

The lamp light was then gradually attenuated by 1, 2, up to 5 identical attenuators. Each distribution is well fitted by a Polya function with an offset. With less than 3 attenuators, the number of events is stable, which means for those sample all the flashes are detected with a mean frequency close to 45 Hz (27,000 events divided by 600 s the run time). In fact the probability not to detect the flash is very low, 2.5% with 2 attenuators, as expected by the Poisson distribution with the mean value given by the fit. With more attenuators the mean frequency is decreasing because the probability of producing zero photoelectrons per flash is increasing exponentially (**Figure 12**—top).

The number of photoelectrons per flash given by the fit decreases asymptotically to one as expected. The mean number of extracted electron per flash decrease exponentially with a reduction factor close to 3 as expected (**Figure 12**—bottom), even with a large number of attenuators, pointing to a negligible inefficiency to detect the single photelectrons.

An example of a good fit using Polya distribution is the case of 3 attenuators giving a mean value of 1.7 photoelectrons per flash as shown in **Figure 13**.

The various distributions for all used attenuators are shown in **Figure 14**. We observe a decrease of the mean value of each distribution up to about 3 attenuators. After that the mean value is not decreasing any more assuring we are at the single electron level.

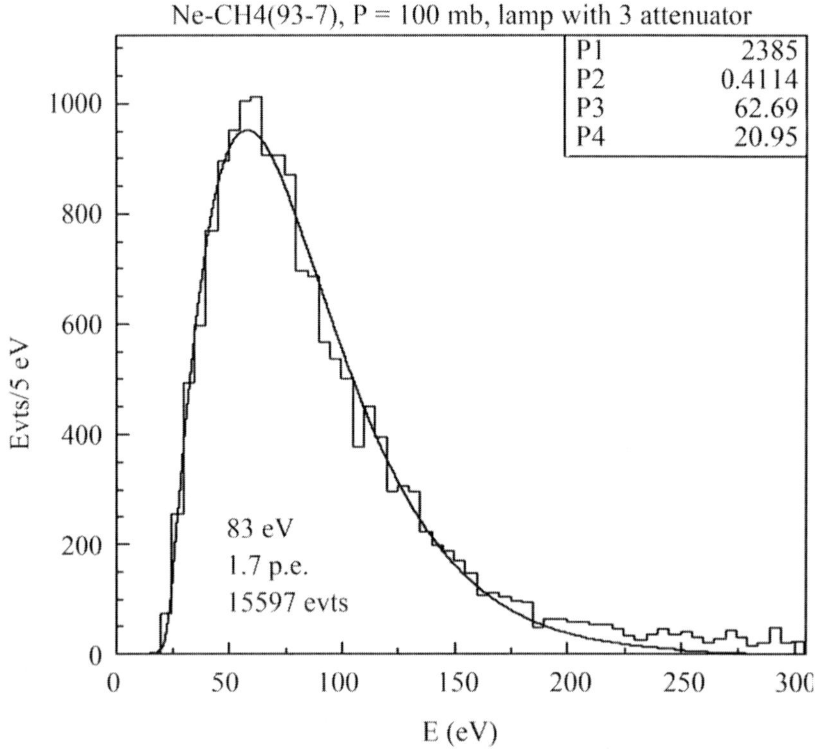

Figure 13. Energy distribution in the case of 3 attenuators.

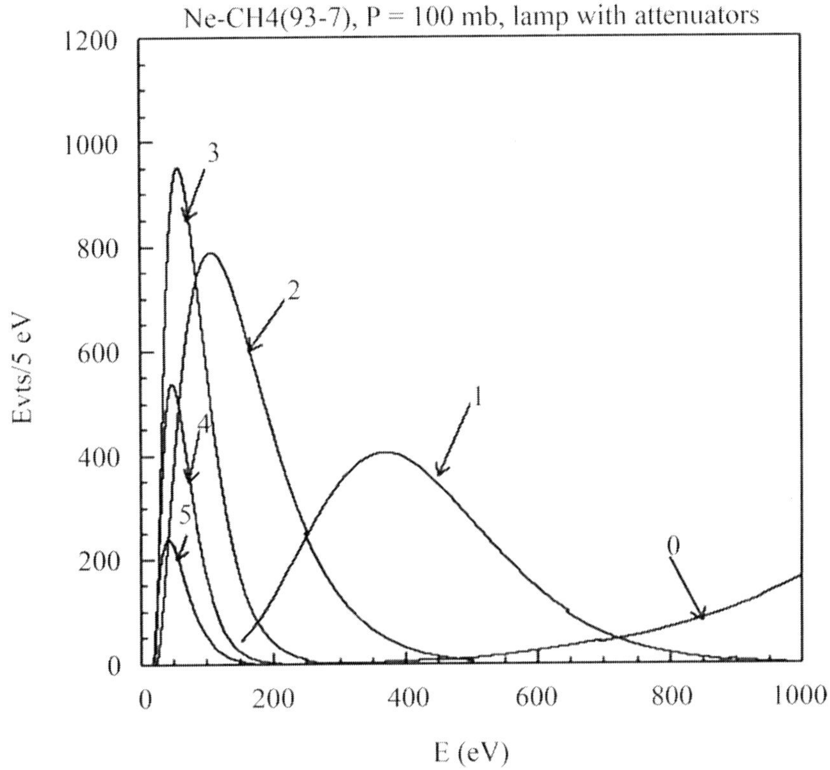

Figure 14. Fitted energy spectrum for several attenuators.

5. FUTURE DEVELOPMENTS AND APPLICATIONS

The counter can be applied for low-background experiments, which require a low energy threshold. An example is the detection of the coherent neutrino-nucleus scattering producing sub keV ion recoils. A spherical detector of radius 4 m and employing Xe gas at a pressure of 10 Atm will detect about 1000 events for a typical supernova explosion at 10 kpc. A world wide network of several such simple, stable and low cost supernova detectors is proposed.

All results that are shown here were taken in the lab using the spherical detector of 1.3 m diameter that is made by raw materials. For underground tests another similar detector of the same diameter has been installed in the LSM laboratory in Modane (Frejus) under 1700 m of rock (4800 meters water equivalent) providing protection from cosmic rays. Comparisons between measurements in the subkeV energy region taken with both detectors are carried out in order to understand the background level and optimize the detector in terms of sensitivity and noise background. The development of a new spherical prototype 70 cm in diameter made of low radioactivity materials is in progress and will be installed in the LSM laboratory. By using a mixture of gas containing ^{3}He, it will be possible to measure not only the thermal neutron component [26] but also the fast neutron energy spectrum by taking profit of the 1 to few barn cross section of capture above 100 keV neutron energy. By adding a lead and Polyethylene shields, it will also be possible to measure the residual background at low energy and assess potential sensitivity to dark matter particle search and coherent neutrino-nucleus scattering.

6. CONCLUSION

We have developed a new detector with large mass and low sub-keV energy threshold. This is a new type of radiation detector based on the radial geometry with spherical proportional amplification read-out that combines simplicity, robustness and low cost. Several applications are open arising from low energy neutrino physics, dark matter search, Supernova detection to neutron background measurement.

ACKNOWLEDGEMENTS

E. Bougamont, P. Colas, J. Derre, I. Giomataris, G. Gerbier, M. Gros, P. Magnier, X. F. Navick, P. Salin, I. Savvidis, G. Tsiledakis, J. D. Vergados We would like to thank Dr. T. Papaevangelou for exciting discussions.

REFERENCES

1. K. Nakamura, et al., "Review of Particle Physics," Journal of Physics G: Nuclear and Particle Physics, Vol. 37, No. 7A, 2010, pp. 1-1422.
2. Giomataris and J. D. Vergados, "Neutrino Properties studied with a Triton Source and a Large Spherical TPC," Nuclear Instruments and Methods in Physics Research Section A, Vol. 530, No. 3, 2004, pp. 330-358.
3. J. I. Collar and Y. Giomataris, "Possible Low-Background Applications of Micromegas Detector Technology, Nuclear Instruments and Methods in Physics Research Section A, Vol. 471, No. 1-2, 2001, pp. 254-259.
4. R. Bernabei, et al., "New Results from DAMA/LIBRA," The European Physical Journal C, Vol. 67, No. 1-2, 2010, pp. 39-49.
5. Dedes, I. Giomataris, K. Suxho and J. D. Vergados, "Searching for Secluded Dark Matter via Direct Detection of Recoiling Nuclei as Well as Low Energy Electrons," Nuclear Physics B, Vol. 826, No. 1-2, 2010, pp.148-173.

6. J. Monroe and P. Fisher, "Neutrino Backgrounds to Dark Matter Searches," Physical Review D, Vol. 76, No. 3, 2007, Article No: 033007.
7. D. Santos, et al., "Ionization Quenching Factor Measurement of 4He," 2008. arXiv: 0810.1137 [astro-ph].
8. C. Grignon, et al., "A Prototype of a Directional Detector for Non-Baryonic Dark Matter Search: MIMAC (MicroTPC Matrix of Chambers)," JINST, Vol. 4, 2009, Article No. P11003.
9. S. Ahlen, et al., "The Case for a Directional Dark Matter Detector and the Status of Current Experimental Efforts," International Journal of Modern Physics A, Vol. 25, No. 1, 2010, pp. 1-51.
10. D. Z. Freedman, et al., "The Weak Neutral Current and Its Effects in Stellar Collapse," Annual Review of Nuclear and Particle Science, Vol. 27, 1977, pp. 167-207.
11. Drukier and L. Stodolsky, "Principles and Applications of a Neutral-Current Detector for Neutrino Physics and Astronomy," Physical Review D, Vol. 30, No. 11, 1984, pp. 2295-2309.
12. J. Horowitz, K. J. Coakley and D. N. McKinsey, "Supernova Observation via Neutrino-Nucleus Elastic Scattering in the CLEAN Detector," Physical Review D, Vol. 68, No. 2, 2003, p. 023005.
13. Y. Giomataris and J. D. Vergados, "A Network of Neutral Current Spherical TPCs for Dedicated Supernova Detection," Physics Letters B, Vol. 634, No. 1, 2006, pp. 23-29.
14. J. D. Vergados, F. T. Avignone III and I. Giomataris, "Coherent Neutral Current Neutrino-Nucleus Scattering at a Spallation Source: A Valuable Experimental Probe," Physical Review D, Vol. 79, No. 11, 2009, Article No: 113001.
15. K. Scholberg, "Prospects for Measuring Coherent Neutrino-Nucleus Elastic Scattering at a Stopped-Pion Neutrino Source," Physical Review D, Vol. 73, No. 3, 2006, Article No: 033005.
16. B. R. Davis, P. Vogel, F. M. Mann and R. E. Schenter, "Reactor Antineutrino Spectra and Their Application to Antineutrino-Induced Reactions," Physical Review C, Vol. 19, No. 6, 1979, pp. 2259-2266.
17. T. Araki, et al., "Experimental Investigation of Geologically Produced Antineutrinos with KamLAND," Nature, Vol. 436, No. 7050, 2005, pp. 499-503.
18. S. Dye, "Testing Geological Models with Terrestrial Antineutrino Flux Measurements," 2009. arXiv:0912.2775 [physics.geo-ph].
19. P. Ila, et al., "Considerations for a Dedicated Geoneutrino Detector for Geosciences," 2009. arXiv:0902.3607 [physics.ins-det].
20. K. A. Hochmuth, et al., "Probing the Earth's Interior with the LENA Detector," Earth Moon Planets, No. 99, No. 1-4, 2006, pp. 253-264.
21. H. T. Wong, "Ultra-Low-Energy Germanium Detector for Neutrino-Nucleus Coherent Scattering and Dark Matter Searches," Modern Physics Letters A, Vol. 23, No. 17-20, 2008, pp. 1431-1442.
22. C. E. Aalseth, et al., "Results from a Search for LightMass Dark Matter with a P-Type Point Contact Germanium Detector," 2010. arXiv:1002.4703 [astro-ph.CO].
23. S. Aune, et al., "NOSTOS: A Spherical TPC to Detect Low Energy Neutrinos," AIP Conference Proceedings, Vol. 785, 2005, pp. 110-118.
24. Giomataris, et al., "NOSTOS Experiment and New Trends in Rare Event Detection," Nuclear Physics B— Proceedings Supplements, Vol. 150, 2006, pp. 208-213.
25. Giomataris, et al., "A Novel Large-Volume Spherical Detector with Proportional Amplification Read-Out," JINST, Vol. 3, 2008, Article No: P090007.
26. Giomataris, et al., "Spherical TPC Development and Trends," Journal of Physics: Conference Series, Vol. 179, 2009, Article No: 012003.

CHAPTER 11

Finite Temperature Asymmetric Nuclear Matter Using Pion Dressing

Saroj Kumar Sahu

Department of Physics, Banki College (Autonomous), Cuttack, India

ABSTRACT

We study the asymmetric nuclear matter in a nonperturvative manner at finite temperatures using thermofield dynamics method. The nucleon-meson interaction is taken to examine the binding energy (E_B), pressure (P) for various proton fractions.

Keywords: Binding Energy, Equation of State, Pion Dressing, Hot Dense Nuclear Matter

1. INTRODUCTION

The properties of hot dense nuclear matter are very important in the context of neutron stars [1] as well as in high-energy heavy-ion collisions experiments. In those experiments, nuclei undergo collisions [2] and produce states of nuclear matter at different temperatures at conditions far from those normally encountered in low energy collisions. At low temperatures, this can be attained in medium energy heavy-ion reactions; no such exotic state can be produced but there is the possibility of interesting liquid-gas phase transitions leading to the breakup of heated nuclei into small clusters or droplets of nucleons [3] .

Such phase transitions have been identified in multifragmentation experiments and in the crusts of neutron stars [4] [5] . The possibilities of such phase transitions were previously considered by several authors using different approaches [6] - [8] . Walecka [9] in a attempt, developed a method consisting of interactions of nucleons with a neutral scalar field σ and ρ mesons as well as neutral vector meson ω. Variations of the same model have also been considered including cubic and quartic terms in the σ fields to reproduce correct bulk modulus of nuclear matter [10] . These calculations however use meson fields as classical, and use a σ-field which is not observed.

A model for infinite nuclear matter consisting of interacting nucleons and pions was considered in [11] . The Pion Nucleon interaction was taken with off mass shell of pion. The scalar isoscalar pion condensates simulated the effects of σ mesons [12] with the short distance repulsion arising from composite structure of nucleons and/or through vector meson exchanges. This appeals aesthetically as classical σ fields arise from quantum mechanical structures and also has a stronger phenomenological appeal as σ mesons have not been found in nature. With a similar approach we shall reconsider nuclear matter at finite temperatures with pion dressing [13] with off mass shell. The methods of thermofield dynamics [14] fit naturally for this purpose because here statistical average is done through an expectation value over a "thermal vacuum" [11] in an extended Hilbert space.

The article is organised as follows. In Section 2, we review the thermofield dynamics to consider hot and dense nuclear matter and obtain expressions for temperature-dependant pressure P, binding energy B and nuclear density ρ. In Section 3, we evaluate numerically the above applying variational technique, study their characteristics and discuss the results so obtained.

2. FORMALISM

We consider the effective Hamiltonian for pion nucleon interaction at zero temperature [15] -[17] as

$$\mathcal{H}(x)=\mathcal{H}_N(x)+\mathcal{H}_{int}(x)+\mathcal{H}_M(x), \tag{1}$$

where $\mathcal{H}_N(x)$, $\mathcal{H}_{int}(x)$ and $\mathcal{H}_M(x)$ are the Hamiltonians for the free nucleon part, the pion-interaction part and the free meson part respectively. Here $\mathcal{H}_N(x)$ is given by

$$\mathcal{H}_N(\boldsymbol{x}) = \psi_I^\dagger(\boldsymbol{x})\,\varepsilon_x\,\psi_I(\boldsymbol{x}), \tag{2}$$

and the effective Hamiltonian $\mathcal{H}_{int}(\boldsymbol{x})$ for pion nucleon πN interaction

$$\mathcal{H}_{int}(\boldsymbol{x}) = \psi_I^\dagger(\boldsymbol{x})\left[-\frac{iG}{2\epsilon_x}\boldsymbol{\sigma}\cdot\boldsymbol{p}\varphi + \frac{G^2}{2\epsilon_x}\varphi^2\right]\psi_I(\boldsymbol{x}). \tag{3}$$

We have taken $\epsilon_x = \left(M^2 - \nabla_x^2\right)^{1/2}$, with M as the mass of the nucleon. The free meson part of the Hamiltonian is given by

$$\mathcal{H}_M(\boldsymbol{x}) = \frac{1}{2}\left[\dot{\varphi}_i^2 + (\nabla\varphi_i)\cdot(\nabla\varphi_i) + m_\pi^2\varphi_i^2\right], \tag{4}$$

where m_π denotes the mass of the meson and $\varphi = \tau_i\varphi_i$. The pion field expansion in Equation (4) in terms of annihilation and creation operators is given by

$$\varphi_i(x) = \sqrt{\frac{1}{2\omega_x}}\left(a_i(x) + a_i^\dagger(x)\right)$$

$$\text{and}\quad \dot{\varphi}_i(x) = i\sqrt{\frac{\omega_x}{2}}\left(a_i^\dagger(x) - a_i(x)\right), \tag{5}$$

with $\omega_x = \left(m^2 - \nabla_x^2\right)^{1/2}$. We now study asymmetric nuclear matter at finite temperature using thermofield dynamics method. Here the thermal average of a quantum operator $\hat{O}$ given in [13] [14] with $\beta = 1/kT$, as

$$\left\langle \hat{O} \right\rangle_\beta = \frac{Tr\left(e^{-\beta H}\hat{O}\right)}{Tr\left(e^{-\beta H}\right)} \tag{6}$$

The above reduces to ground state expectation, in zero temperature limit value for the operator $\hat{O}$, shown below as

$$\lim_{\beta\to\infty}\left\langle \hat{O}\right\rangle_{\beta} = \lim_{\beta\to\infty}\frac{\langle 0|\hat{O}|0\rangle e^{-\beta\epsilon_0}+\langle 1|\hat{O}|1\rangle e^{-\beta\epsilon_1}+\cdots}{e^{-\beta\epsilon_0}+e^{-\beta\epsilon_1}+\cdots}$$
$$= \lim_{\beta\to\infty}\frac{\langle 0|\hat{O}|0\rangle+\langle 1|\hat{O}|1\rangle e^{-\beta(\epsilon_1-\epsilon_0)}+\cdots}{1+e^{-\beta(\epsilon_1-\epsilon_0)}+\cdots}=\langle 0|\hat{O}|0\rangle, \tag{7}$$

where $|0\rangle$ is the lowest energy state. In thermofield method, one generalises [13] to the case of finite temperature and for some $|0(\beta)\rangle$, the relationship holds

$$\left\langle \hat{O}\right\rangle_{\beta} = Tr\left(e^{-\beta H}\hat{O}\right)Tr\left(e^{-\beta H}\right)\equiv\langle 0(\beta)|\hat{O}|0(\beta)\rangle, \tag{8}$$

where $|0(\beta)\rangle$ is defined as the "thermal vacuum". Here corresponding to physical operator a, a "tilde" operator $\tilde{a}$ is introduced to denote thermal modes. At finite temperature the ground state is replaced by $|0(\beta)\rangle$ given as [13]

$$|0(\beta)\rangle\equiv U_B(\beta)|vac\rangle=\exp\left(\int\theta_B(\boldsymbol{k},\beta)\left(a(\boldsymbol{k})^{\dagger}\tilde{a}(-\boldsymbol{k})^{\dagger}-h.c.\right)d\boldsymbol{k}\right)|vac\rangle, \tag{9}$$

where $\tilde{a}(-\boldsymbol{k})^{\dagger}$ in the above corresponds to the extra Hilbert space. The function $\theta_B(k,\beta)$ is calculated through minimization of thermodynamic potential density given as

$$\Omega=\left(E(\beta)-\frac{1}{\beta}S_B+\mu N\right) \tag{10}$$

where μ, the chemical potential, N, number of nucleons and the entropy density

$$S_B=-\frac{3}{(2\pi)^3}\int d\boldsymbol{k}\left[\sinh^2\theta_B(\boldsymbol{k},\beta)\ln\left(\sinh^2\theta_B(\boldsymbol{k},\beta)\right)-\cosh^2\theta_B(\boldsymbol{k},\beta)\ln\left(\cosh^2\theta_B(\boldsymbol{k},\beta)\right)\right].$$

For zero chemical potential and for free fields, extremization of the free energy then yields

$$\sinh^2\theta_B = \frac{1}{e^{\beta\omega(\boldsymbol{k},\beta)}} - 1 \tag{11}$$

with Hamiltonian density as $\int a^{\dagger}(z)\omega_z a(z)\mathrm{d}z\ \mathcal{H}^0 =$ so that

$$E = \frac{1}{(2\pi)^3}\int \sinh^2\theta_B\,\omega(\boldsymbol{k})\,\mathrm{d}\boldsymbol{k}$$

If we substitute this value the free energy density becomes

$$\mathcal{F} = E(\beta) - \frac{1}{\beta}S_B = \frac{1}{\beta}\frac{1}{(2\pi)^3}\int \ln\left(1 - e^{-\beta\omega(\boldsymbol{k})}\right)\mathrm{d}\boldsymbol{k}$$

Similarly for the fermionic sector the thermal vacuum is

$$|0(\beta)\rangle \equiv U_F(\beta)|vac\rangle = \exp\left(\int \theta_F(\boldsymbol{k},\beta)\left(\psi_I(\boldsymbol{k})^{\dagger}\tilde{\psi}_I(-\boldsymbol{k})^{\dagger} - h.c.\right)\mathrm{d}\boldsymbol{k}\right)|vac\rangle, \tag{12}$$

where $\tilde{\psi}_I^{\dagger}$ corresponds to the creation of the fermionic thermal modes. The entropy is given as

$$S_F = -\frac{\gamma}{(2\pi)^3}\int \mathrm{d}\boldsymbol{k}\left[\sin^2\theta_F(\boldsymbol{k},\beta)\ln\left(\sin^2\theta_F(\boldsymbol{k},\beta)\right) + \cos^2\theta_F(\boldsymbol{k},\beta)\ln\left(\cos^2\theta_F(\boldsymbol{k},\beta)\right)\right],$$

and $\theta_F(\boldsymbol{k},\beta)$ is given as [13]

$$\sin^2\theta_F = \frac{1}{e^{\beta(\omega(\boldsymbol{k},\beta)-\mu)}} + 1, \tag{13}$$

where μ is the chemical potential corresponding to baryon number conservation and $\omega(\boldsymbol{k},\beta) = \sqrt{k^2 + M^2}$ for free fermions of mass M. In this methodology the correct distribution function with extra thermal modes can be obtained and this enables us to have the temperature dependent background off-shell pion pair configuration given as

$$|f,\beta\rangle = U_I(\beta)|f\rangle = U_I(\beta)U|vac\rangle \tag{14}$$

with

$$U_I(T) = \exp\left(B_I^\dagger(T) - B_I(T)\right) \tag{15}$$

and

$$B_I(T)^\dagger = \frac{1}{2}\int \theta_B(\boldsymbol{k},T) b_i(\boldsymbol{k})^\dagger \tilde{b}_i(-\boldsymbol{k})^\dagger \,\mathrm{d}\boldsymbol{k} \tag{16}$$

In the above,

$$b_i(\boldsymbol{k})^\dagger = U a_i(\boldsymbol{k}) U^\dagger \tag{17}$$

In nucleon sector for fermions

$$U_{II}(T) = \exp\left(B_{II}^\dagger(T) - B_{II}(T)\right) \tag{18}$$

with

$$B_{II}(T)^\dagger = \frac{1}{2}\int \theta_F(\boldsymbol{k},T) \psi_I(\boldsymbol{k})^\dagger \tilde{\psi}_I(-\boldsymbol{k})^\dagger \,\mathrm{d}\boldsymbol{k} \tag{19}$$

where $\theta_F(\boldsymbol{k},T)$ will be determined later. We then have the nuclear matter density

$$\rho = \langle vac|U_{II}(T)^\dagger \psi_\alpha(\boldsymbol{x})^\dagger \psi_\alpha(\boldsymbol{x}) U_{II}(\beta)|vac\rangle = \frac{\gamma}{(2\pi)^{-3}}\int \mathrm{d}\boldsymbol{k}\, \sin^2\theta_F. \tag{20}$$

Here the thermodynamic potential [18] is given by

$$\mathcal{F} = \epsilon - \mu_p \rho_p - \mu_n \rho_n - TS, \tag{21}$$

where μ_p and μ_n are chemical potentials of proton and neutron respectively, ρ_n and ρ_p are neutron and proton densities and S is the entropy of the asymmetric nuclear matter $(S = S_p + S_n)$. Clearly, with $\sin^2\theta_F = \Theta(k_f - k)$, Equation (13) gives $\rho = \gamma k_f^3/6\pi^2$ of zero temperature. $\sin^2\theta_F$ for the interacting system will be determined here from the construction of the thermodynamic potential. We thus have for the nucleon kinetic term temperature dependent as

$$\epsilon_f = \gamma(2\pi)^{-3}\int \mathrm{d}\boldsymbol{k}\,\frac{k^2}{2M}\sin^2\theta_F \tag{22}$$

The temperature dependent kinetic energy due to the mesons is given by

$$\epsilon_k = 3(2\pi)^{-3}\int \mathrm{d}\boldsymbol{k}\,\omega(\boldsymbol{k})\left[\sinh^2\tilde{f}(\boldsymbol{k})\cosh 2\theta_B(\boldsymbol{k},\beta) + \sinh^2\theta_B(\boldsymbol{k},\beta)\right] \tag{23}$$

where $\omega(\boldsymbol{k}) = \sqrt{\boldsymbol{k}^2 + m^2}$ and $f(\boldsymbol{k})$ is pion dressing function [13] which extremises energy, pressure and entropy etc of nuclear matter. Then the interaction energy density becomes

$$\epsilon_{int} = \frac{G^2\rho}{2M}J \tag{24}$$

where

$$J = \frac{3}{(2\pi)^3}\int \frac{\mathrm{d}\boldsymbol{k}}{\omega(\boldsymbol{k})}\left(\frac{\sinh 2\tilde{f}(\boldsymbol{k})\cosh 2\theta_B}{2} + \sinh^2\tilde{f}(\boldsymbol{k})\cosh 2\theta_B + \sinh^2\theta_B\right). \tag{25}$$

we shall now assume a phenomenological term corresponding to meson repulsion due to composite structure of mesons given as

$$\epsilon_m^R(\beta) = 3a(2\pi)^{-3}\int\left(\sinh^2\tilde{f}(\boldsymbol{k})\cosh 2\theta_B + \sinh^2\theta_B\right)e^{R_\pi^2 k^2}\,\mathrm{d}\boldsymbol{k} \tag{26}$$

Finally, the nucleon repulsion term given as

$$\epsilon_R = \lambda \rho^2 \tag{27}$$

where ρ is as given in Equation (20). Thus the energy density is given by

$$E = \frac{\left(\epsilon_f + \epsilon_m + \epsilon_R\right)}{\rho} \tag{28}$$

where

$$\epsilon_m = \epsilon_k + \epsilon_m^R + \epsilon_{int}$$

as before.

The thermodynamic potential density Ω is given by

$$\Omega = E\rho - \frac{S}{\beta} - \mu_p \rho_p - \mu_n \rho_n \tag{29}$$

where the last term corresponds to nucleon number conservation with μ_p and μ_n as the chemical potentials of proton and neutron respectively. The temperature dependent pressure is given as $P = -\Omega$. The entropy density above is $S = S_F + S_B$ with S_F being the entropy in fermion sector given as

$$S_F = -\frac{\gamma}{(2\pi)^3}\int d\boldsymbol{k}\left[\sin^2\theta_F(\boldsymbol{k},\beta)\ln\left(\sin^2\theta_F(\boldsymbol{k},\beta)\right) + \cos^2\theta_F(\boldsymbol{k},\beta)\ln\left(\cos^2\theta_F(\boldsymbol{k},\beta)\right)\right].$$

and similarly the meson sector contribution S_B is given as

$$S_B = -\frac{3}{(2\pi)^3}\int d\boldsymbol{k}\left[\sinh^2\theta_B(\boldsymbol{k},\beta)\ln\left(\sinh^2\theta_B(\boldsymbol{k},\beta)\right) - \cosh^2\theta_B(\boldsymbol{k},\beta)\ln\left(\cosh^2\theta_B(\boldsymbol{k},\beta)\right)\right].$$

Thus the thermodynamic potential density now is a functional of $\theta_F(\boldsymbol{k},T)$, $\theta_B(\boldsymbol{k},T)$ as well as the pion dressing function $\tilde{f}(\boldsymbol{k})$ which will of course depend upon temperature. Extremisation of Equation (29) with respect to $\tilde{f}(\boldsymbol{k})$ yields

$$\tanh 2\tilde{f}(\boldsymbol{k}) = -\frac{G^2\rho}{2M}\cdot\frac{1}{\omega^2(\boldsymbol{k}) + \frac{G^2\rho}{2M} + a\omega(\boldsymbol{k})\mathrm{e}^{R_\pi^2 k^2}} \tag{30}$$

which is of the same form as [16] for zero temperature. Similarly minimising the thermodynamic potential with respect to $\theta_B(\boldsymbol{k},T)$ we get

$$\sinh^2\theta_B = \frac{1}{\mathrm{e}^{\beta\omega'} - 1} \tag{31}$$

where

$$\omega' = \left(\omega + \frac{G^2\rho}{2M\omega} + a\mathrm{e}^{R_\pi^2 k^2}\right)\cosh 2\tilde{f}(\boldsymbol{k}) + \frac{G^2\rho}{2M\omega}\sinh 2\tilde{f}(\boldsymbol{k}) \tag{32}$$

Once we substitute the optimised dressing as in Equation (30), the above simplifies to

$$\omega' = \left(\omega + \frac{G^2\rho}{M\omega} + a\mathrm{e}^{R_\pi^2 k^2}\right)^{1/2}\left(\omega + a\mathrm{e}^{R_\pi^2 k^2}\right)^{1/2} \tag{33}$$

which is different from ω due to interactions. Further, minimising the thermodynamic potential with respect to $\theta_F(\boldsymbol{k},T)$ we have the solution

$$\sin^2\theta_F = \frac{1}{\mathrm{e}^{\beta(\epsilon_F - \mu)} + 1} \tag{34}$$

with

$$\epsilon_F = \frac{G^2}{2MJ} + 2\lambda\rho + \frac{k^2}{2M} \tag{35}$$

where J is given in Equation (25). We may note that the change in ϵ_F above from $k^2/2M$ is also due to interaction.

3. RESULTS AND DISCUSSIONS

The parameters a, R_π, λ_ω and λ_ρ [15] [16] are determined variationally. The first three of these four parameters are evaluated by constraining the binding energy per nucleon $E_B \equiv (\epsilon/\rho - M)$, the pressure P and the compressibility K of the symmetric nuclear matter i.e. $E_{B_{sat}} = -16\ \text{MeV}$, $P_{sat} = 0$ and $K_{sat} = 270\ \text{MeV}$ [19] .

The fourth parameter λ_ρ of our calculation is evaluated by fixing the value of E_{sym} to the standard value of 31 MeV.

The pion-nucleon coupling constant $G^2/4\pi = 14.6$ and a = 115.264 MeV, $R_\pi = 1.061\ \text{fm}$, $\lambda_\omega = 3.164\ \text{fm}^2$ and $\lambda_\rho = 0.650\ \text{fm}^2$.

In Figure 1 we study the nature of variation of E_B with different baryon densities at different temperatures. The saturation binding energy increases from −16 MeV at zero temperature to higher values with rise of temperature. This clearly shows that the temperature has a significant effect on symmetric nuclear matter.

In Figure 2, the variation of pressure P with nucleon densities have been studied at different temperatures (T = 0, 4, 8, 12, 16, 20 MeV). It shows that at the saturation density, the pressure is zero for zero temperature as usual. But pressure becomes negative if one goes towards lower density and it becomes zero at zero nuclear density. This peculear behaviour is first order phase transition. This trend disappears at higher temperatures.

In Figure 3, the variation of pressure with nucleon densities for different Proton fraction at different temperatures. It shows that at Proton fraction $y_p = 0.1$ the pressure becomes positive. If one goes to higher y_Ps the Pressuer becomes negative at different lower densities reaching zero Pressure at zero nucleon densities. So the asymmetric nuclear matter behaves differently at different nucleon densities.

In Figure 4, the slope $(\partial P/\partial \rho)_{T,y_p}$ at different relative nucleon densities ρ/ρ_0 are plotted at two different temperatures for different proton fractions. It shows that the slope becomes negative as the relative nucleon densities increases to a certain value and then increases at higher values. The trend is same at different higher temperatures but terminates to a certain value of slope at certain lower densities.

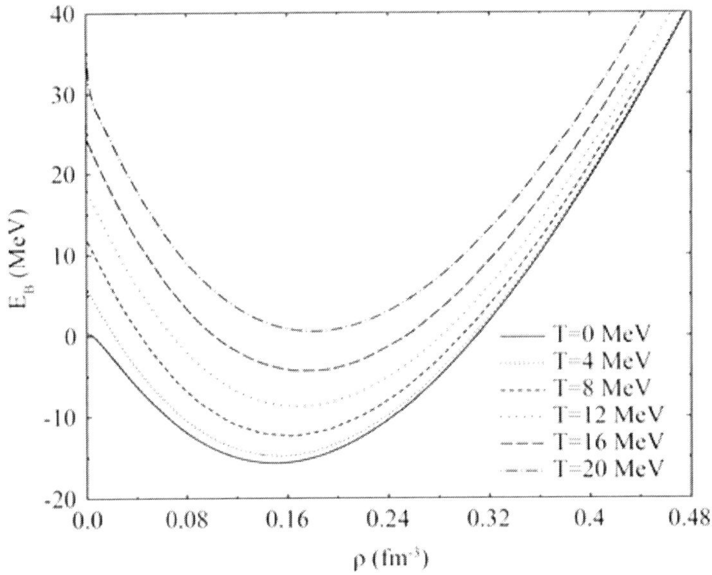

Figure 1. The binding energy E_B as a function of nucleon density ρ.

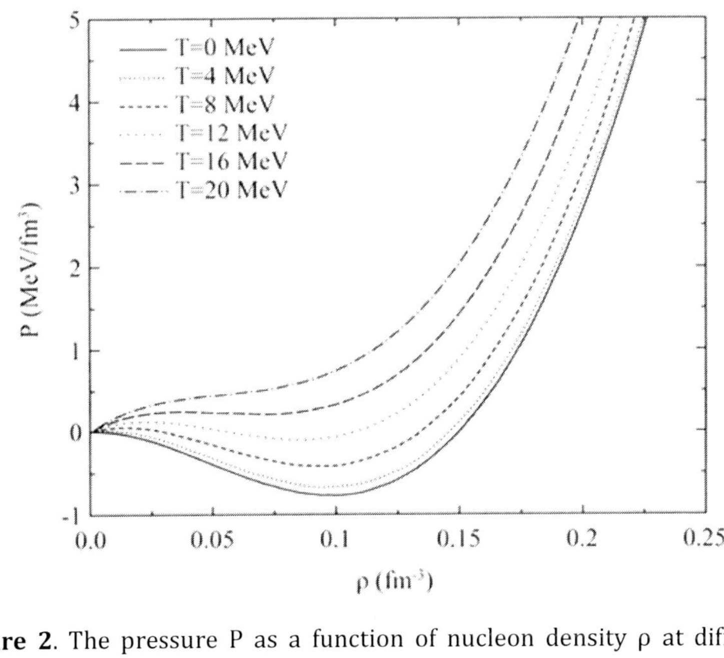

Figure 2. The pressure P as a function of nucleon density ρ at different temperature.

4. SUMMARY

In this paper, we study the warm equation of state (EOS) of asymmetric nuclear matter taking pion-nucleon interaction with repulsive effect due to ρ and ω mesons. We observe here that our results are comparable with the Non-Linear Walecka (NLWM) and Quark Meson Coupling (QMC) model [20] . The binding energy changes from negative to positive value with increase of temperature around saturation density. The profile of pressure variation with density at different temperatures shows that there is a liquid-gas phase transition and the transition is continuous above the critical temperature T_c at which $(\partial P/\partial \rho)_{T,y_p}=0$. We also observe that the pressure gradient becomes more negative due to increase of proton fraction leading to a possibility of instability.

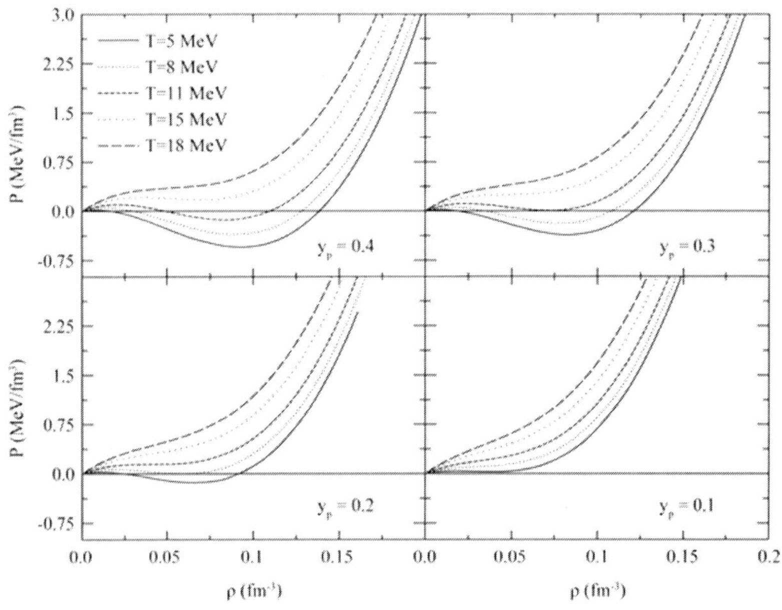

Figure 3. The pressure P as a function of nucleon density ρ for different proton fraction.

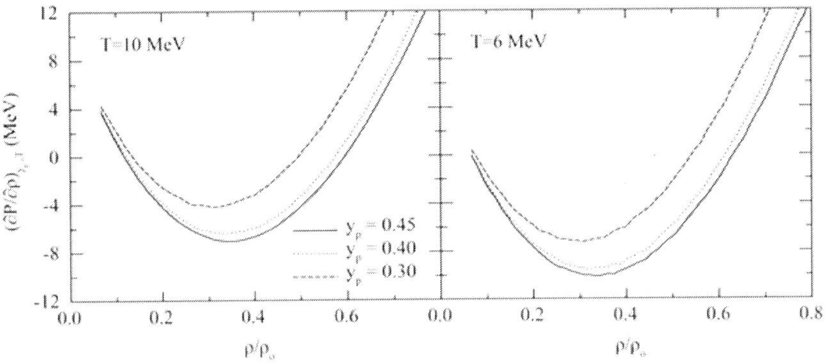

Figure 4. The pressure variation $(\partial P/\partial \rho)_{T,y_p}$ as a function of relative nucleon density ρ/ρ_0 for different proton fraction.

REFERENCES

1. Prakash, M., Bombaci, I., Prakash, M., Ellis, P.J., Lattimer, J.M. and Knorren, R. (1977) Composition and Structure of Protoneutron Stars. Physics Report, 280, 1-77.
2. Abreu, M.C., et al. (2000) Evidence for Deconfinement of Quarks and Gluons from the J/ψ Suppression Pattern Measured in Pb-Pb Collisions at the CERN-SPS. Physics Letters B, 477, 28-36.
3. Hauger, J.A., et al. (1996) Dynamics of the Multifragmentation of 1A GeV Gold on Carbon. Physical Review Letters, 77, 235.
4. Pethick, C.J., Ravenhall, D.G. and Lorenz, C.P. (1995) The Inner Boundary of a Neutron-Star Crust. Nuclear Physics A, 584, 675-703.
5. Lorenz, C.P., Ravenhall, D.G. and Pethick, C.J. (1993) Neutron Star Crusts. Physical Review Letters, 70, 379.
6. Barranco, M. and Buchler, J.R. (1980) Thermodynamic Properties of Hot Nucleonic Matter. Physical Review C, 22, 1729.
7. Müller, H. and Serot, B.D. (1995) Phase Transitions in Warm, Asymmetric Nuclear Matter. Physical Review C, 52, 2072.
8. Menezes, D.P. and Providencia, C. (1999) Droplet Formation in Cold Asymmetric Nuclear Matter. Nuclear Physics A, 650, 283-198.
9. Menezes, D.P. and Providencia, C. (1999) Nucleation Process in Asymmetric Hot Nuclear Matter. Physical Review C, 60, Article ID: 024313
10. Walecka, J.D. (1974) A Theory of Highly Condensed Matter. Annals of Physics, 83, 491-529.
11. Serot, B.D. and Walecka, J.D. (1986) Advances in Nuclear Physics, 16, 1. Gmuca, S. (1991) Relativistic Mean-Field Fit to Microscopic Results in Nuclear Matter. Journal of Physics G, 17, 1115.
12. Gambhir, Y.K. and Ring, P. (1988) Relativistic Description of Deformed Rare Earth Nuclei. Physics Letters B, 202, 5-9.
13. Mishra, A., Mishra, H. and Misra, S.P. (1990) Nuclear Matter with Constituent Meson Quanta. International Journal of Modern Physics A, 5, 3391-3399.
14. Eisenberg, J.M. (1981) A Heuristic Model for Nucleon Effective Mass Features in the Presence of Pion Condensation. Physics Letters B, 104, 353-356.

15. Jung, H., Beck, F. and Miller, G.A. (1989) Relativistic Pion-Ring Series for Nuclear Matter. Physical Review Letters, 62, 2357-2360.
16. Mishra, H., Misra, S.P., Panda, P.K. and Parida, B.K. (1992) Hot Nuclear Matter: A Variational Approach. International Journal of Modern Physics E, 1, 405-419.
17. Umezawa, H., Matsumoto, H. and Tachiki, M. (1982) Thermo Field Dynamics and Condensed States. North Holand, Amsterdam.
18. Sarangi, S., Panda, P.K., Sahu, S.K. and Maharana, L. (2008) Asymmetric Nuclear Matter: A Variational Approach. International Journal of Modern Physics B, 22, 4524-4537.
19. Sarangi, S., Panda, P.K., Sahu, S.K. and Maharana, L. (2010) Study of the Asymmetric Nuclear Matter with Pion Dressing. Indian Journal of Physics, 84, 431-447.
20. Sahu, S.K. (2015) Low Density Instability in Asymmetric Nuclear Matter Using Pion Dressing. Journal of Modern Physics, 6, 1350-1359.
21. Fetter, A.L. and Walecka, J.D. (1971) Quantum Theory of Many Particle System. McGraw-Hill Book Company, New York, 34.
22. Steiner, A.W., Prakash, M., Lattimer, J.M. and Ellis, P.J. (2005) Isospin Asymmetry in Nuclei and Neutron Stars. Physics Reports, 441, 325-375.
23. Santos, A.M., Providecia, C. and Panda, P.K. (2009) Low Density Instabilities in Asymmetric Nuclear Matter within the Quark-Meson Coupling (QMC) Model with the δ Meson. Physical Review C, 79, Article ID: 045805.

CHAPTER 12

5D World-Universe Model. Multicomponent Dark Matter

Vladimir S. Netchitailo

Biolase Inc., Irvine, USA

ABSTRACT

5D World-Universe Model (WUM) is based on the decisive role of the Medium of the World composed of massive particles: protons, electrons, photons, neutrinos, and Dark Matter Particles (DMP). The model forecasts the masses of DMP, discusses the possibility of all macroobject cores consisting of DMP (galaxy clusters, galaxies, star clusters, extrasolar systems, and planets), and explains the diffuse cosmic gamma-ray background radiation as the sum of contributions of multicomponent dark matter annihilation. The signatures of DMP annihilation with expected masses of 1.3 TeV, 9.6 GeV, 70 MeV, 340 keV, and 3.7 keV, are found in spectra of the diffuse gamma-ray background and the emission of various macroobjects in the World. The correlation between different emission lines in spectra of macroobjects is connected to their structure, which depends on the composition of the cores and surrounding shells made up of DMP. Consequently, the diversity of Very High Energy (VHE) gamma-ray sources in the World has a clear explanation.

Keywords: 5D World-Universe Model, Medium of the World, Dark Matter Particles, Cores of Macroobjects, Gamma-Ray Background Radiation, Pioneer Anomaly

1. INTRODUCTION

We can't solve problems by using the same kind of thinking we used when we created them.

Albert Einstein

In the World-Universe Model (WUM), we introduce the basic unit of mass m_0 that equals to

$$m_0 = \frac{h}{ac} = 70.025267 \text{ MeV}/\text{c}^2$$

where h is Planck constant, c is the electrodynamic constant, $a = 2\pi a_0$, and a_0 is the classical electron radius. m_0 plays a key role when the masses of Dark Matter Particles (DMP) are discussed in the next Section.

The Fine-structure constant (FSC) α is a fundamental physical constant that has several physical interpretations. α is the rest mass of an electron m_e measured in terms of basic unit m_0. FSC plays a central role in WUM.

According to WUM, all stable particles are created in the 3-sphere World due to the surface energy of the 4-ball Nucleus of the World provided by the 4-dimensional Universe. The World consists of the Medium (protons, electrons, photons, neutrinos, and DMP) and Macroobjects (Galaxy clusters, Galaxies, Star clusters, Extrasolar systems, planets, etc.) made of these particles. There is no empty space or dark energy in WUM. The role of the Intergalactic plasma consisting of protons, electrons, and photons as part of the Medium of the World is analyzed in [1] .

This paper discusses the Multicomponent Dark Matter and its decisive role in the Medium and Macroobjects of the World. DMP includes three Majorana fermions (Neutralinos, WIMPs, and Sterile neutrinos) with spin of 1/2 and two spin-0 bosons (named DIRACs and ELOPs in the World-Universe Model), as detailed below. Multicomponent dark matter models consisting of both bosonic and fermionic components were analyzed in literature (for example, see [2] - [10] and references therein).

2. DARK MATTER PARTICLES

Dark Matter (DM) is among the most important open problems in both cosmology and particle physics. There are three prominent hypotheses on nonbaryonic DM, namely Hot Dark Matter (HDM), Warm Dark Matter (WDM), and Cold Dark Matter (CDM).

A neutralino with mass m_N in $100 \Leftrightarrow 10000\ \mathrm{GeV}/c^2$ range is the leading CDM candidate. Light DMP that are heavier than WDM and HDM but lighter than neutralinos is DM candidates too. Subsequently, we will refer to the light DMP as WIMPs. Their mass m_{WIMP} falls into $1 \Leftrightarrow 10\ \mathrm{GeV}/c^2$ range. It is known that a sterile neutrino with mass m_{ν_s} in $1 \Leftrightarrow 10\ \mathrm{keV}/c^2$ range is a good WDM candidate. In our opinion, a tauonic neutrino is a good HDM candidate.

In addition to fermions discussed above, we offer another type of DMP—spin-0 bosons, consisting of two fermions each. There exist two types of DM bosons which we called DIRACs and ELOPS.

DIRACs are magnetic dipoles with mass m_0, consisting of two Dirac monopoles with mass $\frac{m_0}{2}$ and charge $\mu = \frac{e}{2\alpha}$, where e is an electron charge. Dissociated DIRACs can only exist at nuclear densities or at high tem-

peratures. In our opinion, Dirac monopoles are the smallest building blocks of constituent quarks and hadrons (mesons and baryons).

The second spin-0 boson is the ELOP (named by analogy to an ELectron-nortisOP dipole). ELOP weighs $\frac{2}{3}m_e$ and consists of two preons with mass $m_{pr} = \frac{1}{3}m_e$ and charge $e_{pr} = \frac{1}{3}e$. ELOPs break into two preons at

nuclear densities or at high temperatures. In particle physics, preons are postulated to be "point-like" particles, conceived to be subcomponents of quarks and leptons [11] .

We did not take into account the binding energies of DIRACs and ELOPs, and thus the values of the masses of monopoles and preons are approximate. They have negligible electrostatic and electromagnetic charges because the separation between charges is very small. The signatures of these bosons' annihilation in gamma-ray spectra will be discussed in Section 6.

WUM postulates that masses of DMP are proportional to multiplied by different exponents of α and can be expressed with the following formulae:

CDM particles (neutralinos and WIMPs):

$$m_N = \alpha^{-2} m_0 = 1.3149950\ \mathrm{TeV}/c^2 \tag{2.1}$$

$$m_{\mathrm{WIMP}} = \alpha^{-1} m_0 = 9.5959823\ \mathrm{GeV/c^2} \tag{2.2}$$

DIRACs:

$$m_{\mathrm{DIRAC}} = 2\alpha^0 \frac{m_0}{2} = 70.025267\ \mathrm{MeV/c^2} \tag{2.3}$$

ELOPs:

$$m_{\mathrm{ELOP}} = 2\alpha^1 \frac{m_0}{3} = 340.66606\ \mathrm{keV/c^2} \tag{2.4}$$

WDM particles (sterile neutrinos):

$$m_{\nu_s} = \alpha^2 m_0 = 3.7289402\ \mathrm{keV/c^2} \tag{2.5}$$

These values fall into the ranges estimated in literature. The roles of those particles in macroobject cores built up from fermionic dark matter, in gamma-ray spectra of the diffuse gamma-ray background, and the emission of various macroobjects in the World will be discussed in Sections 3, 4 and 6 respectively.

Our Model holds that the energy densities of all types of DMP are proportional to the proton energy density ρ_p in the World's Medium [1] :

$$\rho_p = \frac{2\pi^2\alpha}{3}\rho_{cr} \tag{2.6}$$

where ρ_{cr} is a critical energy density of the World:

$$\rho_{cr} = 3\rho_0 \times Q^{-1} \tag{2.7}$$

$$\rho_0 = \frac{hc}{a^4} \tag{2.8}$$

ρ_0 is a basic unit of energy density and a dimensionless time-varying quantity Q equals to the ratio of the size of the World R at cosmological time τ to the Worlds' size a at the beginning:

$$Q = \frac{R}{a} \tag{2.9}$$

In all, there are 5 different types of DMP. Then the total energy density of DM is

$$\rho_{DM} = 5\rho_p = 0.24007327\rho_{cr} \tag{2.10}$$

which is close to the measured DM energy density: $\rho_{DM} \cong 0.268\rho_{cr}$ [12] . Note that one of outstanding puzzles in particle physics and cosmology relates to so-called cosmic coincidence: the ratio of dark matter density in the World to baryonic matter density in the Medium of the World $\cong 5$ [10] [13] .

Neutralinos, WIMPs, and sterile neutrinos are Majorana fermions, which partake in the annihilation interaction with strength equals to α^{-2} , α^{-1} , and α^{2} respectively (see Section 3). The signatures of DMP annihilation with expected masses of 1.3 TeV, 9.6 GeV, 70 MeV, 340 keV, and 3.7 keV are found in spectra of the diffuse gamma-ray background and the emission of various macroobjects in the World (see Section 6).

3. MACROOBJECT CORES BUILT UP FROM FERMIONIC DARK MATTER

In this section, we discuss the possibility of all macroobject cores consisting of DMP introduced in Section 2. The first phase of stellar evolution in the history of the World may be dark stars, powered by Dark Matter heating rather than fusion. Neutralinos and WIMPs, which are their own antiparticles, can annihilate and provide an important heat source for the stars and planets in the World.

In our view, all macroobjects of the World (including galaxy clusters, galaxies, star clusters, extrasolar systems, and planets) possess the following properties:

- Macroobject cores are made up of DMP;
- Macroobjects consist of all particles under consideration, in the same proportion as they exist in the World's Medium;
- Macroobjects contain other particles, including DM and baryonic matter, in shells surrounding the cores.

Taking into account the main principle of the World-Universe Model (all physical parameters can be expressed in terms of α, Q, small integer numbers, and π) we modify the published theory of Fermionic Compact Stars (FCS) developed by G. Narain, et al. [14] as follows. We take a scaling solution for a free Fermi gas consisting of fermions with mass m_f in accordance with following equations:

Maximum mass:

$$M_{max} = A_1 M_F \tag{3.1}$$

Minimum radius:

$$R_{min} = A_2 R_F \tag{3.2}$$

Maximum density:

$$\rho_{max} = A_3 \rho_0 \tag{3.3}$$

where

$$M_F = \frac{M_P^3}{m_f^2}; \quad R_F = \frac{M_P}{m_f}\frac{L_{Cf}}{2\pi}; \quad \rho_0 = \frac{hc}{a^4} \tag{3.4}$$

and M_P is Planck mass, L_{Cf} is a Compton length of the fermion. A_1, A_2, and A_3 are parameters. Let us choose π as the value of A_2 (instead of $A_2 = 3.367$ taken by G. Narain, et al. [14]). Then diameter of FCS is proportional to the fermion Compton length L_{Cf}. We use $\pi/6$ as the value of A_1 (instead of $A_1 = 0.384$ taken by G. Narain, et al. [14]). Then A_3 will equal to

$$A_3 = \left(\frac{m_f}{m_0}\right)^4 \tag{3.5}$$

Table 1 summarizes the parameter values for FCS made up of various fermions:

Table 1. Parameter values of Compact Stars built up from fermionic dark matter particles.

Fermion	Fermion relative mass m_f/m_0	Macroobject relative mass M_{max}/M_0	Macroobject relative radius R_{min}/L_g	Macroobject relative density ρ_{max}/ρ_0
Sterile neutrino	α^{2}	α^{-4}	α^{-4}	α^{8}
Preon	$3^{-1}\alpha^{1}$	$3^{2}\alpha^{-2}$	$3^{2}\alpha^{-2}$	$3^{-4}\alpha^{4}$
Electron-proton (white dwarf)	α^{1}, β	β^{-2}	$(\alpha\beta)^{-1}$	$\alpha^{3}\beta$
Monopole	2^{-1}	2^{2}	2^{2}	2^{-4}
WIMP	α^{-1}	α^{2}	α^{2}	α^{-4}
Neutralino	α^{-2}	α^{4}	α^{4}	α^{-8}
Interacting WIMPs	α^{-1}	β^{-2}	β^{-2}	β^{4}
Interacting neutralinos	α^{-2}	β^{-2}	β^{-2}	β^{4}
Neutron (star)	$\approx \beta$	β^{-2}	β^{-2}	β^{4}

where

$$M_0 = \frac{4\pi m_0}{3} \times Q^{3/2} \tag{3.6}$$

$$L_g = a \times Q^{1/2} \tag{3.7}$$

$$\beta = \frac{m_p}{m_0} \tag{3.8}$$

and m_p is the mass of a proton. A maximum density of neutron stars equals to the nuclear density:

$$\rho_{\max} = \beta^4 \rho_0 \tag{3.9}$$

which is the maximum possible density of any macroobject in the World.

A Compact Star made up of heavier particles—WIMPs and neutralinos—could in principle have a much higher density. In order for such a star to remain stable and not exceed the nuclear density, WIMPs and neutralinos must partake in an annihilation interaction whose strength equals to α^{-1} and α^{-2} respectively.

Scaling solution for interacting WIMPs can also be described with Equations (3.1), (3.2), (3.3) and the following values of A_1, A_2 and A_3:

$$A_{1\max} = \frac{\pi}{6}(\alpha\beta)^{-2} \tag{3.10}$$

$$A_{2\min} = \pi(\alpha\beta)^{-2} \tag{3.11}$$

$$A_{3\max} = \beta^4 \tag{3.12}$$

The maximum mass and minimum radius increase about two orders of magnitude each and the maximum density equals to the nuclear density. Note that parameters of a FCS made up of strongly interacting WIMPs are identical to those of neutron stars.

In accordance with the paper by G. Narain, et al. [14] , the most attractive feature of the strongly interacting Fermi gas of WIMPs is practically constant value of FCS minimum radius in the large range of masses M_{WIMP} from

$$M_{\text{WIMPmax}} = \frac{\pi}{6}(\alpha\beta)^{-2} M_F = \frac{1}{\beta^2} M_0 \tag{3.13}$$

down to

$$M_{\text{WIMPmin}} = \alpha^4 M_{\text{WIMPmax}} \tag{3.14}$$

M_{WIMPmin} is more than eight orders of magnitude smaller than M_{WIMPmax}. It makes strongly interacting WIMPs good candidates for stellar and planetary cores of extrasolar systems with Red stars (see Section 4).

When the mass of a FCS made up of WIMPs is much smaller than the maximum mass, the scaling solution yields the following equation for parameters A_1 and A_2:

$$A_1 A_2^3 = \pi^4 \tag{3.15}$$

Compare $\pi^4 \cong 97.4$ with the value of 91 used by G. Narain, et al. [14] .

Minimum mass and maximum radiuses take on the following values:

$$A_{1\min} = \frac{\pi}{6}\sqrt{6}(\alpha\beta)^2 \tag{3.16}$$

$$A_{2\max} = \pi\sqrt[6]{6}(\alpha\beta)^{-2/3} \tag{3.17}$$

It follows that the range of FCS masses $(A_{1\min} \Leftrightarrow A_{1\max})$ spans about three orders of magnitude, and the range of FCS core radii $(A_{2\min} \Leftrightarrow A_{2\max})$—one order of magnitude. It makes WIMPs good candidates for brown dwarf cores too (see Section 4).

Scaling solution for interacting neutralinos can be described with the same Equations (3.1), (3.2), (3.3) and the following values of A_1^*, A_2^* and A_3^*:

$$A^*_{1\max} = \frac{\pi}{6}\left(\alpha^2\beta\right)^{-2} \tag{3.18}$$

$$A^*_{2\min} = \pi\left(\alpha^2\beta\right)^{-2} \tag{3.19}$$

$$A^*_{3\max} = \beta^4 \tag{3.20}$$

In this case, the maximum mass and minimum radius increase about four orders of magnitude each and the maximum density equals to the nuclear density. Note that parameters of a FCS made up of strongly interacting neutralinos are identical to those of neutron stars.

Practically constant value of FCS minimum radius takes place in the huge range of masses M_N from

$$M_{N\max} = \frac{\pi}{6}\left(\alpha\beta\right)^{-2}\alpha^2 M_F = \frac{1}{\beta^2}M_0 \tag{3.21}$$

down to

$$M_{N\min} = \alpha^8 M_{N\max} \tag{3.22}$$

$M_{N\min}$ is more than seventeen orders of magnitude smaller than $M_{N\max}$. It makes strongly interacting neutralinos good candidates for stellar and planetary cores of extrasolar systems with Main-sequence stars (see Section 4).

When the mass of a FCS made up of neutralinos is much smaller than the maximum mass, the scaling solution yields the following equation for parameters A_1^* and A_2^*:

$$A_1^* A_2^{*3} = \pi^4 \tag{3.23}$$

Minimum mass and maximum radiuses take on the following values:

$$A^*_{1\min} = \frac{\pi}{6}\sqrt{6}\left(\alpha^2\beta\right)^2 \tag{3.24}$$

$$A^*_{2\max} = \pi\sqrt[6]{6}\left(\alpha^2\beta\right)^{-2/3} \tag{3.25}$$

It means that the range of FCS masses $\left(A^*_{1\min} \Leftrightarrow A^*_{1\max}\right)$ is about twelve orders of magnitude, and the range of FCS core radiuses $\left(A^*_{2\min} \Leftrightarrow A^*_{2\max}\right)$ is about four orders of magnitude. The numerical values for FCS masses and radii will be given in Section 4.

Fermionic Compact Stars have the following properties:

The maximum potential of interaction $U_{\max}$ between any particle or macroobject and FCS made up of any fermions

$$U_{\max} = \frac{GM_{\max}}{R_{\min}} = \frac{c^2}{6} \tag{3.26}$$

does not depend on the nature of fermions;

The minimum radius of FCS made of any fermion

$$R_{\min} = 3R_{SH} \tag{3.27}$$

equals to three Schwarzschild radii and does not depend on the nature of the fermion;

FCS density does not depend on $M_{\max}$ and $R_{\min}$ and does not change in time while $M_{\max} \propto \tau^{3/2}$ and $R_{\min} \propto \tau^{1/2}$.

4. MACROOBJECTS OF THE WORLD

According to WUM, all macroobjects of the World (galaxies, stars, planets) possess cores consisting of DMP. The theory of fermion compact stars made up of DMP is well developed. Scaling solutions are derived for a free and an interacting Fermi gas in Section 3. Table 2 describes the numerical values for masses and radii of FCS made up of different fermions:

Table 2. Numerical values for parameters of Compact Stars built up from fermionic dark matter particles.

Fermion	Fermion mass m_f, MeV/c^2	Macroobject mass M_{max}, kg	Macroobject radius R_{min}, m	Macroobject density ρ_{max}, kg/m^3
Sterile neutrino	3.73×10^{-3}	1.2×10^{41}	5.4×10^{14}	1.8×10^{-4}
Preon	≥ 0.17	5.9×10^{37}	2.6×10^{11}	7.8×10^{2}
Monopole	≥ 35	1.4×10^{33}	6.2×10^{6}	1.4×10^{12}
Interacting WIMPs	9596	1.9×10^{30}	8.6×10^{3}	7.2×10^{17}
Interacting neutralinos	1315×10^{3}	1.9×10^{30}	8.6×10^{3}	7.2×10^{17}
Electron; proton (white dwarf)	0.511; 938.3	1.9×10^{30}	1.6×10^{7}	1.2×10^{8}
Neutron (star)	939.6	1.9×10^{30}	8.6×10^{3}	7.2×10^{17}

The calculated parameters of FCS show that

- White Dwarf Shells (WDS) around the nuclei made of strongly interacting WIMPs or neutralinos compose cores of stars in extrasolar systems;
- Shells of dissociated DIRACs to monopoles around the nuclei made of strongly interacting WIMPs or neutralinos form cores of globular clusters;
- Shells of dissociated ELOPs to preons around the nuclei made of strongly interacting WIMPs or neutralinos constitute cores of galaxies;

Ÿ Shells of sterile neutrinos around the nuclei made of strongly interacting WIMPs or neutralinos make up cores of galaxy clusters.

Although there are no free Dirac's monopoles and preons in the World, they can arise in the cores of FCS as the result of DIRACs and ELOPs gravitational collapse with density increasing up to the nuclear density $\left(\sim 10^{17}\ \mathrm{kg/m^3}\right)$ and/or at high temperatures, with subsequent dissociation of dipoles to monopoles and preons.

4.1. Galaxies and Galaxy Clusters

A number of non-traditional models explaining the supermassive dark objects observed in galaxies and galaxy clusters, formed by self-gravitating non-baryonic matter composed of fermions and bosons, are widely discussed in literature [2] - [10] .

Dark matter can be, in principle, achieved also through extended theories of gravity. It has been shown, for example, that in the framework of R^2gravity and in the linearized approach, it is possible to obtain spherically symmetric and stationary galaxy states which can be interpreted like an approximated solution of the Dark Matter problem [15] [16] .

According to WUM, the heaviest macroobjects include a high-density preon plasma shell around their cores:

- Macroobjects with a cold preon shell emit strong radio waves. Such objects are good candidates for the compact astronomical radio sources at centers of galaxies like Sagittarius A* in the Milky Way Galaxy;
- Red Giants are macroobjects with hot preon shells;
- Blazars are members of a larger group of active galaxies that host active galactic nuclei (AGN). They are macroobjects with hot preon and sterile neutrinos shells;

- Quasars are the most energetic and distant members of AGN. They are macroobjects with very hot preon and sterile neutrinos shells;
- Seyfert galaxies are one of the two largest groups of AGN, along with quasars. They have quasar-like nuclei, but unlike quasars, their host galaxies are clearly detectable. Seyfert galaxies account for about 10% of all galaxies.

Note that the temperature of the preon and sterile neutrinos shells depends on the composition of the macroobject core. Macroobjects whose cores are made up of WIMPs and preons remain cold. Macroobjects with cores made up of WIMPs and WDS produce hot preon and sterile neutrino shells. Macroobjects whose cores consist of neutralinos and WDS have very hot preon and sterile neutrino shells.

The mass of an AGN is about 7-11 orders of magnitude larger than the mass of the Sun. The radius of an AGN is about 4 - 7 orders of magnitude larger than the radius of WDS (see Table 2). The area of the closed spherical surface around the AGN is 8 - 14 orders of magnitude greater than the surface area of WDS. Luminosity of the AGN is then 8 - 14 orders of magnitude higher than the luminosity of the largest star. This take on AGNs explains the fact that the most luminous quasars radiate at a rate that can exceed the output of average galaxies, equivalent to two trillion suns.

To summarize, macroobjects of the World have cores made up of DM particles. The cores are surrounded by shells made up of DM and baryonic matter. Every macroobject consists of all particles under consideration that are present in the same proportion as they exist in the World's Medium. No compact stars are made up solely of fermionic DMP, for instance.

4.2. Extrasolar Systems

There are two primary types of stars: main-sequence stars and red stars. They differ in their surface temperatures and radii:

- Red stars have cool surface temperatures: 3500 ⇔ 4500 K for Hypergiants, Supergiants, Giants, lower for Red dwarfs (2300 ⇔ 3800 K), and significantly lower for Brown dwarfs (300 ⇔ 1000 K). These stars have enormous range of radii: from $1650R_{Sun}$ for Hypergiants down to $0.08R_{Sun}$ for Red dwarfs, and lower still for Brown dwarfs.
- Main-sequence stars have surface temperatures in the range of 3000 ⇔ 45,500 K, and radii in the range from $35R_{Sun}$ for the most massive known star R136a1 down to $0.1R_{Sun}$ for least heavy stars.

In our opinion, the difference between main-sequence stars and red stars lies in composition of stellar cores. Main-sequence stars cores are made up of neutralinos, while red star cores consist of WIMPs. As we have shown in Section 3, in both cases the cores' maximum mass and minimum radius equals to that of a neutron star. The fermions, however, have drastically different interaction strength of annihilation: α^{-1} in case of WIMPs and α^{-2} in case of neutralinos.

The Core temperature is therefore much higher in main-sequence stars whose cores are made up of neutralinos. Ignition of proton-proton chain reaction with the interaction strength equal to $\beta \approx 13.4$ developing in the surrounding WDS happens much more efficiently in these stars.

The developed star model explains the very low power production density produced by fusion inside of the Sun. Wikipedia humorously notes that the power output of the Sun more nearly approximates reptile metabolism than a thermonuclear bomb. In our Model, the core made up of strongly interacting neutralinos is the supplier of proton-electron pairs into WDS and igniter of the proton-proton chain reaction developing in the surrounding WDS with small interaction strength $\beta \cong 13.4$.

New neutralinos freely penetrate through the entire stellar envelope, get absorbed into the core and support neutralino annihilation and proton fusion in the WDS. An important consequence for Solar system, and in fact for all other stars in the World, is that they will never burn their "fuel" out. On the contrary, stars accumulate more fuel with time, and output more power.

Enormous radii of Hypergiants (up to $1650 R_{Sun} \cong 10^{12}$ m) and huge luminosity of giant stars can be explained by an additional shell of preons—particles whose charge equals to $\frac{1}{3}e$. They compose hot high-density plasma with surface temperature in the range of 3500 ⇔ 4500 K. The minimum radius of preon shell $R_{min} \cong 2.6 \times 10^{11}$ m (see Table 2).

Brown dwarfs are sub-stellar objects whose masses range from 13 to 80 Jupiter masses. In our opinion, Brown dwarfs differ from red stars in that the density of their cores is smaller than nuclear density. Consequently, WIMPs annihilation takes place less efficiently.

4.3. Extrasolar System Formation

The Nebular Hypothesis is the most widely accepted model of planetary formation. It holds that 4.6 Billion years ago, the Solar System was formed during a gravitation collapse of a giant molecular cloud, some light years across. The most significant criticism of the hypothesis is its inability to explain the Sun's relative lack of angular momentum when compared to the planets [17] .

According to WUM, extrasolar systems arise from clouds of all particles under consideration with mass M_{cl}. As a result of gravitational instability, gravitational collapse takes place and one third of M_{cl} is concentrating at the center of the cloud, increasing the density of the core up to the nuclear density.

The heaviest particles—neutralinos or WIMPs are the first in this stream of matter. When their density achieves the nuclear density, self-annihilation process ignites. As the result, the Stellar Nucleus (SN) grows up to 10^4 for neutralinos and 10^2 times for WIMPs taking additional mass of neutralinos and WIMPs from oncoming stream.

The next heaviest particles—protons, joined by electrons—will follow neutralinos or WIMPs during the gravitational collapse, and form the White Dwarf Shell (WDS) around the SN made of strongly interacting WIMPs or neutralinos.

Expansion of the hot Stellar Core (SC), consisting of SN with WDS, is progressing. Drops of the SC are ejected from the equatorial bulges of an overspinning SC (outward centrifugal forces exceed the inward gravitational force) and give birth to the cores of planets.

The following facts support the creation picture of extrasolar systems outlined above:

- The analysis of a mass-radius ratio for compact stars made of strongly interacting fermions shows that the radius remains approximately constant for a wide range of compact stars masses;
- The analysis of a mass-radius ratio for the lowest mass white dwarfs shows the same behavior—radius does not depend on mass. It happens because at the low mass end the Coulomb pressure (which is characterized by constant density $\propto M/r^3$ and thus $r \propto M^{1/3}$) starts to compensate the degeneracy: $r \propto M^{-1/3}$. The two effects nearly cancel each other out, so $r \propto M^0$—no dependency at all;
- Recent analysis of the Solar and Heliospheric Observatory (SOHO) mission data favors a faster rotation rate in the solar core (below 0.2 solar radius) than in the rest of the radiative zone [18] ;
- By analyzing the minute changes in travel times and wave shapes for earthquake doublets, the researchers of [19] concluded that the Earth's inner core is rotating faster than its surface by about 0.3 - 0.5 degrees per year;

- The authors of [20] found that Earth's inner core, made up of solid iron, "superrotates" in an eastward direction—meaning it spins faster than the rest of the planet—while the outer core, comprising mainly molten iron, spins westwards at a slower pace.

In our opinion, the Earth's inner core is made up of neutralinos, while the outer core is the WDS. The cores of the Sun and the planets comprising the Solar System are not rotating with the same speed as their surfaces. When analyzing the angular momentum distribution of the entire Solar System, one must consider these additional angular momentums. Moreover, the remainder of the original particle cloud weighing $\frac{2}{3}M_{cl}$ may possess additional angular momentum.

As discussed above, the minimum radius of the hot neutralinos and WIMPs core $R_{\min} \cong 8.6\ \text{km}$, and it remains essentially constant whether the core belongs to a star or to a planet. The masses of planets formed around red stars and main-sequence stars differ:

- The smallest possible mass of planets formed around red stars is 8 orders of magnitude smaller than maximum star mass M_0;
- Planets formed around main-sequence stars may be 17 orders of magnitude lighter than the maximum star mass.

Consequently, all round objects in hydrostatic equilibrium, down to Mimas in Solar system, contain hot neutralinos cores with WDS and should be considered planets. Planets can arise only around main sequence and red stars. Due to the less violent nature of their formation, brown dwarfs do not create planets.

4.4. Pioneer Anomaly

According to WUM, the macroobject energy E_{MO} enclosed in surface S_{MO} is proportional to the area of that surface:

$$E_{MO} = \sigma_0 S_{MO} \tag{4.4.1}$$

where σ_0 is the basic unit of surface energy density: $\sigma_0 = \rho_0 a$. It is natural to define surface S_{MO} as the boundary between macroobject and surrounding environment. In case of our Solar system, such a surface is named Heliosphere. We will refer to such a surface as Macroobject Boundary (MOB). According to the developed Model, Macroobjects have cores made up of fermionic DMP possessing minimum radii $R_{\min}$ described in Table 1 and Table 2. In case of extrasolar systems, the cores are made up of interacting neutralinos or WIMPs surrounded with White Dwarf Shells (WDS).

The cores are surrounded by the transitional region. In this region, the density decreases rapidly to the point of the zero level of the fractal structure [21] characterized by radius R_f and energy density ρ_f that satisfy the following equation for $r \geq R_f$:

$$\rho(r) = \frac{\rho_f R_f}{r} \tag{4.4.2}$$

According to Yu. Baryshev: For a structure with fractal dimension D = 2 the constant $\rho_f R_f$ may be actually viewed as a new fundamental physical constant [21] . In our Model, it is natural to connect this constant with the constant σ_0:

$$\rho_f R_f = 4\sigma_0 \tag{4.4.3}$$

The value of 4 above follows from the ratio for all Macroobjects of the World: 1/3 of the total energy is in the central macroobject (for example, star in extrasolar system) and 2/3 of the total energy is in the fractal structure around it. Taking the radius of a Macroobject Boundary R_{MOB} we find the macroobject energy:

$$E_{MO} = 4\pi R_{MOB}^2 \sigma_0 \tag{4.4.4}$$

The energy in the fractal structure E_{FS} at $R_{MOB} \gg R_f$ is:

$$E_{FS} = \int_{R_f}^{R_{MOB}} \frac{4\sigma_0}{r} \times 4\pi r^2 dr \approx 8\pi R_{MOB}^2 \sigma_0 \tag{4.4.5}$$

and the total energy E_{tot} equals to: $E_{tot} = 12\pi R_{MOB}^2 \sigma_0$.

It allows us to explain the so-called "Pioneer anomaly". Wikipedia describes this effect the following way: The Pioneer anomaly is the observed deviation from predicted accelerations of the Pioneer 10 and Pioneer 11 spacecraft after they passed about 20 astronomical units $(3\times10^9 \text{ km}; 2\times10^9 \text{ mi})$

on their trajectories out of the Solar System. An unexplained force appeared to cause an approximately constant sunward acceleration of $a_P = 8.74 \pm 1.33 \times 10^{-10}\ \mathrm{m/s^2}$ for both spacecraft. The magnitude of the Pioneer effect a_P is numerically quite close to the product of the speed of light c and the Hubble constant H_0 hinting at cosmological connection.

Let us calculate a deceleration a_P at the distance $r_P \gg R_f$ due to the additional mass of the fractal structure $M_{FS}(r_P) \propto r_P^2$ with the following equation for the gravitational parameter G [1] :

$$G = \frac{c^4}{8\pi\sigma_0 R_0} \tag{4.4.6}$$

$$a_P = \frac{GM_{FS}}{r_P^2} = \frac{c^4}{8\pi\sigma_0 R_0} \times \frac{8\pi\sigma_0}{c^2} = \frac{c^2}{R_0} = cH_0 = 6.68 \times 10^{-10}\ \mathrm{m/s^2} \tag{4.4.7}$$

which is in good agreement with the experimentally measured value (R_0 and H_0 are the values of the World's size R and Hubble's parameter H at the current time t). It is important to notice that the calculated deceleration does not depend on r_P and equals to cH_0 for all objects around the macroobject at the distance $r > R_f$.

Mass of the fractal structure around Sun M_V at distances $R_V \gg R_f$ is

$$M_V = 8\pi R_V^2 \sigma_0 / c^2 \tag{4.4.8}$$

At distance $R_V = 1.8 \times 10^{13}\ \mathrm{m}$ away from the Sun (approximate distance to Voyager 1 [22]),

$$M_V \cong 3.3 \times 10^{27}\ \mathrm{kg} \tag{4.4.9}$$

that is $\sim 0.15\% M_{\mathrm{Sun}}$. Note that the distances traveled by Voyagers are much smaller than the radius of the MOB: $R_V \ll R_{MOB} \sim 10^{15}\ \mathrm{m}$.

5. X RAYS AND GAMMA RAYS

All "elementary" particles of the World are fermions and they possess masses. Bosons such as photons, X- quants, and Gamma-quants are composite particles and consist of two fermions. Gamma rays are usually distinguished from X rays by their origin: X rays are emitted by electrons outside the nucleus, while gamma rays are emitted by the nucleus. A better way to distinguish the two, in our opinion, is the type of fermions composing the core of X-quants and Gamma-quants.

Super-soft X rays possess energies in the 0.09 2.5 keV range, whereas soft Gamma rays have energies in the 10 5000 keV range. We assume that X-quants are composed of two interacting neutrinos. New Physics with the dineutrinos in the Rare Decay $B \rightarrow K\nu\bar{\nu}$ is actively discussed in literature in recent years (for example, see [23] [24]).

Soft Gamma-quants are composed of two sterile neutrinos (3.7 keV each). Hard and super-hard Gamma-quants may be composed of two preons ($\gtrsim$ 0.17 MeV each), which are ELOPs in our Model, two Dirac monopoles ($\gtrsim$35 MeV each) which are, in fact, DIRACs.

We propose that Super-soft gamma rays (<10 keV) can arise as the result of sterile neutrino annihilation in the low energy case. Two or three super-soft gamma-quants with the energy < 3.7 keV are created. Similarly,

- ELOP annihilation produces hard gamma rays with energies < 340 keV;
- DIRAC annihilation produces hard gamma rays with energies < 70 MeV;
- WIMP annihilation produces super-hard gamma rays with energies < 9.6 GeV;
- Neutralino annihilation produces super-hard gamma rays with energies < 1.3 TeV.

Diffuse cosmic gamma-ray background is the sum of the contributions of the multicomponent self-interacting dark matter annihilation.

6. DARK MATTER SIGNATURES IN GAMMA-RAY SPECTRA

Large number of papers has been published in the field of X-ray and gamma-ray astronomy. The X-ray and gamma-ray background from $\lesssim$0.1 keV to $\gtrsim$10 TeV has been studied using high spectral and spatial resolution data from different spectrometers. Numerous papers were dedicated to

Dark Matter searches with astroparticle data (see reviews [25] - [34] and references therein). Dark Matter annihilation is proportional to the square of the DM density and is especially efficient in places of highest concentration of dark matter, such as compact stars built up from fermionic dark matter particles (see Section 3).

The models of DM annihilation and decay for various types of macroobjects (galaxy clusters, blazars, quasars, Seyfert galaxies) are well-developed. Physicists working in the field X-ray and gamma-ray astronomy attempt to determine masses of DM particles that would fit the experimental results with the developed models.

Recall that no macroobjects are made up of just a single type of DM particles, since other DM particles as well as baryonic matter are present in the shells. It follows that macroobjects cannot irradiate gamma rays in a single spectral range. On the contrary, they irradiate gamma-quants in different spectral ranges with ratios of fluxes depending on structure of a given macroobject.

WUM forecasts existence of DM particles with 1.3 TeV, 9.6 GeV, 70 MeV, 340 keV, and 3.7 keV masses. We will look for signs of annihilation of these particles in the observed gamma-ray spectra. We connect gamma-ray spectra with the structure of macroobjects (core and shells composition).

C. Boehm, P. Fayet, and J. Silk have this to say about Light and Heavy Dark Matter Particles:

It has recently (2003) been pointed out that the 511 keV emission line detected by Integral/SPI from the bulge of our galaxy could be explained by annihilations of light Dark Matter particles into e^+e^-. If such a signature is confirmed, then one might expect a conflict with the interpretation of very high energy gamma rays if they also turn out to be due to Dark Matter annihilations.

They proposed a way to reconcile the low and high energy signatures, even if both of them turn out to be due to Dark Matter annihilations. One would be a heavy fermion, for example, the lightest neutralino (>100 GeV) and the other one a possibly light spin-0 particle (~100 MeV). Both of them would be neutral and also stable as a result of two discrete symmetries (say R and M-parities) [9] [35].

According to our Model, the two couples of co-annihilating DMP are: a heavy fermion—neutralino with mass 1.3 TeV and a light spin-0 boson—DIRAC with mass 70 MeV; a heavy fermion—WIMP with mass 9.6 GeV and a light spin-0 boson?ELOP with mass 340 keV.

6.1. Neutralino 1.3 TeV

J. Holder has this to say about TeV Gamma-ray Astronomy: In leptonic scenarios, a population of electrons is accelerated to TeV energies, typically through Fermi acceleration by shocks in the AGN jet. These electrons then cool by radiating X-ray synchrotron photons. TeV emission results from inverse Compton interactions of the electrons with either their self-generated synchrotron photons, or an external photon field. The strong correlation between X-ray and TeV emission which is often observed provides evidence for a common origin such as this, although counter examples do exist [36] .

In our opinion, the TeV blazar emission should be classified as extremely-hard X rays and not gamma rays, since by definition: X rays are emitted by electrons outside the nucleus, while gamma rays are emitted by the nucleus.

R.C.G. Chaves, et al. have found that a significant fraction of the Galactic VHE (Very High Energy) gamma-ray sources (from the observed approximately 100 VHE γ-ray sources [38] - [42]) do not appear to have obvious counterparts at other wavelengths [37] .

This correlation between keV emission and TeV emission can be easily explained by the annihilation of the sterile neutrinos (3.7 keV) in the shell around the core of AGN made of neutralinos (1.3 TeV). Lack of the counterpart in gamma-ray spectra means the absence of sterile neutrino shell.

A detailed global analysis on the interpretation of the latest data of PAMELA, Fermi-LAT, AMS-02, H.E.S.S, and other collaborations in terms of dark matter annihilation and decay in various propagation models showed that for the Fermi-LAT and H.E.S.S. data favor DMP mass is $m_\chi \approx 1.3$ TeV [43] - [46] . The obtained data in [47] - [55] require DM mass m_χ to be around 1 to 1.5 TeV which is in good agreement with the predicted mass of a neutralino (1.3 TeV). Pulsars are the most natural candidates for such sources [41] .

The presence of spectral break at 1.3 TeV in VHE spectra was measured for different blazars [56] - [58] . Some nearby sources, e.g. Vela, Cygnus Loop, and Monogem Supernova Remnant (SNR) have unique signatures in the electron energy spectrum in the TeV region: broken power-law at ~1.3 TeV [59] . The DM interpretations of the $e^\pm$ excesses observed by PAMELA, Fermi and ATIC suggest the DMP mass of 1.3 TeV [60] .

As we mentioned above, pulsars are the most natural candidates for such VHE gamma-ray sources. According to WUM, FCS made up of strongly interacting neutralinos and WIMPs have maximum mass and minimum size which is exactly equal to parameters of neutron stars (see Table 1 and Table 2). It follows that pulsars might be in fact rotating Neutralino stars and WIMP stars with different shells around them.

The cores of such pulsars may also be made up of the mixture of neutralinos (1.3 TeV) and WIMPs (9.6 GeV) surrounded by shells composed of the other DMP: DIRACs (70 MeV), ELOPs (340 keV), and sterile neutrinos (3.7 keV). Annihilation of those DMP can give rise to any combination of gamma-ray lines. Thus the diversity of VHE gamma-ray sources in the World has a clear explanation in frames of the World-Universe Model.

In our opinion, results obtained by the CALET program are the closest to the ultimate discovery of the first confirmed dark matter particle—neutralino with mass 1.3 TeV [59] .

6.2. WIMP 9.6 GeV

Dan Hooper summarized and discussed the body of evidence which has accumulated in favor of dark matter in the form of approximately 10 GeV particles, including the spectrum and angular distribution of gamma rays from the Galactic Center, the synchrotron emission from the Milky Way's radio filaments, the diffuse synchrotron emission from the Inner Galaxy (the "WMAP Haze") and low-energy signals from the direct detection experiments DAMA/LIBRA, CoGeNT and CRESST-II. Dan Hooper finds that gamma-ray signal observed from the Galactic Center is consistent with 7 - 12 GeV dark matter particles annihilating mostly to leptons [61] [62] .

Based on EGRET observations, P. Sreekumar, et al. attribute the high-energy gamma ray emissions to blazars: Most of the measured spectra of individual blazars only extend to several GeV and none extend above 10 GeV, simply because the intensity is too weak to have a significant number of photons to measure [63] . WUM proposes that cores of blazars are composed of annihilating WIMPs (9.6 GeV), explaining why no observed radiation extends above 10 GeV. The results of gamma-ray emission between 100 MeV to 10 GeV detected from 18 globular clusters in our Galaxy are also in a good correlation with the predicted mass of WIMPs [64] [65] .

The DAMA/LIBRA, CoGeNT, CRESST-II, CDMS-II collaborations conduct direct detections of DMP by nuclear recoils due to the elastic scattering of DMP. An 8.6 GeV DMP is deemed most probable [66] .

Based on its core assumptions, WUM analytically predicts WIMPs to possess the mass of 9.6 GeV. A large number of experimental results seem to converge to a number in the neighborhood of 10 GeV, providing additional support to WUM.

6.3. DIRAC 70 MeV

S. D. Hunter, et al. discuss a peak at 67.5 MeV: Below about 100 MeV, gamma rays produced via electron bremsstrahlung are the dominant component of the observed spectrum, whereas, above about 100 MeV, the gamma-rays from π^0 decay, which form the broad "pion bump" centered at 67.5 MeV, are the dominant component of the spectrum. The "pion bump", clearly visible in this spectrum, is the only spectral feature in the diffuse gamma ray emission in the EGRET energy range [67] .

70 MeV peak in EGRET data was discussed by Golubkov and Khlopov [68] . They explained this peak by the decay of π^0 -mesons, produced in nuclear reactions. B. Wolfe, et al. said that gamma rays at 70 MeV are notably detectable by GLAST and EGRET [69] . R. Yamazaki, et al. attribute the 70 MeV peak in the emission spectrum from an old supernova remnant (SNR) to π^0 -decay [70] [71] .

Note that whenever the 70 MeV peak appears in gamma-ray spectra, it is always attributed to pion decay. We claim that π^0 decay produces a 67.5 MeV peak, while DIRAC annihilation is responsible for 70 MeV peak. To find out the source of the observed broad peak about 70 MeV, we suggest utilization of exponentially cutoff power-law for analysis of experimental data for gamma-ray energies < 70 MeV. A better fit of experimental data will be evidence of DIRACs annihilation.

In our opinion, the DIRAC may indeed be the so-called U boson, target of intense search by the scientific community. Note that the mass of DIRAC proposed by WUM—0.07 GeV/c^2—falls into the mass range of U boson: $M_U = 0.02 - 0.1\ \mathrm{GeV/c^2}$ [72] - [77] .

6.4. ELOP 340 keV

An ELOP is a spin-0 boson with 340 keV mass. Existence of DMP of similar masses $(m_\chi < 0.42\ \mathrm{MeV})$ has been discussed by Y. Rasera, et al. [78] . The experimental 100 - 400 keV "bump" [79] is in good agreement with the theoretical analysis in [78] and with annihilating ELOPs with mass 340 keV proposed in our Model.

D. E. Gruber, et al. describes a wide gamma-ray extragalactic background spectrum between 1 keV and 10 GeV: Above 60 keV selected data sets included the HEAO 1 A-4 (LED and MED), balloon, COMPTEL, and EGRET data. The fit required the sum of three power laws [80] .

According to our Model, the fit of the total diffuse spectrum in the range between 3 keV and 10 GeV should be performed based on three exponentially cutoff power-laws with injection spectral $J(E) \propto E^{-\gamma} \exp\{-E/E_{cut}\}$ with the spectral index γ and E_{cut} being the cutoff energy of the source spectra. For values of E_{cut}, we should use

- 9.6 GeV (annihilating WIMPs) in the 9.6 GeV - 70 MeV range;
- 70 MeV (annihilating DIRACs) in the 70 MeV - 340 keV range;
- 340 keV (annihilating ELOPs) in the 340 keV - 3.7 keV range.

The fit in the range between 9.6 GeV and 1.3 TeV should be done with $E_{cut} = 1.3\ \text{TeV}$, which equals to the mass of a neutralino.

6.5. Sterile Neutrino 3.7 keV

The very first signature of the emission around 3.7 keV was found in 1967 by P. Gorenstein, R. Giacconi, and H. Gursky. They analyzed the counting rate in the 2 - 5 keV range and found that the sources GX-10.7, +9.1, +13.5, and +16.7 are qualitatively different from Sco X-1, Cyg X-1 or Cyg X-2 in that the highest number of net counts is recorded in the bin centered at 3.75 keV [81].

An important result was obtained by S. Safi-Harb and H. Ogelman in 1997. They reported that the observations of the X-ray lobes of the large Galactic source W50 [are] associated with the two-sided jets source SS 433. A broken power-law model gives the best fit. The power-law indices are 1.9 and 3.6, with the break occurring at 3.7 keV [82].

T. Itoh analyzed the broad-band (3.0 - 50 keV) spectra of NGC 4388 and found line-like residual around 3.7 keV at the high confidence level [83].

A. Bykov, et al. investigated the nature of the extended hard X-ray source XMMU J061804.3 + 222732 and its surroundings using XMM-Newton, Chandra, and Spitzer observations. The X-ray emission consists of a number of bright clumps embedded in an extended structured non-thermal X-ray nebula larger than 30″ in size. Some clumps show evidence for line emission at ~1.9 keV and ~3.7 keV at the 99% confidence level. A feature at 3.7 keV was found in the X-ray spectrum of Src 3 at the 99% confidence level [84].

In our opinion, the line emission ~3.7 keV corresponds to the annihilation of sterile neutrinos and the line ~1.9 keV corresponds to their decay.

R. Fukuoka, et al. observed the South End of the Radio Arc and found the line-like residual at ~3.7 keV with ~3σ significance [85]. In 2012, A.

Moretti, et al. measured the diffuse gamma-ray emission at the deepest level and with the best accuracy available today. An emission line around 3.7 keV is clearly visible in the obtained spectrum [86] .

7. CONCLUSIONS

Ÿ Emission lines of 1.3 TeV, 9.6 GeV, 70 MeV, 340 keV, and 3.7 keV, can be found in spectra of the diffuse gamma-ray background radiation and various macroobjects of the World in different combinations depending on their structure.

- The diffuse cosmic gamma-ray background radiation in the <1.3 TeV range is the sum of the contributions of multicomponent dark matter annihilation.
- The total cosmic-ray radiation consists of gamma-ray background radiation plus X-ray radiation from the different highly ionized chemical elements in the hot areas of the World and is due to various electron processes such as synchrotron radiation, electron bremsstrahlung, and inverse Compton scattering.

ACKNOWLEDGEMENTS

I am very grateful to referees for the important critical remarks. Many thanks to Felix Lev for our frequent discussions of history and philosophy of Physics. Special thanks to my son Ilya Netchitailo who questioned every aspect of the paper and helped shape it to its present form.

REFERENCES

1. Netchitailo, V.S. (2015) 5D World-Universe Model. Space-Time-Energy. Journal of High Energy Physics, Gravitation and Cosmology, 1, 25.
2. Arrenberg, S., et al. (2013) Complementarity of Dark Matter Experiments. http://www-public.slac.stanford.edu/snowmass2013/docs/CosmicFrontier/Complementarity-27.pdf
3. Heeck, J. and Zhang, H. (2013) Exotic Charges, Multicomponent Dark Matter and Light Sterile Neutrinos.
4. Aoki, M., et al. (2012) Multi-Component Dark Matter Systems and Their Observation Prospects.
5. Kusenko, A., Loewenstein, M. and Yanagida, T. (2013) Moduli Dark Matter and the Search for Its Decay Line Using Suzaku X-Ray Telescope. Physical Review D, 87, Article ID: 043508.
6. Feldman, D., Liu, Z., Nath, P. and Peim, G. (2010) Multicomponent Dark Matter in Supersymmetric Hidden Sector Extensions.

7. Feng, J.L. (2010) Dark Matter Candidates from Particle Physics and Methods of Detection.
8. Zurek, K.M. (2009) Multi-Component Dark Matter.
9. Boehm, C., Fayet, P. and Silk, J. (2003) Light and Heavy Dark Matter Particles.
10. Feng, W.Z., Mazumdar, A. and Nath, P. (2013) Baryogenesis from Dark Matter.
11. D'Souza, I.A. and Kalman, C.S. (1992) Preons: Models of Leptons, Quarks and Gauge Bosons as Composite Objects. World Scientific, Singapore.
12. NASA's Planck Project Office (2013) Planck Mission Brings Universe into Sharp Focus.https://www.nasa.gov/mission_pages/planck/news/planck20130321.html#.VZ4k5_lViko
13. Feng, W.Z., Nath, P. and Peim, G. (2012) Cosmic Coincidence and Asymmetric Dark Matter in a Stueckelberg Extension.
14. Narain, G., Schaffner-Bielich, J. and Mishustin, I.N. (2006) Compact Stars Made of Fermionic Dark Matter.
15. Corda, C., Cuesta, H.J.M. and Gomez, R.L. (2012) High-Energy Scalarons in R2 Gravity as a Model for Dark Matter in Galaxies. Astroparticle Physics, 35, 362-370.
16. Corda, C. (2009) Interferometric Detection of Gravitational Waves: The Definitive Test for General Relativity. International Journal of Modern Physics D, 18, 2275-2282.
17. Woolfson, M.M. (1984) The Evolution of Rotation in the Early History of the Solar System. Philosophical Transactions of the Royal Society A, 313, 5-18.
18. García, R.A., Turck-Chieze, S., Jimenez-Reyes, S.J., et al. (2007) Tracking Solar Gravity Modes: The Dynamics of the Solar Core. Science, 316, 1591-1593.
19. Zhang, J., Song, X.D., Li, Y.C., et al. (2005) Inner Core Differential Motion Confirmed by Earthquake Waveform Doublets. Science, 309, 1357-1360.
20. Livermore, P.W., Hollerbach, R. and Jackson, A. (2013) Electromagnetically Driven Westward Drift and Inner-Core Superrotation in Earth's Core. Proceedings of the National Academy of Sciences, 110, 15914-15918.
21. Baryshev, Y.V. (2008) Field Fractal Cosmological Model as an Example of Practical Cosmology Approach.
22. Agle, D.C. and Brown, D. (2012) Data from NASA's Voyager 1 Point to Interstellar Future. http://www.nasa.gov/mission_pages/voyager/voyager20120614.html
23. Altmannshofer, W., Buras, A.J., Straub, D.M., et al. (2009) New Strategies for New Physics Search in Decays.
24. Del Amo Sanchez, P., et al., The BABAR Collaboration (2011) Search for the Rare Decay .
25. Strigari, L.E. (2012) Galactic Searches for Dark Matter.
26. Bechtol, K. (2011) The Extragalactic Gamma-Ray Background. A Census of High Energy Phenomena in the Universe.http://astro.fnal.gov/events/Seminars/Slides/Bechtol%20120611.pdf
27. Buckley, J., Byrum, K., Dingus, B., et al. (2008) The Status and Future of Ground-Based TeV Gamma-Ray Astronomy. A White Paper Prepared for the Division of Astrophysics of the American Physical Society.
28. Jeltema, T. (2012) Observational Cosmology and Astroparticle Physics. http://physics.ucsc.edu/~joel/12Phys205/Feb6-Jeltema.pdf
29. Aharonian, F.A. (2004) Very High Energy Cosmic Gamma Radiation. A Crucial Window on the Extreme Universe. http://www.worldscientific.com/worldscibooks/10.1142/4657
30. Totani, T. (2009) The Cosmic Gamma-Ray Background Radiation. AGNs, and More?http://www-conf.kek.jp/past/HEAP09/ppt/1day/Totani_HEAP09.pdf
31. Johnson, R.P. and Mukherjee, R. (2009) GeV Telescopes: Results and Prospects for Fermi. New Journal of Physics, 11, Article ID: 055008.
32. Giovannelli, F. and Sabau-Graziati, L. (2012) Multifrequency Behavior of High Energy Cosmic Sources. A Review. Memorie Della Societa Astronomica Italiana, 83, 17.
33. Essig, R., Kuflik, E., McDermott, S.D., et al. (2013) Constraining Light Dark Matter with Diffuse X-Ray and Gamma-Ray Observations.
34. Porter, T.A., Johnson, R.P. and Graham, P.W. (2011) Dark Matter Searches with Astroparticle Data.
35. Boehm, C., Hooper, D., Silk, J., et al. (2003) MeV Dark Matter: Has It Been Detected?
36. Holder, J. (2012) TeV Gamma-Ray Astronomy: A Summary.

37. Chaves, R.C.G., for the H.E.S.S. Collaboration (2009) Extending the H.E.S.S. Galactic Plane Survey.
38. Tibolla, O., Chaves, R.C.G., de Jager, O., et al. (2009) New Unidentified H.E.S.S. Galactic Sources.
39. Hoppe, S., de Ona-Wilhemi, E., Khélifi, B., et al. (2009) Detection of Very-High-Energy Gamma-Ray Emission from the Vicinity of PSR B1706-44 with H.E.S.S.
40. Tam, P.H.T., Wagner, S.J., Tibolla, O., et al. (2009) A Search for VHE Counterparts of Galactic Fermi Bright Sources and MeV to TeV Spectral Characterization.
41. Tibolla, O., Chaves, R.C.G., Domainko, W., et al. (2009) New Unidentified Galactic H.E.S.S. Sources.
42. Tam, P.H.T., Wagner, S., Tibolla, O. and Chaves, R. (2010) A Search for VHE Counterparts of Galactic Fermi Sources.
43. Aleksic, J., Ansoldi, S., Antonelli, L.A., et al. (2013) Optimized Dark Matter Searches in Deep Observations of Segue 1 with MAGIC.
44. Moralejo, A. (2013) http://projects.ift.uam-csic.es/multidark/images/moralejoalcala.pdf.
45. Abramowski, A., Acero, F., Aharonian, F., et al. (2013) Search for Photon Line-Like Signatures from Dark Matter Annihilations with H.E.S.S.
46. Jin, H.B., Wu, Y.L. and Zhou, Y.F. (2013) Implications of the First AMS-02 Measurement for Dark Matter Annihilation and Decay.
47. Abdo, A.A., et al., Fermi/LAT Collaboration (2009) Measurement of the Cosmic Ray e+ + e- Spectrum from 20 GeV to 1 TeV with the Fermi Large Area Telescope.
48. Adriani, O., Barbarino, G.C., Bazilevskaya, G.A., et al. (2011) The Cosmic-Ray Electron Flux Measured by the PAMELA Experiment between 1 and 625 GeV.
49. He, X.G. (2009) A Brief Review on Dark Matter Annihilation Explanation for e± Excesses in Cosmic Ray.
50. Cholis, I. and Goodenough, L. (2010) Consequences of a Dark Disk for the Fermi and PAMELA Signals in Theories with a Sommerfeld Enhancement.
51. Morselli, A. (2011) Indirect Detection of Dark Matter, Current Status and Recent Results. Progress in Particle and Nuclear Physics, 66, 208-215.
52. Abazajian, K.N. and Harding, J.P. (2011) Constraints on WIMP and Sommerfeld-Enhanced Dark Matter Annihilation from HESS Observations of the Galactic Center.
53. Kawanaka, N., Ioka, K., Ohira, Y., et al. (2010) TeV Electron Spectrum for Probing Cosmic-Ray Escape from a Supernova Remnant.
54. Aharonian, F.A., Akhperjanian, A.G., de Almeida, U.B., et al. (2008) Energy Spectrum of Cosmic-Ray Electrons at TeV Energies. Physical Review Letters, 101, Article ID: 261104.
55. Ibarra, A., et al. (2010) Extragalactic Diffuse Gamma-Rays from Dark Matter Decay.http://calet.phys.lsu.edu/Science/DGR.php
56. Orr, M. and Krennrich, F. (2011) Constraining the Extragalactic Background Light in the Near-Mid IR with the Cherenkov Telescope Array (CTA). 32nd International Cosmic Ray Conference, Beijing.http://www.ihep.ac.cn/english/conference/icrc2011/paper/ proc/v8/v8_1156.pdf
57. Orr, M., Krennrich, F. and Dwek, E. (2011) Strong New Constraints on the Extragalactic Background Light in the Near- to Mid-IR.
58. Madhavan, A. (2013) The VHE γ-Ray Spectra of Several Hard-Spectrum Blazars from Long-Term Observations with the VERITAS Telescope Array. PhD Thesis, Iowa State University, Ames.
59. Torii, S., for the CALET Collaboration (2014) The CALorimetric Electron Telescope (CALET): A High Energy Cosmic-Ray Observatory on the International Space Station.http://www.crlab.wise.sci.waseda.ac.jp/eng/wp-content/uploads/downloads/2014/09/VHEPU2014-CALET_final.pdf
60. Papuccia, M. and Strumia, A. (2009) Robust Implications on Dark Matter from the First FERMI Sky Gamma Map.
61. Hooper, D. (2012) The Empirical Case For 10 GeV Dark Matter.
62. Hooper, D. and Goodenough, L. (2010) Dark Matter Annihilation in the Galactic Center as Seen by the Fermi Gamma Ray Space Telescope.

63. Sreekumar, P., Bertsch, D.L., Dingus, B.L., et al. (1998) EGRET Observations of the Extragalactic Gamma Ray Emission. The Astrophysical Journal, 494, 523-534.
64. Abdo, A.A., et al. (1997) A Population of Gamma-Ray Emitting Globular Clusters Seen with the Fermi Large Area Telescope.
65. Tam, P.H.T., Kong, A.K.H., Hui, C.Y., et al. (1997) Gamma-Ray Emission from Globular Clusters.
66. Frandsen, M.T., Kahlhoefer, F., McCabe, C., et al. (2013) The Unbearable Lightness of Being: CDMS versus XENON.
67. Hunter, S.D., Bertsch, D.L., Catelli, J.R., et al. (1997) EGRET Observations of the Diffuse Gamma-Ray Emission from the Galactic Plane. The Astrophysical Journal, 481, 205-240.
68. Golubkov, Y.A. and Khlopov, M.Y. (2000) Antiprotons Annihilation in the Galaxy as a Source of Diffuse Gamma Background.
69. Wolfe, B., Melia, F., Crocker, R.M., et al. (2008) Neutrinos and Gamma Rays from Galaxy Clusters.
70. Yamazaki, R., Kohri, K., Bamba, A., et al. (2006) TeV Gamma-Rays from Old Supernova Remnants.
71. Nakamori, T. (2012) Fermi Observations of Galactic Sources.www.heap. phys. waseda.ac.jp/cnf1203/Files/Oral/Nakamori.pdf
72. Agakishiev, G., Balanda, A., Belver, D., et al. (2013) Searching a Dark Photon with HADES.
73. Merkel, H., Achenbach, P., Gayoso, C.A., et al., A1 Collaboration (2011) Search for Light Gauge Bosons of the Dark Sector at the Mainz Microtron. Physical Review Letters, 106, Article ID: 251802.
74. Abrahamyan, S., Ahmed, Z., Allada, K., et al., APEX Collaboration (2011) Search for a New Gauge Boson in Electron-Nucleus Fixed-Target Scattering by the APEX Experiment. Physical Review Letters, 107, Article ID: 191804.
75. Drees, R.M., Waltham, C., Bernasconi, T., et al., SINDRUM I Collaboration (1992) Measurement of the π0 Electromagnetic Transition form Factor. Physical Review D, 45, 1439-1447.
76. Adlarson, P., et al., WASA-at-COSY Collaboration (2013) Search for a Dark Photon in the π0→e+e-γ Decay. Physics Letters B, 726, 187-193.
77. Babuski, D., Badoni, D., Balwierz-Pytko, I., et al., KLOE-2 Collaboration (2013) Limit on the Production of a Light Vector Gauge Boson in φ Meson Decays with the KLOE Detector. Physics Letters B, 720, 111-115.
78. Rasera, Y., Teyssier, R., Sizun, P., et al. (2006) Soft Gamma-Ray Background and Light Dark Matter Annihilation.
79. Zdziarski, A.A. (1996) Contributions of AGNs and SNe Ia to the Cosmic X-Ray and Gamma-Ray Backgrounds. Monthly Notices of the Royal Astronomical Society, 281, L9-L13.
80. Gruber, D.E., Matteson, J.L. and Peterson, L.E. (1999) The Spectrum of Diffuse Cosmic Hard X-Rays Measured with HEAO-1.
81. Gorenstein, P., Giacconi, R. and Gursky, H. (1967) The Spectra of Several X-Ray Sources in Cygnus and Scorpio. The Astrophysical Journal, 150, L85.
82. Safi-Harb, S. and Ogelman, H. (1997) ROSAT and ASCA Observations of W50 Associated with the Peculiar Source SS 433. The Astrophysical Journal, 483, 868-881.
83. Itoh, T. (2007) Suzaku Studies of Time Variable X-Ray Spectra of Edge-On Active Galactic Nuclei. PhD Thesis, University of Tokyo, Tokyo. http://www.astro.isas. jaxa.jp/suzaku/bibliography/phd/titoh_dron_print080220.pdf
84. Bykov, A.M., Krassilchtchikov, A.M., Uvarov, Y.A., et al. (2009) Isolated X-Ray—Infrared Sources in the Region of Interaction of the Supernova Remnant IC 443 with a Molecular Cloud.
85. Fukuoka, R., Koyama, K., Ryu, S.G., et al. (2008) Suzaku Observation Adjacent to the South End of the Radio Arc.
86. Morretti, A., Vattakunnel, S., Tozzi, P., et al. (2012) Spectrum of the Unresolved Cosmic X-Ray Background: What Is Unresolved 50 Years after Its Discovery.

CHAPTER 13

The Conservation of Energy Space-Time Metric for Space Outside Matter

V. N. E. Robinson

ETP Semra Pty Ltd., Canterbury, Australia

ABSTRACT

By using experimentally determined measurements of potential energy together with the principle of conservation of energy and solving directly, the space-time geometry equation for space outside matter is obtained. That equation fits all the experimental observations that support the accepted Schwarzschild metric, yet predicts there isn't a singularity at the Schwarzschild radius. The accepted Schwarzschild metric is the first approximation of the conservation of energy space-time metric. No observation yet made can distinguish between the predictions of the two metrics.

Keywords: Gravity; Photons; Redshift; Conservation of Energy; Space-Time Metric; No Black Holes

1. INTRODUCTION

With the passage of time, Einstein's gravitational field equations [1-4], now often written as

$$R_{\mu\nu} - \frac{1}{2} g_{\mu\nu} R = \frac{8\pi G}{c^4} T_{\mu\nu} \qquad (1)$$

where $R_{\mu\nu}$ is the Ricci curvature tensor, $g_{\mu\nu}$ is the metric tensor, R is the scalar curvature, G is Newton's universal gravitational constant, c is the speed of light and $T_{\mu\nu}$ is the stress energy tensor, remain the only theory of gravity to correctly predict observations against which they were tested [The cosmological constant term $g_{\mu\nu}\Lambda$ is often included on the left hand side of Equation (1)]. Among the features Einstein predicted from his theory were:

- Photons leaving the sun would be redshifted by the time they reached Earth;
- The orbit of the planet Mercury would undergo a precession due to the sun's mass curving space-time;
- The sun's mass would deflect the path of photons that travelled close to it;
- Photons leaving the surface of a star or planet would be redshifted.

His subsequent predictions of gravity waves and gravitational lensing have also been confirmed.

Schwarzschild [5] was the first to attempt to solve them. Later solutions [6-11] yielded the now accepted exact solution [12]

$$\begin{aligned} ds^2 &= -dt^2\left(1-\frac{\alpha}{r}\right)+dr^2\left(1-\frac{\alpha}{r}\right)^{-1} \\ &\quad +r^2\left(d\theta^2+\sin^2 d\varphi^2\right) \end{aligned} \qquad (2)$$

where s is the space-time geometry co-ordinate, t is time, r is distance from the centre of mass, θ and φ are angles and $\alpha = \frac{2GM}{c^2}$ where M is the mass of the body that is distorting space-time, G is Newton's universal gravitational constant and c is the speed of light. It is rather obvious that Equation (2) behaves badly at r = α and this is the source of the concept of black holes with an event horizon at radius r = α.

Einstein's gravitational field equations were based upon mass distorting space-time. The success of the general theory of relativity in predicting gravitational effects has been highly significant. As well as explaining the effects mentioned above, Einstein also predicted that gravity waves would be produced by rotating non symmetric objects, gravitational lensing and moving mass dragging space-time with it. These were subsequently observed. A prediction from Equation (2) was that anything originating at a distance $r < \alpha$ from a massive object could not pass through the barrier at $r = \alpha$ because it would need to travel at faster than c, which led to the concept of black holes. Subsequent studies have suggested that black holes lost all properties except mass and rotation. Kerr [13] studied the properties of rotating black holes and was able to predict a spreading of photon wavelengths when seen by outside observers.

The developments of the general theory of relativity have been very good at explaining most planetary and stellar gravitational effects as well as many galactic phenomena and some inter galactic effects. There is no doubt as to the validity of the derivations of the spacetime metric of Equation (2) from Einstein's field Equations (1). Einstein showed the relationship between his work and Newtonian gravity. Newton [14] unified potential energy with stellar behavior and gravitational effects on planet Earth. His theory successfully described the behavior of objects under the influence of large masses such as planets and stars. In particular it enabled gravitational effects to be calculated mathematically, explaining all effects observed to that time.

Despite its success, the general theory of relativity still has its critics, particularly those aspects dealing with the predictions of an event horizon associated with collapsed matter forming black holes. From Equation (2) it is obvious that a black hole will form with an event horizon at $r = \alpha$, beyond which even light cannot travel. In the case of matter collapsing to a point singularity the event horizon requires gravity to have an action at a distance. There is no physical principle in which action at a distance can occur. One of the early criticisms of Newton's theory of gravity was that it required action at a distance for two objects to attract each other across space. Despite that criticism, Newton's calculations dominated all gravitational studies for over two hundred years. Einstein's general relativity theory overcame the problem of action at a distance by showing that mass distorted space-time and in turn, space-time determined how mass moved. That has allowed general relativity to successfully describe all situations against which it has been tested since its introduction.

All theories are attempts to describe reality. That does not mean that a theory is reality even if it successfully describes all situations against which it has been tested. When it comes to gravitational effects, the theory of general relativity gives the best description, having successfully predicted the outcome of all measurements against which it has been tested. Despite almost a century passing since it was first published, its format is still regarded as entirely mathematical. Some of its predictions, such as a black hole having an event horizon, are seen as predictions from the mathematics of an accurate theory for which no physical explanation is invoked. This lack of a physical explanation and the complexity of the mathematics has led many to question the integrity of general relativity, despite the accuracy of its experimentally tested predictions. General relativity has its supporters and its detractors. None of its supporters appear to have been able to offer a physical explanation for its effect. None of its detractors appear to have come up with an alternative that can satisfy all the measurements that general relativity has successfully predicted.

Reality is the result of gravity determined by measuring such properties as potential energy. It is beneficial for measurements on planet Earth that Newton's theory of gravity gives a means of calculating potential energy. With or without Newton's theory, potential energy is still a measurable property associated with mass, which effect can be calculated by using Newton's universal gravitational constant G. In his derivation of general relativity, Einstein still used the same universal gravitational constant associated with mass. Gravity and potential energy are inseparably associated with mass. It is an aim of this manuscript to show that gravitational effects can be calculated from the well known physical principle of conservation of energy.

2. THE CONSERVATION OF ENERGY METRIC

Potential energy (PE) is associated with mass through the relationship

$$PE = \int_{r_1}^{r_2} \frac{GMm}{r^2} dr \tag{3a}$$

where r_1 and r_2 are the different distances r from the centers of mass of an object of mass m from another object of mass M. When $\Delta r = r_2 - r_1 \ll r$, this simplifies to

$$PE = mg\Delta r \tag{3b}$$

where g is the average acceleration due to gravity over the distance Δr. No experiment or observation has yet been made in which Equation (3a) does not hold. As such it is a suitable starting point for a study of the behavior of photons in space outside matter. A photon is a particle of energy E = hν, where h is Planck's constant and ν is its frequency. From Einstein's special theory of relativity [15,16] it is known that $E = mc^2$, which yields the mass m of a photon as $m = h\nu/c^2$ [17-21]. It is known from experiment that an object of mass m with kinetic energy (KE) moving up against a gravitational field will lose kinetic energy as it gains potential energy. This is illustrated in **Figure 1**, for a photon of mass $h\nu/c^2$ moving distance Δr against a gravitational field of strength g. This was recognized by Einstein [22,23] when he discussed the effect of gravity upon the propagation of light. He expressed the relationship between the frequencies of light at different vertical positions by the equation

$$\nu_2 = \nu_1\left(1+\frac{GM\Delta r}{r^2c^2}\right) \tag{4}$$

where ν_1 and ν_2 are the frequency of the light at vertical distances r_1 and r_2, separated by height $\Delta r = r_2 - r_1$, in a gravitational field of strength $g = -GM/r^2$, as illustrated in **Figure 1**. With a photon being a constant velocity particle, a loss in kinetic energy can only occur by a loss of mass and hence frequency. Equation (4) was used by Einstein [22,23] and was subsequently verified by the work of Pound, Rebka and Snider [24-26]. Equation (4) is based upon experimental observation and has been adequately calculated using Newtonian mechanics. To date no observation has been made in which Equations (3) and (4) do not hold true.

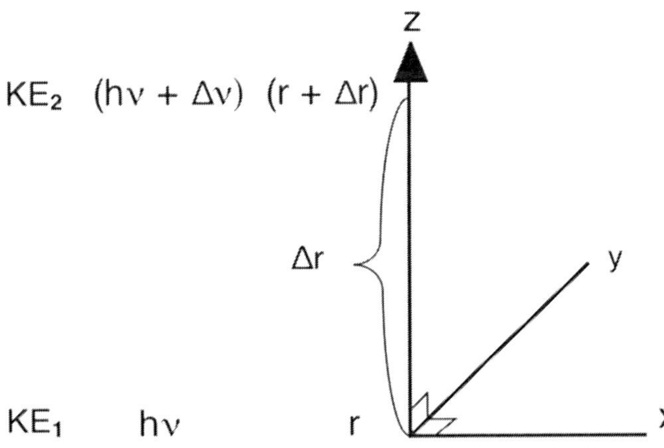

Figure 1. Illustration of the gain in potential energy when a photon travels a distance Δr against gravity.

The conservation of energy as the light travels away from the centre of mass, as illustrated in **Figure 1**, yields

$KE_2 = KE_1 + PE$ which becomes

$$h(\nu + \Delta\nu) = h\nu - h\nu \frac{GM\Delta r}{r^2c^2} \tag{5}$$

Setting the limit as Δr tends to zero in Equation (4), while allowing that G and c are constants and M is the mass of a gravitationally attracting body, enables Equation (5) to be rearranged to give

$$\frac{d\nu}{\nu} = \frac{-GMdr}{r^2c^2} \tag{5a}$$

for which there is no experimental evidence to suggest that it is not applicable for all values of r from the centre of mass of M. Newton [14] showed that mass distributed throughout space behaves gravitationally as if all the mass were concentrated at the centre of mass, when observed from outside the mass, allowing r to be used as the distance from the center of mass M.

Introducing a constant $\alpha = \frac{2GM}{c^2}$ and integrating from r_1 to r_2 ($>r_1$) gives

$$\ln\nu_2 - \ln\nu_1 = -\frac{\alpha}{2r_2} + \frac{\alpha}{2r_1}$$

or

$$\frac{\nu_2}{\nu_1} = \frac{e^{\frac{\alpha}{2r_1}}}{e^{\frac{\alpha}{2r_2}}} \tag{6}$$

Equation (6) is an exact expression for the variation of frequency of light when the energy of an individual photon remains constant as it moves away from the centre of mass M. An observer at a distance from M will see the light coming away from the object as being red shifted. To such an observer, it would appear as if time was slowing down the closer the observation was made to the centre of mass. If we apply this to the case of the red shift from the sun, r_1 = 700,000 km, $r_2 = 1.49 \times 10^8$ km and α = 2.95 km, a photon leaving the surface of the sun would be red shifted by an amount equal to

$$e^{\frac{2.95}{1,400,000}} - e^{\frac{2.95}{298,000,000}} \tag{6a}$$

by the time it reaches the surface of planet Earth. This equates to 2×10^{6}, which agrees with the result reported by Einstein [22,23]. Such a change in frequency has been observed [27,28]. It should be noted that others have used Newtonian mechanics in the same manner to calculate the redshift of photons from the sun, reaching the same answer [29,30]. Those authors did not continue the study in the manner reported in the following work.

From knowledge that c is constant to all observers and the relationship

$$\lambda = \frac{c}{v} \tag{7}$$

where λ is the wavelength of the photon, it follows that this change of frequency will generate an inverse change in wavelength. The only method we have of measuring the properties of distant objects is with photons. If the frequency of photons has changed, the distance measured using photons will also change. Equation (7) can now be substituted into Equation (6) to determine the corresponding variation of length with distance, yielding

$$\frac{\lambda_1}{\lambda_2} = \frac{e^{\frac{\alpha}{2r_2}}}{e^{\frac{\alpha}{2r_1}}} \tag{8}$$

Equations (6) and (8) can now be used to define the exact nature of the variation of time and length with distance from the centre of mass of a body. This is equivalent for each of the coordinates, x, y and z enabling the exact nature of the space-time continuum to be calculated in any direction radially out from the centre of mass of a massive object M. We can now go directly to the Minkowski space-time equation [31,32], which is applicable to flat space-time that is well away from significant mass, which using the most common convention is expressed as

$$s^2 = ct^2 - x^2 - y^2 - z^2$$

Following the work of Schwarzschild [5], when mass is involved, we need to change the form to take into account the distortion of space, giving

$$ds^2 = Fdt^2 - H\left\{dx^2 + dy^2 + dz^2\right\}$$
$$-J\left\{xdx + ydy + zdz\right\} \tag{9}$$

where F, H and J are functions of r when $r^2 = x^2 + y^2 + z^2$.

Equation (9) can be transformed from Cartesian to polar co-ordinates to yield

$$ds^2 = Fdt^2 - \left(H + Jr^2\right)dr^2$$
$$-Jr^2\left(d\theta^2 + \sin^2\theta d\phi^2\right). \tag{10}$$

Allowing that the distortion of space and time is measured by variations to the frequency and wavelength of photons, it is apparent from Equation (7) that the product of $Fx\left(H + Jr^2\right) = 1$. From Equations (6) and (8) this gives

$$F = e^{\frac{-\alpha}{r}}, \left(H + Jr^2\right) = e^{\frac{\alpha}{r}}$$

and $J = 1$ yielding

$$ds^2 = dt^2 e^{\frac{-\alpha}{r}} - dr^2 e^{\frac{\alpha}{r}} - r^2\left(d\theta^2 + \sin^2\theta d\varphi^2\right). \tag{11}$$

Equation (11) describes how space-time varies around an object of mass

$$M = \frac{\alpha c^2}{2G}$$

based upon the principle of conservation of energy. The only way we have of observing space-time is the detection of photons that have traveled through it.

The distortion of space-time is both a measure of and is measured by the distortion of the photons that have traveled through it.

3. COMPARISON WITH OTHER METRICS

From his field equations for space outside matter, Einstein derived his approximate solutions of $dx_1 = 1 - \alpha/2r$ and $dx_4 = 1 + \alpha/2r$. He then used those approximations to solve for the deviation of a photon by the gravitational field of the sun and the precession of the orbit of the planet Mercury. Schwarzschild [5] provided a solution to them, which solution had the form

$$\mathrm{d}s^2 = \mathrm{d}t^2\left(1-\frac{\alpha}{R}\right) - \mathrm{d}R^2\left(1-\frac{\alpha}{R}\right)^{-1}$$
$$-R^2\left(\mathrm{d}\theta^2 + \sin^2\theta \mathrm{d}\phi^2\right)$$

$$R = \left(r^3 + \alpha^3\right)^{1/3}.$$

It should also be noted that Schwarzschild introduced a co-ordinate change from r to R to simplify his calculations. In his answer, the singularity at R = α equates to a singularity at r = 0.

Equation (2) is an exact solution to Einstein's field Equations. Equation (11) is an exact derivation of the space-time geometry equation using the principle of conservation of energy. Equations (2) and (11) are different. Although Einstein was aware of Equation (5) and used it in his first paper to derive the effect of mass upon space-time, his further work that led to the general relativity theory, was based upon the effect of the gravitational field upon space-time. This has led some general relativity practitioners to conclude that his field equations hold for all values of α/r, i.e., all gravitational field strengths, while Newtonian mechanics, which Einstein showed was a first approximation of general relativity, was only a weak field solution and as such could not be extended to high field strengths. This author contends that the principle of conservation of energy applies at all values of α and r, i.e., all gravitational field strengths. Until experimental evidence shows that Equations (3) and (4) do not hold, Equation (11) must be considered to be valid.

Further it should be noted that $e^{\frac{-\alpha}{r}}$ can be expressed by the Maclaurin/Taylor series as:

$1-\frac{\alpha}{r}+\frac{\alpha^2}{r^2 2!}-\frac{\alpha^3}{r^3 3!}+\cdots\frac{(-\alpha)^n}{r^n n!}$ making $1-\frac{\alpha}{r}$ the first approximation to $e^{\frac{-\alpha}{r}}$.

The only mechanism for separating those two solutions is to make measurements well within the difference between the two terms, namely approximately $0.1(\alpha/r)^2$. This can be done either by performing measurements on planet Earth that are accurate to better than 1 part in 10^{21}, or making independent observations of the Schwarzschild radius α, the radius r and measuring the distortions of space-time at distances r less than 10 α to an accuracy of better than 1:100. Many experiments have been forwarded as verifying Equation (2). Each of those observations and subsequent verifications has been based upon only knowledge of Equation (2) only and not of Equation (11). To date no observation has been made that can distinguish between the predictions of Equations (2) and (11) [33].

It should be noted that Eddington [30] pointed out that light had mass and that using Newtonian mechanics as the basis of calculation, the deflection of light rays by the sun would be half that due to Einstein's space-time distortion, the same as was originally calculated by Einstein [22,23]. The other half of the distortion comes about because a change in frequency automatically means a change in time as well as length. Eddington [34] spoke of light pulses and accepted that they behaved according to what is now called the Schwarzschild metric given in Equation (2). It is equally clear that his original work on the influence of gravity upon the propagation of light, Einstein [22,23] considered the effect of Newtonian calculations of gravity upon particles of light, now called photons. Under the Einstein-de Broglie model of photons, they are considered as particles that have relativistic mass given by $m = \frac{h\upsilon}{c^2}$.

4. CALCULATING SOME GRAVITATIONAL EFFECTS

The ability of photons to be treated in the conservation of energy as particles of constant velocity and variable mass makes it somewhat easier to grasp the mechanism involved in the general theory of relativity. The redshift of photons leaving a massive object is done in the same manner as was done by Einstein [22,23] and gives the same answer, namely $\alpha/2r = 2\times10^{-6}$ for photons leaving the sun, as shown in the derivation of Equation (6).

Consider the situation shown in **Figure 2**, in which a photon passes at distance r_0 from the origin O of an object of mass M. From Equation (4) the photon experiences a distortion of time at distance r given by $\alpha/2r$ as it travels from $-\infty$ to $+\infty$. The total time distortion it experiences is given by summing all distortions, namely

$$\int_{\frac{-\pi}{2}}^{\frac{\pi}{2}} \frac{\alpha}{2r}\cos\theta d\theta = \frac{-\alpha}{r_0}$$

where r_0 is the distance of closest approach. From Equations (7) and (11) it is obvious that there is a similar term for the distortion of space, but with opposite sign. Equation (11) shows that the total distortion of space-time is obtained by subtracting these two, which gives a spacetime distortion that shows up as a deflection given by

$$\acute{e} = \frac{-2\alpha}{r} \tag{12}$$

with the negative sign indicating that the distortion is towards the massive object. Using α of the sun = 2.95 km and the radius of the sun as r_0 = 700,000 km, gives é = 1.74 arc seconds which agrees with Einstein's prediction [1,3,4] and was subsequently observed [34,35].

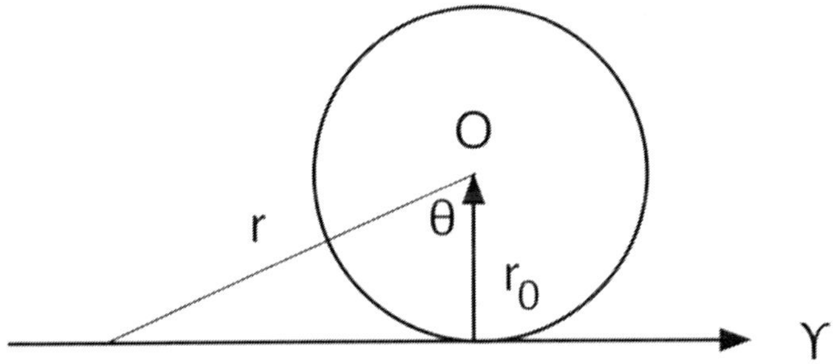

Figure 2. Illustration of the passage of a photon Υ close to the surface of a massive object, making its closest approach at distance r_0.

A study of the derivation of the precession of the orbit of planet Mercury, also predicted by Einstein [2] may help to illustrate the easier understanding of this "relativistic" effect.

Mercury is the innermost planet and has an elliptical orbit, moving between distances of 46.6 million kilometers and 69.8 million kilometers from the sun, giving it an orbit eccentricity e of 0.2056, taking approximately 88 Earth days to travel once around the sun, meaning that it makes 415 orbits per Earth century.

From Equation (12), we know that a photon passing by the sun at the orbit of Mercury, will be deflected by an amount, é, equal to $2\alpha/r'_M$, where r'_M is the corrected radius for Mercury's orbit. Under this presentation é is the distortion of space-time at this distance, as viewed from infinity. On planet Earth, from where these observations are made, there is a time distortion of $e^{\frac{-\alpha}{2r_E}}$ where r_E is radius of planet Earth's orbit, see Equation (6). This is illustrated in **Figure 3**, which shows the time dilation, vertical scale, plotted against distance. When measured from Earth, this must be included in the spacetime distortion Δs at Mercury's orbit, to give

$$\Delta s = \frac{2\alpha}{r_M} - e^{\frac{\alpha}{2r_E}}$$

which simplifies to

$$\Delta s = \frac{2\alpha}{r_M} - \frac{\alpha}{2r_E}$$

when r'_M and $r_E \gg \alpha$.

Note that this calculation only takes planet Earth's time distortion into account and not the space-time distortion because all measurements are made by photons travelling from Mercury to Earth, which, to a first approximation, is at a fixed distance r_E from the sun. We need to consider the distortion at those distances for the whole circle of radius r. **Figure 4** illustrates photons moving at a distance r from mass M.

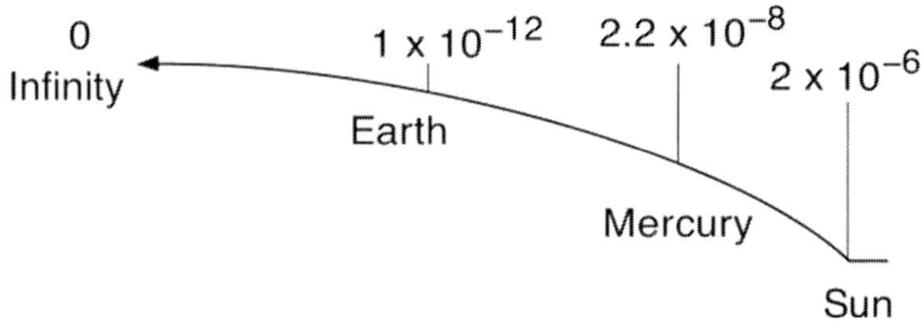

Figure 3. Schematic illustration of the relationship between time dilation and distance for the sun and planets Mercury and Earth.

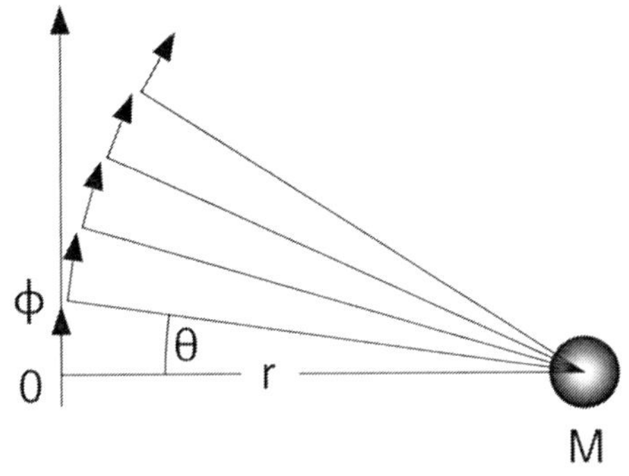

Figure 4. Schematic illustration of the passage of a photon past a massive object centered at point M. Summing the series of those segments yields the complete trajectory.

In each case a photon passing at distance r will be distorted Δs to move in a slightly different direction. Another photon passing at the new point will be distorted by the same Δs, this time starting at a different point. To determine the space-time distortion over a complete orbit, we must consider the deflection of the photon over angle φ. This is given by

$$\int_0^{\varphi}\left[\frac{2\alpha}{\left(1-e^2\right)r_M}-\frac{\alpha}{2r_E}\right]\cos\theta\, d\theta$$

$$=\left[\frac{2\alpha}{\left(1-e^2\right)r_M}-\frac{\alpha}{2r_E}\right]\sin\varphi$$

where e is the eccentricity of Mercury's orbit and r_M is its semi-major axis. To obtain the space-time distortion at Mercury's orbit, we set the limit of φ tends to zero, yielding sinφ = φ, and integrate φ from 0 to 2π, which gives

$$\varepsilon=\left[\frac{2\alpha}{\left(1-e^2\right)r_M}-\frac{\alpha}{2r_E}\right]\int_0^{2\pi}\varphi\, d\varphi$$

yielding

$$\varepsilon=2\pi\left[\frac{2\alpha}{\left(1-e^2\right)r_M}-\frac{\alpha}{2r_E}\right] \qquad (13)$$

where ε is the distortion of space-time that leads to the precession of Mercury's orbit in its direction of rotation. Using the above information, α = 2.95 km and $r_E = 149 \times 10^6$ km, gives the precession ε of Mercury's orbit as ε = 42.15 arc seconds per century.

This is slightly less than that predicted from Einstein's calculation, which gives ε = 42.98 arc seconds per century. Part of these differences can be attributed to the uncertainties in e, α, r_E and r_M. It should also be noted that the total precession of Mercury's orbit is 5599.7 arc seconds per century [36]. Of this, 5560.26 arc seconds per century is explained by other factors [37], leaving 39.44 arc seconds per century unexplained. At 42.98 arc seconds per century, Einstein's explanation is too large by about 3.5 arc seconds per century. At 42.15 arc seconds per century, this derivation leaves about 2.7 arc seconds per century unexplained.

The ease of understanding this approach could be compared to the more complex approaches used by others. For example Trencèvski et al. [38] used flat Minkowski space to calculate their equivalent to Equation (13), yielding

$$\varepsilon = \frac{3\pi\alpha}{\left(1 - e^2\right) r_M}$$

which equates to 35.6 arc seconds per century when applied to the precession of the orbit of the planet Mercury. It should be noted that when Einstein derived his explanation for the motion of Mercury's perihelion, he also derived that same expression. Trencèvski et al. [38] calculation of the equivalent of Equation (12) gave the same answer as Equation (12).

The accuracy of Einstein's field equations from his general theory of relativity in predicting gravitational effects should not be overlooked. Einstein [39] made a further prediction that non-symmetrical rotating massive bodies would radiate gravity waves and hence loose energy. That loss of energy would be detected by the slowing down of the rotating body. Hulse and Taylor [40] reported a binary pulsar, PSR B1913 + 16, two neutron stars rotating about each other, which seemed to fit the situation of a non-symmetrical rotating body. Observation by Weisberg and Taylor [41-43] showed it was indeed slowing down according to Einstein's predictions. Consequently it is generally believed that Einstein's general theory of relativity been verified to an accuracy of approximately 1 part in 10^{14}. The decay rate is independent of whether the metric used is Equation (2) or Equation (11). The same situation applies for other verifications of Einstein's field equations. Unless independent measurements of r and α are made to the accuracy required to separate the two metrics, there is no experimental evidence to suggest that Equation (2) is accurate and Equation (11) is not.

5. DISCUSSION

The above has demonstrated that treating photons as particles of variable mass and constant velocity while calculating the effect of conservation of energy upon them, has produced the space-time metric for space outside matter. These calculations, based upon the experimenttally verified principles of conservation of mass/energy and the existence of potential energy associated with massive objects, were done without any approximations. The relationship between the general relativity and conservation of energy metrics is illustrated in **Figure 5**.

Figure 5(A) suggests that mass generates potential energy, leading to gravity, which in turn distorts spacetime, as derived in this presentation. **Figure 5**(B) suggests that mass distorts space-time, causing gravity, which in turn leads to potential energy. In both cases, the central theme is mass, from which potential energy, gravity and space-time distortion cannot be separated, as suggested in **Figure 5**(C). The pathway for the calculation of gravity from mass can be either conservation of energy or space-time distortion. In either pathway the answers should always be the same because their effect, namely gravity, is the same. The important question is "which pathway offers the best means of calculating gravitational effects?" General relativity predicts a singularity at the Schwarzschild radius, which requires "action at a distance" when the Schwarzschild radius is larger than the radius of the massive body. This conservation of energy metric does not use any approximations and does not require "action at a distance" to explain any gravitational effect. Without independent determinations of r and α at $r < 10\alpha$ and the measurement of gravitational effects to an accuracy of $\approx$ 1%, there is no experimental evidence to suggest that the Schwarzschild metric better describes gravitational effects than those described by this conservation of energy metric. Lasky [33] has shown that current observations are several orders of magnitude away from being able to verify the accuracy of general relativity predictions under all conditions. Until measurements are made that can distinguish between the two metrics, there is no experimental evidence to suggest that the predictions of gravitational effects using general relativity are better than those obtained using conservation of energy.

The relationship between the three descriptions of gravity, Newtonian mechanics, general relativity and conservation of energy was shown above. It is contended that the conservation of energy metric will accurately predict all gravitational phenomena for space-time outside matter.

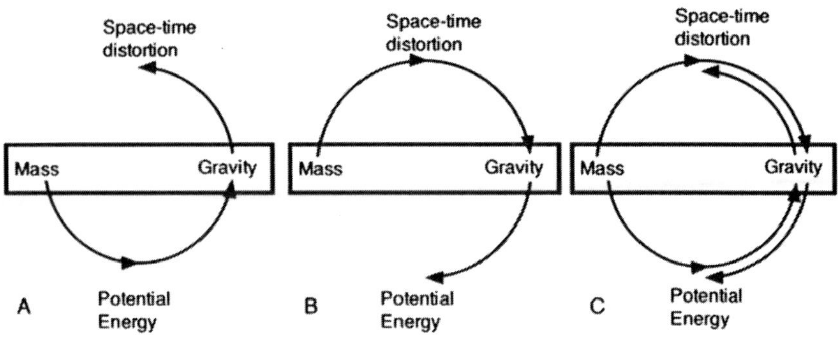

Figure 5. Schematic illustration of the possible relationships between mass, potential energy, gravity and space-time distortion. A illustrates the pathway that gravity arises from mass by potential energy, with gravity distorting space-time, as shown in this presentation. B illustrates the pathway that mass distorts space-time, which leads to gravity that generates potential energy. C suggests that both pathways will lead to the same result.

The Schwarzschild metric from general relativity is the first approximation to the conservation of energy metric in which $e^{\frac{-\alpha}{r}}$ is approximated to $1-\frac{\alpha}{r}$. Newtonian mechanics is based upon the speed of light being infinite and the mass of a photon being zero. It accurately describes local situations where experimental error is not sufficient to distinguish between c and ∞ or measure the change in mass (and hence frequency) of photons. As this presentation shows, giving photons a mass and a finite velocity enables Newtonian mechanics to predict relativistic effects when conservation of energy is applied to the photons. The differences are between the predictions of general relativity and the conservation of energy metrics.

Figure 6 illustrates the difference expected when photons of frequency ν_0 are emitted at different distances r from a massive object at O and detected by an observer at infinity. The conservation of energy metric predicts that a photon emitted at r will have a frequency at infinity given by

$$\nu_\infty = \nu_0 e^{\frac{-\alpha}{2r}} \tag{14}$$

where r is the distance from the centre of mass. The Schwarzschild metric predicts that the photon will have a frequency given by

$$\nu_\infty = \nu_0 \sqrt{1 - \frac{\alpha}{r}} \quad (15)$$

For photons emitted at distances $r \gg \alpha$, the predictions of the two metrics will be the same for observations at infinity. As the emission distance approaches the Schwarzschild radius the predictions become significantly different. The Schwarzschild metric predicts that photons emitted at r = α will have no frequency left when they reach infinity. The conservation of energy metric predicts that photons emitted at r = α will have a redshift of ≈ 0.6065 and still be detected. Until observations are made at distances r < 10α to an accuracy of ≈ 1%, this conservation of energy metric matches all observations that have confirmed the accuracy of general relativity predictions.

Equations (14) and (15) make different predictions for the detected frequency of photons leaving the surface of a massive compact object when seen by an observer in free space. It is apparent that these differences are only significant for distances of r close to α, a situation that can only apply for very massive objects, such as neutron stars. A neutron star of solar mass $M_\oplus$ will have α = 2.95 km and a physical radius r_0 = 13.4 km based upon an average neutron star density of $2\times10^{17}\,\text{kg/m}^3$ [44], giving it $\alpha/r_0 = 0.220$. The redshift of spectral lines emitted from the surface of non rotating neutron stars gives an opportunity to determine the accuracy of Equation (11), provided the density of the neutron star, and hence r_0, is accurately known. The Schwarzschild radius α increases linearly with mass while the physical radius of the neutron star, r_N, increases with the cube root of mass.

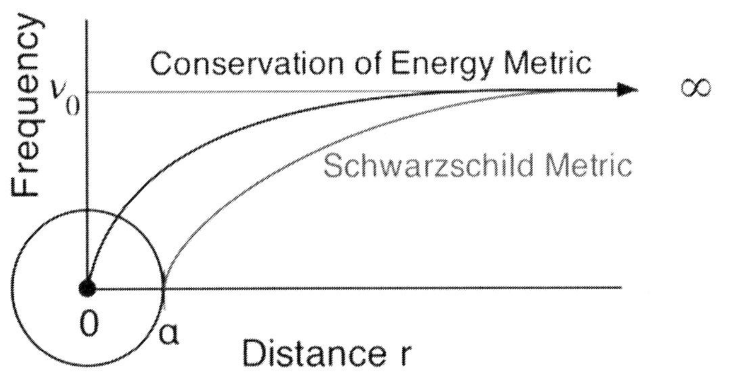

Figure 6. Illustration of the differences predicted by the conservation of energy and Schwarzschild metrics for photons of frequency ν_0 emitted at distances r from an object, as detected by an observer at infinity.

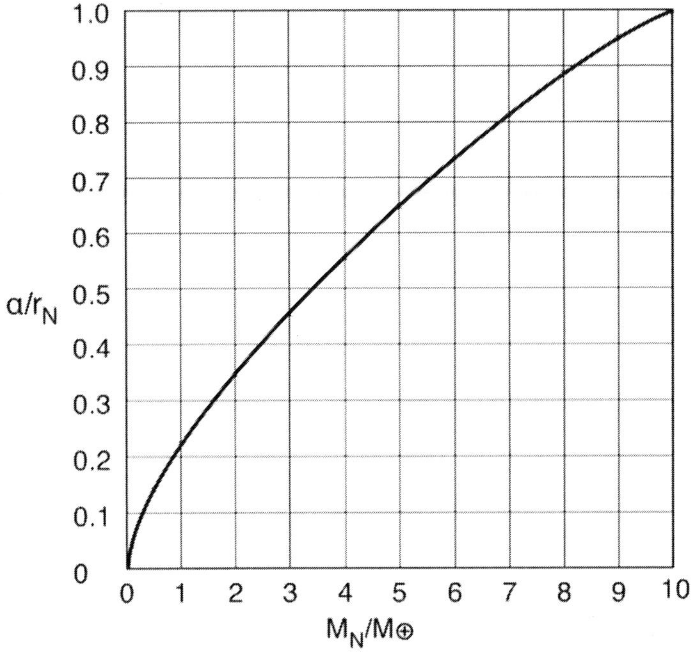

Figure 7. Plot of Schwarzschild radius α divided by the neutron star radius r_N, versus the mass of the neutron star expressed as a proportion of the solar mass $M_\oplus$.

Figure 7 shows a plot of α/r_N for neutron star masses less than 10 $M_\oplus$. Using Equations (14) and (15) for the relationship of frequency versus distance for the conservation of energy and Schwarzschild metrics, **Figure 8** shows a plot of the variation of frequency of photons emitted from the surface of a non rotating neutron star when plotted under these metrics. As can be seen from **Figure 8**, there is a region above ≈ $M_N \approx 0.5$ $M_\oplus$ where there are significant differences between the frequencies of photons emitted from the surface of the neutron star as predicted by the different metrics. This still requires measurements to be made of the M_N, which determines α, and either the density or radius r_N of the neutron star. If this can be done, it provides one means of discriminating between the two metrics. It is postulated that when observations are made with the required accuracy, the predictions from the conservation of energy metric will match all gravitational measurements made in space outside matter.

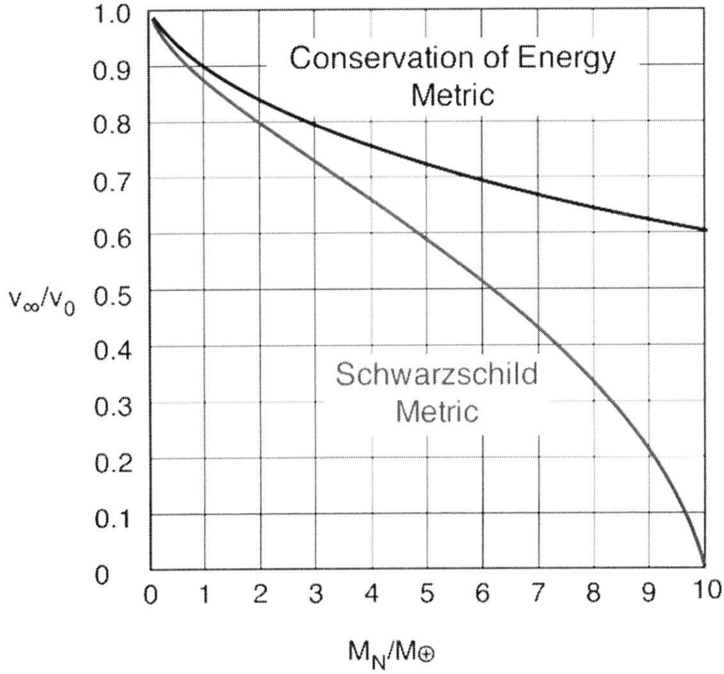

Figure 8. Plot of the detected frequency variation expected from photons emitted from the surface of a neutron star as calculated from the Schwarzschild metric and the conservation of energy metric, as indicated, for neutron star masses up to 10 $M_{\oplus}$.

6. CONCLUSION

This paper has shown that gravitational effects can be calculated from the physical principal of conservation of energy. Potential energy is a measurable properties associated with mass, which, in the case of photons, is given by $m = \frac{h\nu}{c^2}$. It can be calculated using Newtonian mechanics and Newton's universal gravitational constant G. The resultant space-time metric matches all experimental V. N. E. Robinson observations that currently support the predictions of the general relativity theory. It predicts that there is no singularity at the Schwarzschild radius and hence no black holes. It also overcomes the problem of gravitational action at a distance associated with the Schwarzschild radius being outside the mass of a collapsed object.

REFERENCES

1. Einstein, "Zur Allgemeinen Relativätstheorie," Sitzungsberichte der Königlich Preussischen Academie der Wissenschaften, Part 2, 1915, pp. 778-786.
2. Einstein, "Erklärung der Perihelbewegung des Merkur aus der Allgemeinen Relativätstheorie," Sitzungsberichte der Königlich Preussischen Academie der Wissenschaften, Part 2, 1915, pp. 831-839.
3. Einstein, "The Foundation of the General Theory of Relativity," In: H. A. Lorentz, A. Einstein, H. Weyland and H. Minkowski, Eds., The Principle of Relativity, Dover Publications, New York, 1952, pp. 109-164.
4. Einstein, Annalen der Physik, Vol. 42, 1916, pp. 769- 822.
5. K. Schwarzschild, "Auf das Gravitationsfeld Eines Massenpunktes Entsprechend Einsteinschen Theorie," Sitzungsberichte der Königlich Preussischen Academie der Wissenschaften, PK1, 1916, pp. 189-196.
6. J. Droste, General Relativity and Gravitation, Vol. 34, 2002, pp. 1545-1563 (Reprinted from Köninklije Nederland Acadamie Wetenschapen, Series A, Vol. 19, 1917, pp. 197-221.).
7. D. Hilbert, "Die Grundlagen der Physik (Zweite Mitteilung)," Königliche Gesellschaft der Wissenschaftenzu Gottingen, K1, 1917, pp. 53-76.
8. H. Weyl, Annalen der Physik, Vol. 54, 1917, pp. 117-123.
9. M. Brillouin, Journal Physic et Radium, Vol. 4, 1923, pp. 43-48.
10. M. D. Kruskal, Physical Review, Vol. 119, 1960, pp. 1743-1745.
11. G. Szekeres, Mathematicae Debrecen, Vol. 7, 1960, pp. 285-301.
12. W. Misner, K. S. Thorne and J. A. Wheeler, "Gravitation," W. H. Freeman and Co., New York, 1973, Chapter 31.
13. R. P. Kerr, Physical Review Letters, Vol. 11, 1963, pp. 237-238.
14. Newton, "Philosophiae Naturalis Principia Mathematica," 1687.
15. Einstein, "On the Electrodynamics of Moving Bodies," In: H. A. Lorentz, A. Einstein, H. Weyl and H. Minkowski, Eds., The Principle of Relativity, Dover Publications, New York, 1952, pp. 37-65.
16. Einstein, Annalen der Physik, Vol. 17, 1905, pp. 891- 921.
17. Einstein, "Does the Inertia of a Body Depend upon its Energy Content?" In: H. A. Lorentz, A. Einstein, H. Weyl and H. Minkowski, The Principle of Relativity, Dover Publications, New York, 1952, pp. 69-71.
18. Einstein, Annalen der Physsik, Vol. 18, 1905, pp. 639- 641.
19. M. Evans and J. P. Vigier, "The Enigmatic Photon," Kluwer Academic Publishers, Dordrecht, 1994, Chapter 8.
20. M. B. Van der Mark and G. W. t'Hooft, "Light Is Heavy," Van A tot Q, NNV, 2000, pp. 1-6. http://nnvaq.tue.nl/atoq/2000/BIJDRAGE-vdMark-tHooft.pdf
21. R. V. Pound, Classical and Quantum Gravity, Vol. 17, 2000, pp. 2303-2311.
22. Einstein, "On the Influence of Gravitation on the Propagation of Light," In: H. A. Lorentz, A. Einstein, H. Weyl and H. Minkowski, The Principle of Relativity, Dover Publications, New York, 1952, pp. 99-108.
23. Einstein, Annalen der Physik, Vol. 35, 1911, pp. 898- 908.
24. R. V. Pound and G. A. Rebka, Physical Review Letters, Vol. 3, 1959, pp. 439-441.
25. R. V. Pound and G. A. Rebka, Physical Review Letters, Vol. 4, 1960, pp. 337-341.
26. R. V. Pound and J. L. Snider, Physical Review Letters, Vol. 13, 1961, pp. 539-540.
27. W. S. Adams, Astrophysical Journal, Vol. 31, 1910, pp. 30-61.
28. M. G. Adam, Monthly Notices of the Royal Astronomical Society, Vol. 108, 1948, pp. 446-464.
29. J. O'Leary, American Journal of Physics, Vol. 32, 1964, pp. 52-55.
30. S. Eddington, "Space Time and Gravitation," Cambridge University Press, Cambridge, 1920, Chapter 7.

31. H. Minkowski, "Space and Time," In: H. A. Lorentz, A. Einstein, H. Weyl and H. Minkowski, Eds., The Principle of Relativity, Dover Publications, New York, 1952, pp. 75-91.
32. H. Minkowski, Jahrescherícht der Deutschen Mathematiker-Vereinigung, Vol. 18, 1909, pp. 75-88.
33. P. Lasky, Australian Physics, Vol. 48, pp. 84-89
34. S. Eddington, "Space Time and Gravitation," Cambridge University Press, Cambridge, 1920, Chapter 8.
35. F. W. Dyson, A. S. Eddington and C. Davidson, Philosophical Transactions of the Royal Society of London, Vol. 220A, 1920, pp. 291-333.
36. T. L. Chow, "Gravity, Black Holes, and the Very Early Universe: An Introduction to General Relativity and Cosmology," Springer, New York, 2008, Chapter 5.
37. R. A. Matzner, "Dictionary of Geophysics, Astrophysics and Astronomy," CRC Press, London, 2001, p. 360.
38. K. Trencèvski, E. G. Celakoska and V. Balan, International Journal of Theoretical Physics, Vol. 50, 2011, pp. 1-26.
39. Einstein, "Über Gravitationswellen," Sitzungsberichte der Königlich Preussischen Academie der Wissenschaften, 1918, pp. 154-167.
40. R. A. Hulse and J. H. Taylor, Astrophysical Journal, Vol. 195, 1975, pp. 51-53
41. J. M. Weisberg and J. H. Taylor, General Relativity and Gravitation, Vol. 13, 1980, pp. 1-6.
42. J. M. Weisberg and J. H. Taylor, Astrophysical Journal, Vol. 576, 2002, pp. 942-949.
43. J. M. Weisberg and J. H. Taylor, "The Relativistic Binary Pulsar B1913+16: Thirty Years of Observation and Analysis," In: F. A. Rasio and I. H. Stairs, Eds., Binary Radio Pulsars, Aspen Conference, ASP Conference Series, Vol. 328, 2005, pp. 25-31.
44. P. Newman, "Tour the ASM Sky," 2011. http://heasarc.gsfc.nasa.gov/ docs/xte/ learning_center/ASM/ns.html

CHAPTER 14

Accelerated Expansion of Space, Dark Matter, Dark Energy and Big Bang Processes

Auguste Meessen

UCL, Louvain-la-Neuve, Belgium

ABSTRACT

The accelerated expansion of our universe results from properties of dark matter particles deduced from Space-Time Quantization. This theory accounts for all possible elementary particles by considering a quantum of length a in addition to c and h. It appears that dark matter particles allow for fusion and fission processes. The resulting equilibrium enables the cosmic dark matter gas to produce dark energy in an adaptive way. It keeps the combined matter-energy density at a constant level, even when space is expanding. This accounts for the cosmological constant Λ and the accelerated expansion of space without requiring any negative pressure. The Big Bang is related to G, c, h and a. It started with a "primeval photon" and led to the cosmic matter-antimatter asymmetry as well as inflation.

Keywords: Accelerated Expansion, Dark Matter, Dark Energy, Space-Time Quantization, Big Bang, Inflation, Matter-Antimatter Asymmetry

1. INTRODUCTION

1998 was said to be "a very good year for cosmology" [1] , because of the unexpected discovery that the expansion of our universe is accelerating [2] [3] [4] . It became also clear that our universe contains about 4% of ordinary matter, 23% of Dark Matter (DM) and 73% of Dark Energy (DE). However, the nature of DM and DE, as well as the cause of the accelerated expansion of our universe are still unknown. We show here that these problems are related to one another and to elementary particle physics. The semi-empirical Standard Model calls itself for explanations and has to be completed to account at least for DM particles.

This is possible, by generalizing Relativistic Quantum Mechanics. The value a of the smallest measurable length is unknown, but the resulting theory of Space-Time Quantization (STQ) is logically consistent. Moreover, it is sufficient that $a \neq 0$, to justify the Standard Model and to complete it in regard to DM particles [5] [6] . Predicted properties of the cosmic DM gas are confirmed by astrophysical observations [7] . Here, we examine other consequences, especially for the accelerated expansion of space and Big Bang processes.

The idea of an expanding universe arose from realizing that our universe should fatally collapse, since all masses attract one another. Einstein's theory of General Relativity (GR) revealed that space itself would contract because of gravity. This resulted from the fact that a homogeneous and isotropic universe allows that all measured distances are proportional to the same scale factor. It could be a function $R(t)$ of universal cosmic time. Since our universe seems to be stable, Einstein assumed that gravitational collapse is prevented by another force. He characterized its strength by the cosmological constant Λ. Lemaître realized that Einstein's equation for $R(t)$ allowed for an expansion of our universe that starts at $R(0)=0$. Although it was then possible to set $\Lambda = 0$, Lemaître considered that this would be arbitrary. He treated Λ as a parameter of unknown value and showed that the expansion of space would eventually become accelerated when $\Lambda > 0$. This has now been established and suggests that the cosmological constant should be related to DE [8] .

Since the underlying physics is not yet known, the discovery of the accelerated expansion of our universe was very important. It led in 2011 to the Nobel Prize in physics [9] , but also to great perplexity. Where could the required energy come from? Does space contain energy [10] or is it provided by some yet unknown substance? It would then have to be present everywhere in the whole universe and has been called quintessence [11] . This refers to the ancient concept of a fifth element, but requires the existence of some hypothetical field and corresponding particles [12] . Actually, there are various models and scenarios, providing equivalent descriptions [13] , since any fluid that is evenly spread out in the whole universe leads to a cosmological constant $\Lambda = -4\pi G(\rho + 3p)$. The pressure p of this fluid depends on its mass-energy density ρ, according to the equation of state $p = w\rho$. Since $\Lambda > 0$, it would be necessary that $w < -1/3$. The "cosmological constant problem" seems then to be reduced to imagining a substance that has great negative pressure. Actually, it is necessary to determine the real value of Λ by means of measurements and to explain it in a physically coherent way.

The cosmologist Michael Turner, who coined the term "dark energy" in 1998, stated in 2002 that it is "the causative agent of the current epoch of accelerating expansion". Since $w \approx 0$ for matter and $w = -1/3$ for photons, DE is "more energy-like than matter-like... The challenge is to understand it" [14] . He considered that this problem is essential for future developments [15] : "Dark energy is just possibly the most important problem in all physics... As a New Standard Cosmology emerges, a new set of questions arises: What is physics underlying inflation? What is the dark-matter particle? How was the baryon asymmetry produced?... What is the nature of the Dark Energy?... The big challenge for the New Cosmology is making sense of dark energy."

The standard Lambda-Cold Dark Matter (ΛCDM) model assumes non-relativistic collision-less DM particles. This yields $w = -1$. However, DM particles do interact with one another [6] [7] [16] . To determine the nature and properties of DM particles and to explain how it is related to DE seems to be of fundamental importance. Actually [17] , "both dark matter and dark energy require extensions of our current understanding of particle physics... The existence of nonbaryonic dark matter implies that there must be new physics beyond the standard model of particle physics." It has even been stated that the scene may be set for a Kuhnian paradigm shift [18] . The historical context suggests, indeed, that "a new theory will emerge, sooner or later" and that it "will radically change our vision of the world" [10] . Since STQ generalizes present-day theories and determines essential properties of DM particles, it is necessary to explore its possible consequences in regard to DE and the accelerated expansion of space. We try to do that by expressing relevant ideas as simply as possible.

In Section 2, we begin with a review of the evolution of ideas concerning the expansion of space, to locate the basic problem before proposing a new solution. It replaces the conventional model of DM and DE by another one, based on the theory of STQ and the resulting properties of DM particles. They allow for fusion and fission processes. This accounts for the production of DE and the accelerated expansion of space. Section 3 recalls some basic results of STQ, to prepare Section 4. It relates the Big Bang to particle physics. Section 5 summarizes results and raises some new questions.

2. THE ACCELERATED EXPANSION OF SPACE

2.1. The Cosmological Constant

The basic problem results from the fact that masses can only attract one another. Newton postulated that these forces are everywhere identical in the whole universe. Their strength is determined by the constant G and they vanish for great separations of the interacting masses, but they are additive. The astronomer Hugo von Seeliger noted in 1895 that this leads to an inconsistency [19] . It is reasonable, indeed, to assume that the average mass density ρ is everywhere identical in the whole universe. A very large sphere of radius r would thus contain a mass $M = \rho V$, where $V = 4\pi r^3/3$. Any object of mass m that is situated on the surface of this sphere is then attracted towards its center by the force

$$F = -\frac{GMm}{r^2} = -\frac{4\pi G\rho m}{3} r$$

Because of the large-scale homogeneity of our universe, the center of this sphere can be arbitrarily chosen. The whole universe should thus collapse. According to classical mechanics, this would even happen with respect to absolute space. However, Einstein realized in 1905 that space and time can only be defined in terms of possible results of measurement, since the velocity c is a universal constant. His theory of Special Relativity (SR) disclosed that mass and energy are equivalent. From now on, we consider thus that ρ defines a mass-energy density (c = 1). Einstein became also aware of the fact that gravitational forces can be replaced by accelerations of the chosen reference frame. His theory of General Relativity (GR) related thus gravitational forces to the local metric of space and time.

This theory was published in 1915 and in 1917, Einstein applied the new concept of gravity to the whole universe [20] . Because of the cosmological principle, stating that our universe is everywhere identical at sufficiently large scales, the coupled equations of GR were reduced to a single one. It can be established in a more direct and conceptually simpler way, by considering again a very large sphere of volume $V = 4\pi r^3/3$. Its mass-energy content $M = \rho V$. According to SR, the test-mass, situated on the surface of this sphere has some rest mass m_o and because of Newtonian gravity, its energy is

$$E = m_o c^2 + \frac{m_o}{2}\dot{r}^2 - \frac{GMm_o}{r} = 0 \tag{1}$$

We consider the case where the kinetic energy is small compared to the rest energy, since this is possible and will be sufficient. The negative potential energy accounts for gravitational attraction towards the center of the sphere. We set $E = 0$, to avoid a blow-up for the total energy of many test-masses. Although (1) implies Euclidean geometry, all distances r could be proportional to the same scale factor $R(t)$. This can be justified by assuming that our 3-D space is the surface of an immense hypersphere in 4-D space. Any value of r is then proportional to the radius R of this sphere. It could even be a function of cosmic time t, which is everywhere identical in the whole universe. Since $\dot{r} = dr/dt$, we get then a differential equation for $R(t)$. The value of m_o is irrelevant, but $M = \rho V$. We get thus two relations:

$$\frac{1}{2}\dot{R}^2 = \frac{GM}{R} + K \qquad \text{and} \qquad V = \frac{4\pi}{3}R^3 \tag{2}$$

K is a constant. The theory of GR yields the same result, but K depends then on the curvature of space. Einstein considered a closed space of constant curvature, as for the surface of a hypersphere. K is then positive, but would be negative for an open, hyperbolic space. The intermediate "flat" space yields Euclidean geometry and $K = 0$. Anyway, derivation of Equation (2) with respect to t will eliminate the constant K, but the resulting equation is then

$$\dot{R}\ddot{R} = -\frac{GM}{R^2}\dot{R} + \frac{G}{R}\dot{M} \tag{3}$$

Considering only ordinary matter, distributed with the same mass-energy density ρ_o in the whole universe, we get $M = M_o = \rho_o V$. Since this mass remains constant when $R = R(t)$, Equation (3) is reduced to a simpler one, which resulted also from Einstein's theory of GR:

$$\ddot{R} = -\frac{GM_o}{R^2} \tag{4}$$

This means that even the new theory of gravity cannot prevent gravitational collapse. The acceleration is greater for small values of R, because of stronger forces. However, it is now attributed to variations of the scale factor $R(t)$ instead of motions in absolute space. Since our universe seems to be stable, Einstein thought that something is missing. His conjecture was that

$$\ddot{R} = -\frac{GM_o}{R^2} + \frac{\Lambda}{3}R \tag{5}$$

The so-called "cosmological constant" Λ can be viewed as defining the strength of a repulsive force that would be opposed to gravity at cosmological scales. It should be noted that (5) results from (3), when we assume that $M = (\rho_o + \rho)V$ and that $\dot{M} = \dot{\rho}V + \rho\dot{V}$, but $\dot{\rho} = 0$. That would be very strange, since vacuum would not only correspond to some ether-like substance. Its mass-energy density ρ would even remain constant when $R = R(t)$. Nevertheless, this assumption was implicit in Equation (5). Einstein circumvented the resulting physical problems, by assuming that $\ddot{R} = 0$ and $\Lambda = 3GM_o/R^3$, but this does not necessarily imply that our universe is stable ($\dot{R} = 0$).

Willem de Sitter noted already in 1917 that variations of $R(t)$ are not excluded, since Equation (5) would even allow for an exponential increase of $R(t)$, when we assume that our universe is empty ($M_o = 0$) and that $R(0)$ is finite. Georges Lemaître recognized that the real universe allows also for an increasing function $R = R(t)$ when we assume that its expansion did start at $R(0) = 0$. It is then sufficient to assume some initial speed $\dot{R}(0) > 0$, to account for the present finite value of $R(t)$. Since the first term on the right side of Equation (5) is predominant for small values of R, the initial expansion would be decelerated. However, the last term becomes predominant for large values of R. While Einstein assumed that $\Lambda > 0$, to get a static universe, it would now mean that the expansion of space will eventually get accelerated.

Lemaître published this theory in 1927 [21] and republished it in 1931 [22] . He translated himself the text from French to English, but dropped some minor parts and added another text. Lemaître insisted always on the assumption that our universe is homogeneous, but his most brilliant idea was that $R(0) = 0$, because of the "arrow of time". Entropy is increasing, since disorder is more probable than order, and structuring yields even more degrees of freedom. Extrapolating backwards, there should have been "fewer and fewer quanta, until we find all the energy of the universe packed in a few or even a unique quantum" [23] . This was a logical, but revolutionary deduction from observable facts.

In 1927, Lemaître had transmitted the initial paper to Einstein, before they met at the fifth Solvay Conference in Brussels. Einstein accepted the mathematical treatment of Equation (5), but rejected the idea of an expanding universe. He thought that such a bold interpretation of (5) is not plausible. He told Lemaître about similar work of Alexander Friedmann, who had studied mathematical physics and published in German. Friedmann considered also a function $R(t)$, but was mainly interested in the mathematical consequences of GR for different curvatures of space. He solved Equation (2) in 1922 for a closed universe [24] and in 1924, he considered the evolution of an open one.

Lemaître ignored this work in 1927. Since he studied engineering, he was concerned with the real world and considered the problem of a possible variation of $R(t)$ also in the context of thermodynamics. The initial state of our universe had then to be the simplest possible one. Lemaître could thus justify his assumption that $R(0)=0$, but he wanted also to know if our universe is really expanding. He knew that the speed of recession of stellar objects can be determined by measuring the red-shift of spectral lines. Vesto Slipher did this already in 1913 for the Andromeda nebula [25] . Edwin Hubble evaluated the distances of neighboring nebulas by means of Cepheid variable stars and established in 1929 that the recession speed of 22 nebulas is proportional to their distance [26] . Lemaître had predicted this relation two years earlier, but the observational confirmation was also essential. The discovery of the expansion of our universe resulted thus from independent, but complementary scientific research.

The term "Big Bang" was introduced by Hoyle in 1949, to ridicule this idea. He preferred a "steady state" cosmology to the concept of a universe that emerged from a single point and could even blow-up when the cosmological constant $\Lambda > 0$. Those who accepted the idea of an expansion of our universe considered that it was sufficient to set $\Lambda = 0$. Equation (4) and the same initial conditions would then yield a function $R(t)$ that passes through a maximum and decrease until $R = 0$. Einstein thought also that the cosmological constant Λ is not needed anymore, while Lemaître considered that instead of postulating that $\Lambda = 0$, we should consider the value of Λ as being unknown [27] .

Today, we know that the Big Bang occurred about 17.8 billion years ago and that the accelerated expansion of space began about 5 billion years ago. Thus $\Lambda > 0$, but an accelerated expansion of space is baffling. It requires a yet unknown energy source. It may be related to the existence of DM and DE, but it is necessary to clarify what these terms do really mean and how this might be possible. In such a situation, the first rational step is to describe what is known in terms of usual concepts.

2.2. The Conventional DM and DE Model

We can assume that DM and DE are substances that have together a mass-energy density ρ. It has thus to be added to the mass-energy density ρ_o of ordinary, baryonic matter. This implies that the mass M in Equation (3) is composed of two parts:

$$M = M_o + U \text{ where } U = \rho V \tag{6}$$

We have thus to determine the value of $\dot{M} = \dot{U}$. This seems to be simple, since energy conservation requires that $\Delta Q = \Delta U + p\Delta V$. Added thermal energy ΔQ increases the internal energy U of any substance, which is enclosed in the volume V, but it could also do work by means of its pressure p. This relation applies to usual gases. The total number of molecules remains then constant, but their average kinetic energy ($3kT/2$) could be modified. Cosmic DM and DE, contained in a huge volume V, are not thermally isolated, but the inflow and the outflow of heat are balanced. Since $\Delta Q = 0$, we expect that $\Delta U = -p\Delta V$. It follows that $\dot{U} = -p\dot{V}$, where $\dot{V} = -p4\pi R^2 \dot{R}.$ Equation (3) leads then to

$$\ddot{R} = -\frac{GM_o}{R^2} - \frac{4\pi G}{3}(\rho + 3p)R \text{ where } p = w\rho \tag{7}$$

This is equivalent to (5) when the cosmological constant

$$\Lambda = -4\pi G(\rho + 3p) = -4\pi G(1 + 3w)\rho \tag{8}$$

Since $\Lambda > 0$, it would be necessary that cosmic DM and DE have everywhere a sufficiently great negative pressure, so that $w < -1/3$. This leaves room for many speculative propositions, but we want to find out if the concept of DM particles, which results from STQ, could be helpful to understand the enigmatic accelerated expansion of space.

We showed that DM particles interact with one another by exchanging N_2 bosons and that this does usually lead to elastic scattering [6] . The cosmic DM gas behaves then like a usual molecular gas. The actual nature and mass of DM particles are irrelevant here. Their average kinetic energy would be $3kT/2$ (since classical Maxwell-Boltzmann statistics apply also to fermions and bosons at low densities). The cosmic DM gas has thus a pressure $p = nkT$, where n is the average density of DM particles. The mass-energy density $\rho = n(mc^2)$, where m is the average mass of DM particles. It follows that $p = w\rho$, where $w = kT/mc^2$. These concepts are confirmed by astrophysical observations [7] , which proved that the cosmic DM gas is cold. We mentioned in the introduction that the ΛCDM implies that $w = -1$, but DM particles interact with one another. Recent cosmological measurements [28] implied that $w = -1.16$ for (8). The local value of the Hubble constant, measured with improved precision, revealed also that the universe is expanding 5% to 9% faster than expected [29] .

2.3. The Model of Adaptive DM and DE

Since $U = \rho V$, the previous relation $\dot{U} = -p\dot{V}$ could be replaced by

$$\dot{U} = \dot{\rho}V + \rho\dot{V} - p\dot{V} \quad \text{where} \quad \dot{\rho} = 0 \tag{9}$$

This may seem to be unbelievable, since it requires that DM and DE have the capacity to keep their mass-energy density ρ at a constant level, even when space is expanding. Nevertheless, this possibility has to be considered. It would imply that $\dot{U} = (\rho - p)\dot{V}$. Replacing p by $p - \rho$ in (8), we get

$$\Lambda = 4\pi G(2\rho - 3p) = 4\pi G(2 - 3w)\rho \tag{10}$$

This yields $\Lambda > 0$ even when $w \approx 0$, but is it possible to prove that $\dot{\rho} = 0$, which seems to be an extraordinary claim? It follows from STQ that DM particles allow also for fusion and fission processes [6] . Fusion liberates energy, while fission requires energy. We have thus to examine the properties of the common mass-energy density ρ. To focus our attention on the essential mechanism, we consider the particular case where the cosmic DM gas contains only two types of particles. Those of mass m are present with a density n, but they can be fused together to constitute compound DM particles of mass $m' = 2m - \Delta m$. The mass defect results from the binding energy $E = \Delta mc^2$, but we set $c = 1$. It is convenient to define a dimensionless binding parameter $b = \Delta m/m$. Thus $m' = (2-b)m$. The density of these compound particles is n', but they can be split, to yield again particles of mass m. Fusion requires the encounter of two particles. The probability is then proportional to n^2, while fission of the particles of mass m' can be spontaneous. This yields the rate equation

$$\dot{n} = -\alpha n^2 + \beta n' \tag{11}$$

The density n would remain constant when $n' = (\alpha/\beta)n^2$. To determine the actual values of n and n', we need a second relation. It results from the fact that the total number of DM particles in the volume V is $(n+n')V$ and their total mass is $(nm + n'm')V$. The ratio defines the average mass $m'' = (nm + n'm')/(n+n')$, which is independent of V. Constancy of m'' implies that $n' = qn$. It follows that $n = q\beta/\alpha = n_1$. Equilibrium is thus possible for this system. Figure 1represents it by means of transitions between energy states, defined by the masses m and m' of two types of particles.

To verify if this equilibrium is stable, we consider a local perturbation for constant values of the parameters α, β as well as q. Thus,

$$n'(t) = qn(t) \quad \text{and} \quad m'' = \frac{1+q(2-b)}{1+q}m \tag{12}$$

The rate Equation (11) is then reduced to

$$\dot{n} = -\alpha n^2 + \gamma n \quad \text{where} \quad \gamma = q\beta \tag{13}$$

Setting $n = u/v$, we get $\dot{u} = \gamma u$ and $\dot{v} = \alpha u$. Thus,

$$n(t) = \frac{\exp(\gamma t)}{A + (\alpha/\gamma)\exp(\gamma t)} \rightarrow n_1 = \frac{\gamma}{\alpha} \text{ when } t \gg 1/\gamma \tag{14}$$

The constant A is determined by $n(0)$, which could be greater or smaller than n_1, but equilibrium would always be restored. It will be reached more rapidly when $\gamma = q\beta$ is great. It could never be reached, of course, if fusion were irreversible ($\beta = 0$). The cosmic DM gas is in a state of homeostasis.

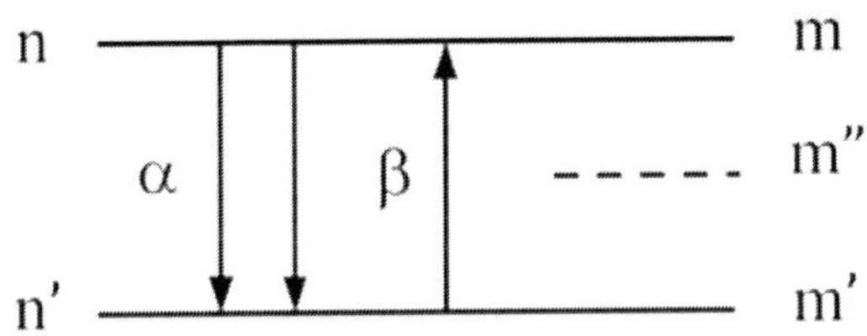

Figure 1. Fusion and fission processes of DM particles yield an equilibrium that accounts for the accelerated expansion of space.

It constitutes an adaptive system, where DM particles produce DE in such a way that the density n of unfused particles and the density n' of fused particles remain constant. The common mass-energy density ρ is invariant when space is expanding. Even the mass-energy density $\rho_{DM} = nm + n'm'$ of DM alone and the mass- energy density $\rho_{DE} = n'\Delta m$ of the total liberated DE remain constant. The ratio χ and the total mass-energy density ρ are also constants:

$$\chi = \frac{\rho_{DE}}{\rho_{DM}} = \frac{qb}{1+q(2-b)} \tag{15}$$

$$\rho = \rho_{DM} + \rho_{DE} = (1+2q)n_1 m \tag{16}$$

According to reported values of ρ_{DE} and ρ_{DM}, the remarkable ratio $\chi = 70/26,\ 71/24,\ 73/23$ or $74/22$. This would yield values between 2.7 and 3.4. Setting $\chi = 3$, it follows from (16) that $q = 3/(4b-6)$, while $m' = (2-b)m$. Since q and m' have to be positive, $3/2 < b < 2$. When $b = 7/4$, for instance, we get $q = n'/n = 3$. It follows that $m' = m/4$ and $m''/m = 7/16 \approx 1/2$. The 2015 results of the Planck measurements of the cosmic background [30] would imply that $\chi = 69/31 = 2.2 \approx 2$. This allows for $b = 5/3$ and $q = n'/n = 2$, which implies that $m' = m/3$ and $m''/m = 5/9$. According to the proposed theory and cosmological observations, there are thus more fused than unfused DM particles. The quotient q is of the order of 2 or 3.

It also appears that fusion of DM particles would be characterized by a high binding energy, so that $m'/2m \approx 1/8$ or $1/6$. This is unaccustomed, but not impossible. Great values of b facilitate the liberation of the enormous amount of energy, which is required to allow for the accelerated expansion of space.

Even equilibrium would be reestablished more rapidly if it were disturbed somewhere. Nevertheless, it is necessary that fused DM particles can be broken up, spontaneously or by collisions. This could result from the excitation of collective oscillations of neutralons inside the compound DM particles.

The accelerated expansion of space is due to the adaptability of the cosmic DM gas. When the volume V increases, fusion and fission processes continue to equilibrate one another by producing more DM and more DE. Since DM particles are electrically neutral, they cannot produce photons and they are not heated or cooled by contact with ordinary matter [6] . Invisible cosmic DM gas is thus isothermal in the whole universe. Even when space is expanding, its density and temperature is regulated everywhere by mutually controlled fusion and fission processes.

Thermal agitation leads to pressure effects. They are important for the constitutions of DM atmospheres [7] , but nearly negligible for the accelerated expansion of space. By measuring Λ, we could determine the average mass-energy density ρ.

Einstein's conjecture (5) was very remarkable, since the existence of DM and DE was totally unknown. The conjecture that the scale factor $R(t)$ had to be finite for the present universe was correct, but it is not constant.

It has been stated and often repeated that Einstein told Gamow that the introduction of the cosmological constant was his "biggest blunder". This is not sure anymore [31] and may result from a misunderstanding. Indeed, Einstein could only regret that he did not realize himself that our universe might be expanding, even when $\Lambda = 0$.

However, this required the additional idea that cosmic expansion could start with $R(0) = 0$. Maybe, Einstein did not consider this possibility because of (5). Anyway, it is interesting to know that Lemaître was fully aware of the physical meaning of the cosmological constant.

He wrote [32] : "Everything happens as though the energy in vacuo would be different from zero". He could not explain how apparently empty space might produce energy, but he did not simply believe that this is impossible.

3. SPACE-TIME QUANTIZATION

3.1. The Positive Energy Content of Our Universe

To prepare the following chapter, we present some essential consequences of the theory of STQ in a short and different way. The basic idea was that Nature could impose a third restriction in addition to those which led to the development of relativity and quantum mechanics. In a nutshell, they are summarized by Einstein's energy momentum relation and de Broglie's redefinition of E and p:

$$\left(E/c\right)^2 = p^2 + \left(m_o c\right)^2 \quad \text{where} \quad E = h\nu \quad \text{and} \quad p = h/\lambda \tag{17}$$

The function $E = E(p)$ applies to free particles in any inertial reference frame. It depends on the rest-mass m_o of these particles. Louis de Broglie discovered that every particle has an "associated wave". It is its wave function, which allows us to express knowledge. It defines the probability distribution for possible positions in space, but provides also information about motions in terms of possible values of p and E. Relativistic Quantum Mechanics (RQM) combines the relations (17) and accounts thus for c and h, but it is assumed that the wavelength λ could be infinitely small. This is equivalent to believing that the energy E and momentum p could have arbitrarily high values. If there did exist a finite limit a for the smallest measurable distance, we would have to accept that $\lambda/2 \geq a$.

The value of a is thus determined by the highest possible momentum $p = h/2a$. Because of (17), it would be obtained when $m_o = 0$ and when the energy E has the highest possible value. This requires a photon and that its energy $E = cp$ cannot be increased. It would thus have to be equal to the total (positive) energy content of the whole universe. Its value would be $E_u = hc/2a$. Although it is gigantic, it could be finite. Einstein's energy momentum relation (17) has to be generalized when $a \neq 0$, but it yields also $E = cp$ when $m_o = 0$. STQ confirms that energies $E \leq hc/2a$, even for material particles [6].

3.2. All Possible Elementary Particles

They can be distinguished from one another by means of their wave functions when the quantum of length $a \neq 0$. Indeed, there are two sets of possible results when the coordinate x is precisely measured along a given reference axis: $x = 0, \pm a, \pm 2a, \cdots$ and $x = \pm a/2, \pm 3a/2$,... The "normal lattice" contains the origin $x = 0$, but a symmetrically intercalated lattice is also possible, since the orientation of the x-axis is arbitrary. The wave function $\psi(x)$ has to be defined for all values of x, but it can have the same or opposite signs on the intercalated lattice with respect to the normal lattice. Although the ψ functions are different, the probability distribution will be unaffected. There are even more degrees of freedom, since $\psi(x)$ can be multiplied everywhere on the intercalated lattice by a "sign-function". It corresponds to a vector of magnitude 1 that can be rotated by an integer number u_x of half turns towards the left or the right, to yield the values ±1 for the sign-function.

Such a modulation of wave functions is possible for any reference frame in our four-dimensional space-time. This yields four quantum numbers (u_x, u_y, u_z, u_{ct}). Each one of them can be a positive or negative integer number, but they are everywhere identical for particles of given type. Every pattern of possible modulations at the smallest possible scale in space and time defines a "particle state", while large-scale variations define possible "states of motion". This is compatible, by multiplexing [6] . Small and large-scale variations can be combined.

The electric charge is always determined (in units e) by $Q = (u_x + u_y + u_z)/3$. The Standard Model of elementary particles accounts for three generations of quarks and leptons. They correspond to $u_{ct} = 0, \pm 1$ and display the same family structure, in terms of triplets (u_x, u_y, u_z). When $u_{ct} = 0$, states of type $(0,1,1)$ correspond to up-quarks $(Q = 2/3)$, while states of type $(0,0,-1)$ correspond to down-quarks $(Q = -1/3)$. In both cases, there are 3 possible permutations, defining different color states (R, G or B). The electron is an elementary particle in the $(-1,-1,-1)$ state, where $Q = -1$. Antiparticles are characterized by opposite signs for all (u_x, u_y, u_z, u_{ct}) quantum numbers.

The Standard Model is not complete, since the u-quantum numbers are not only equal 0 or ±1, but also to ±2, for instance. Moreover, there are states of type $(0,1,-1)$ with 6 possible permutations and two $(0,0,0)$ states, when $u_{ct} = 0$. This octet defines particles and antiparticles of charge $Q = 0$. They are elementary DM particles. Since they behave like neutral quarks, we called them "narks". They are the supersymmetric partners of gluons. Supersymmetry results from the fact that the z-component of the spin vector along a given z-axis is defined by large-scale angular variations of ψ -

functions around this axis. These variations are independent of small-scale variations, defined by u-quantum numbers. Every (u_x, u_y, u_z, u_{ct}) particle state for fermions corresponds thus to a $[u_x, u_y, u_z, u_{ct}]$ state for bosons and vice-versa.

3.3. Conservation Laws for Possible Transformations

Elementary particles can be transformed into one another by means of annihilation and creation processes. However, the sum of u-quantum numbers has to be conserved for every one of the four space-time axes, as well for bosons as for fermions. This accounts for the fact that a quark can change its color by creating or annihilating a gluon. For instance, $(0,1,1) \rightarrow (1,0,1)+[-1,1,0]$. Narks can also create or annihilate gluons, since $(-1,0,1) \rightarrow (0,-1,1)+[-1,1,0]$, for instance. Narks and quarks are thus particles that are subjected to strong interactions. They yield attractive forces that can lead to scattering or binding.

All compound particles have to be "color neural". This means that the three spatial reference axes have to be involved with the same probability. Nucleons are thus constituted of 3 quarks in R, G and B color states. Narks can constitute a greater variety of compound particles. We called them "neutralons", since they are electrically neutral. Nucleons interact with one another by exchanging π mesons, while neutralons interact with one another by exchanging N_2 bosons [6] . The cosmic DM gas is composed of neutralons and compound neutral particles. They interact most frequently by elastic scattering, which leads to pressure effects and explains astrophysical observations [7] . DM particles allow also for fusion and fission processes. They are important for cosmology, but STQ has also other consequences.

4. BIG BANG PROCESSES

4.1. The Primeval Photon

Georges Lemaître justified the idea of an expansion of space, starting at $R(0)=0$, by considering the thermodynamic "arrow of time". Since the initial state of our universe should be the simplest possible one, it would correspond to a unique quantum [24] . He called it the "primeval atom" [33] , which meant only that it should be an elementary particle. Is it one among those, which are possible according to STQ? The best candidate would then be a photon. Its energy-momentum relation $E = E(p)$ is reduced to $E = cp$, but a unique "primeval photon" had the highest possible (positive) energy $E = hc/2a$. Since STQ requires only that $a \neq 0$, to account for all possible elementary particles, the value of the quantum of length could be a function $a(t)$ of cosmic time. Its initial value $a_o = a(0)$ would thus determine the energy of the primeval photon. According to quantum mechanics, it is not possible to specify its position with absolute precision in our 3-D space. However, the probability distribution could be uniformly distributed over the surface of the smallest possible hypersphere. This means that its radius $r_o = R(0) \neq 0$.

The primeval photon had even to be in a quantum mechanical state where all orientations of its momentum vector were equally probable. Their magnitude was defined by $E = cp$ and $p = h/\lambda$, where the wavelength λ was determined by periodic boundary conditions. The available 3-D space was reduced, indeed, to the surface of the hypersphere of radius r_o. The wavelength λ was thus equal to the length $2\pi r_o$ of any great circle of the hypersphere. All possible waves were propagating there with equal amplitude in any direction and the average value of all momenta p was zero, although $p = h/2a_o$. The energy E of the primeval photon was equivalent to a mass M, since

$$E = \frac{hc}{2a_o} = \frac{\hbar c}{r_o} = \underline{M}c^2 \tag{18}$$

The primeval photon was confined by gravity on the surface of the hypersphere of radius r_o. Indeed, the distributed mass $\underline{M}$ was attracted by the equivalent mass $\underline{M}$, situated at the center of the hypersphere. The total energy of the primeval photon was thus

$$\underline{M}c^2 - \frac{G}{r_o}\underline{M}^2 = 0 \text{ and } \underline{M} = \frac{r_o c^2}{G} \tag{19}$$

By combining (18) and (19), we see that the initial radius r_o of the hypersphere was equal to the Planck length, according to its usual definition:

$$r_o^2 = \frac{G\hbar}{c^3} \tag{20}$$

This yields $r_o = 1.6\times10^{-35}$ m and $M = 2.2\times10^{-8}$ kg $\approx 1.2\times10^{19}$ GeV/c^2. Since our universe was in a zero-energy state, it is reasonable to assume that it arose from vacuum fluctuations. The resulting state was stabilized, but gave then rise to an amazing sequence of transformations and a highly astonishing evolution at different levels of complexification. It began with the conversion of the primeval photon into many material elementary particles, according to the conservation law for u-quantum numbers. These particles interacted with one another, which led to the formation of compound particles in negative energy states. The total amount of negative energy increased, as well as the total amount of positive energy, but energy conservation implies that our universe is still in a zero-energy state.

4.2. The Matter-Antimatter Asymmetry

According to STQ any particle state (u_x, u_y, u_z, u_{ct}) for a fermion implies the existence of an antiparticle state $(-u_x, -u_y, -u_z, -u_{ct})$. Nevertheless, there are no (or nearly no) antiparticles in our universe. It is customary to assume that the Big Bang created particles and antiparticles in equal proportions, but that all (or nearly all) antiparticles were annihilated at some early stage. We propose another explanation. Pair production of particles and antiparticles is not required, since the conservation law for u-quantum numbers requires merely that their sum is unchanged for every one of the four space-time axes. For the first generation of elementary particles ($u_{ct} = 0$), we could thus get for instance

$$[0,0,0] \rightarrow (-1,0,0)+(0,-1,\ 0)+(1,1,0)$$

$$[0,0,0] \rightarrow (0,1,1)+(1,0,1)+(0,0,-1)+(-1,-1,-1)+(0,0,0)$$

This would respectively correspond to the production of a neutron, combining (d,d,u) quarks in R, G and B states, as well as the production of a proton, where (uud) quarks are in R, G and B states. We add an electron and an electron neutrino, to preserve the electric charge and to account for weak interactions. There are many other possible primary or secondary transformations, but they had to create compatible fermions. This means that their fields could "coexist" in the extremely compact initial 3-D space.

They did not appear when they could annihilate one another. Pair-production is possible today, since the fields of the resulting particle and its antiparticle can rapidly be separated from one another, but this was not possible on the initial hypersphere.

Elementary particles of the first generation were most frequently produced, since they have lower energies. They had also the greatest chance to survive. Moreover, quarks in R, G and B color states had to be created with equal probabilities, since the 3 spatial reference axes are physically equivalent. DM particles could not be directly created by the primeval photon, but any quark can create a gluon, by changing color. This gluon is annihilated by other quarks or creates two narks in other color states. This allowed for the creation of a compatible ensemble of narks, participating in the general particle-antiparticle asymmetry [6] .

We can also explain why the Big Bang "opted" for particles instead of antiparticles. Their production would have been equally probable, but that does not imply simultaneity. Once the primeval photon started to be transformed into so- called "particles", this process was immediately amplified by stimulated emission. Einstein discovered this process in 1917, when he tried to explain Planck's law for the frequency distribution of black body radiation. Emission of photons can be spontaneous, but also stimulated by the presence of other photons of the adequate type. In quantum mechanics, this is justified by means of great intensities of quantized fields and would also apply to material particles.

4.3. The Initial Inflation

The Big Bang was triggered by the transformation of the primeval photon into quarks and narks. Since they are spin-$1/2$ fermions they could not coexist close to one another in identical states. They darted away, but continued to interact by exchanging gluons. Even very brief binding implied negative energy states. The initial "soup" of interacting particles got thus very hot and the radius $R(t)$ of the hypersphere grew extremely fast in gigantic proportions. Eventually, all available quarks were definitively bound to one another inside nucleons and all narks inside neutralons. Because of the enormous frequency of these fusion processes and the very great binding energy for strong interactions, the resulting "inflation" of space was gigantic and extremely rapid. However, it had to stop when all available elementary particles were definitively bound to one another. Fusion processes became irreversible, but during a very short period, the positive energy content of our universe and the scale factor $R(t)$ grew extremely fast.

Science progresses by successive approximations. Lemaître solved Equation (5) and developed the concept of an expanding universe by adopting the simplest possible hypothesis: $R(0)=0$. He assumed also that the expansion began in a linear way, which defines an initial speed of expansion: $\dot{R}(0)$. Alan Guth improved this theory [34] , since he strongly felt that homogeneity and isotropy of today's extremely vast universe had to be explained. A very rapid and enormous increase of $R(t)$ during the Big Bang would be sufficient. This inflation can now be justified in terms of Bang processes.

5. CONCLUSIONS

Initially, we wanted only to find out if space and time are continuous or not. It appeared that Space-Time Quantization (STQ) is possible and accounts for elementary particle physics [6] . This theory applies also to DM particles and we showed here that they account for the accelerated expansion of space. This results from fusion and fission of DM particles. They liberate and require energy in such a way that the common mass-energy density remains constant, even when space is expanding.

The conventional model called for strong negative pressure. They cannot be justified by elastic collisions of DM particles and are not necessary to account for the cosmological constant Λ and the resulting accelerated expansion of space. It is also possible to account for Big Bang processes and to prove that the initial value a_o of the quantum of length was determined by the initial radius r_o of the 4-D hypersphere. It appeared that r_o is the Planck length and that the smallest possible wavelength was $2a_o = 2\pi r_o$.

It should be noted that the reality of STQ is justified by many remarkable empirical observations, summarized by the Standard Model of elementary particle physics [6] . STQ leads also to the concept of pressure for the cosmic DM gas, which is confirmed by astrophysical observations [7] . STQ accounts even for cosmological processes, but many new questions are emerging. It is necessary, for instance, to clarify the mechanism of fission and fusion processes. Observations of the evolution of Λ and the masses of DM particles in galactic halos may be helpful. This will eventually lead to the development of DM physics, partially similar to nuclear physics. At least DM-electron interactions allowed already for direct detection of DM particles [7] , but we are only beginning to unravel the mysteries of the dark sector of our universe.

Instead of worrying about the ultimate fate of our universe, we have to realize that its accelerated expansion cannot continue forever. This would imply an unphysical divergence, which can only result from an approximation. Actually, the accelerated expansion cannot exceed the capacity of DM to generate more DM. It depends on the transition probabilities α and β, as well as the parameter q. Since fusion of DM particles constitutes an energy source, at least in the form of creating more DM particles, we may wonder if it can be harnessed. At first sight, this seems to be impossible, since they are electrically neutral, but creation of hybrid particles is not excluded [6] . Could much older ET civilizations be using this energy source for interstellar space travel? Anyway, the essential conclusion of this article is that the validity of STQ is confirmed by applying it to cosmology. The intimate connection between cosmology and elementary particle is strengthened and invites to challenging research.

REFERENCES

1. Turner, M.S. (1999) Dark Matter and Dark Energy in the Universe, ASP Conf.
2. Riess, A.G., et al. (1998) The Astronomical Journal, 116, 1009-1038.
3. Perlmutter, S., et al. (1999) The Astrophysical Journal, 517, 565-586.
4. Tonry, J.L., et al. (2003) The Astrophysical Journal, 594, 1-24.
5. Meessen, A. (2011) Space-Time Quantization, Elementary Particles and Dark Matter.
6. Meessen, A. (2017) Journal of Modern Physics, 8, 35-56.
7. Meessen, A. (2017) Journal of Modern Physics, 8, 268-298.
8. Peebles, P.J.E. and Ratra, B. (2003) The Cosmological Constant and Dark Energy, RMP. 75, 559-606.
9. Kungl. Wetenskaps Akademien (2011) The Accelerating Universe, Scientific Background on the Nobel Prize in Physics.
10. Elizalde, E. (2012) Chapter 1: Cosmological Constant and Dark Energy: HistoricalInsight, In: Olmo, G.L., Ed., Open Questions in Cosmology, InTech, 1-30.
11. Steinhardt, P.J. (2011) Philosophical Transactions of the Royal Society of London A, 361, 2497-2513.
12. Tsujikawa, S. (2013) Classical and Quantum Gravity, 30, Article ID: 214003.
13. Bamba, K., et al. (2012) Astrophysics and Space Science, 342, 155-228.
14. Turner, M.S. (2002) Dark Matter and Dark Energy: The Critical Questions.
15. Turner, M. (2001) Dark Energy and the New Cosmology.
16. Spergel, D.N. and Steinhardt, P.J. (2000) Observational Evidence for Self-Inte-racting Cold Dark Matter.
17. Spergel, D.N. (2015) Science, 347, 1100-1102.
18. Horvath, J.E. (2009) Cosmos and History: The Journal of Natural and Social Philosophy, 5 287-303.
19. Von Seeliger, H. (1895) Astronomische Nachrichten, 137, 130-136.
20. Einstein, A. (1917) Sitzungsberichte der K öniglich Preubischen Akademie der Wissenschaften, 1, 142-152.
21. Lemaître, G. (1927) Annales de la Soci ét é Scientifique de Bruxelles, 47A, 49-59.
22. Lemaître, G. (1931) Monthly Notices of the Royal Astronomical Society, 91, 483-490.
23. Lemaître, G. (1931) Letters to Nature, 127, 706,
24. Friedman, A. (1922) Zeitschrift für Physik, 10, 377-386.
25. Slipher, V. (1913) Lowell Observatory Bulletin, 1, 56-57.

26. Hubble, E. (1929) Proceedings of the National Academy of Sciences, 15, 168-173.
27. Lema ître, G. (1949) The Cosmological Constant. In: Schilpp, Ed., Albert Einstein: Philosopher-Scientist II, Harper Torch, New York, 437-456.
28. Scolnic, D. (2014) The Astrophysical Journal, 795, 45.
29. Riess, A.G., et al. (2016) The Astrophysical Journal, 826, 56R.
30. Planck Collaboration (2016) Astronomy & Astrophysics, 594, A13.
31. Weinstein, G. (2013) George Gamow and Albert Einstein: Did Einstein Say That the Cosmological Constant Was the "Biggest Blunder" He Ever Made in His Life?
32. Luminet, J.P. (2014) Georges Lemaître: The Scientist Lemaître's Big Bang (Slide 25/29). https://fys.kuleuven.be/ster/meetings/lemaitre/lemaitre-luminet.pdf
33. Lemaître, G. (1946) L'Hypothése de l'atomeprimitif: Essai de cosmogonie. Griffon & Dunod.
34. Guth, A.H. (1997) The Inflationary Universe: The Quest for a New Theory of Cosmic Origins, Jonathan Cape.

CHAPTER 15

Gravitational Forces Explained as the Result of Anisotropic Energy Exchange between Baryonic Matter and Quantum Vacuum

Stefan L. Hahn

Institute of Radioelectronics, Warsaw University of Technology, Warsaw, Poland

ABSTRACT

Gravitational forces are explained as a result of energy exchange between baryonic matter having the property of mass and the Quantum Vacuum. The derivations are starting with a hypothesis that baryonic matter, particles, atoms and molecules exchange energy with the Quantum Vacuum with zero balance. It is assumed that in absence of an external gravitation field the emission pattern is isotropic. There is no recoil force of radiation. The application of an external gravitation field induces an anisotropy which results in a recoil force of radiation. An ellipsoidal radiation pattern is applied. The eccentricity of the ellipsoid is defined using the maximum possible value of any gravitation field estimated to have the value about 5×10^{12} [m/s^2]. A formula is derived for calculating the power of the isotropic radiation. It was shown that two masses attract due to the fact that gravitation field lowers the energy density of the Quantum Vacuum. Using the results of measurements of a binary neutron star by Taylor and Hulse

(Nobel Prize in Physics 1993) it was shown that possibly gravitational waves carry negative energy.

Keywords: Gravitation, Quantum Vacuum, Negative Energy, Recoil Forces

1. PRELIMINARIES

For convenience, let us recall some basic definitions and notations used in this paper. We apply the SI system of units. The macroscopic Newtonian law defining the force of attraction of two bodies of mass M_1 and M_2 is

$$F_{12} = G\frac{M_1 M_2}{R_{12}^2}\,[\mathrm{N}] \tag{1}$$

where

$$G = 6.67384\times10^{-11}\left[\frac{\mathrm{m}^3}{\mathrm{kg}\cdot\mathrm{s}^2}\right] \tag{2}$$

is the gravitational constant, M_1 and M_2 are masses of the bodies [kg] and R_{12} the distance between their center of mass. Using analogies between gravity and electromagnetism it is convenient to apply the reciprocal constant

$$\gamma = \frac{1}{4\pi G} = 1.92379\times10^{9}\left[\frac{\mathrm{kg}\cdot\mathrm{s}^2}{\mathrm{m}^3}\right]. \tag{3}$$

The gravitational field generated by a hypothetical point mass M_1 is

$$g = \frac{F_{12}}{M_2} = -G\frac{M_1}{R^2}\overline{r} = \frac{-1}{4\pi\gamma}\frac{M_1}{R^2}\overline{r} = \frac{-\rho_M}{\gamma}\overline{r}\left[\mathrm{m/s^2}\right]\mathrm{or}\left[\mathrm{N/kg}\right] \tag{4}$$

is called gravitational acceleration. R is the distance from the center of mass of M_1, $\overline{r}$ a unit vector indicating the direction of the field and r_M is the equivalent surface mass density. The minus sign indicates that acceleration is directed towards the center of mass.

Let us remind that:

1) The gravitational field is a vector quantity.
2) Two opposite gravitational fields of the same modulus cancel. Remark: This cancellation should not be interpreted as annihilation The hypothetical gravitons propagating in opposite directions do not collide (see Figure 9 in Appendix B).
3) Ordinary matter (baryonic matter) is transparent for gravitational fields. Differently to electrostatic fields gravitational screens are unknown.
4) The energy density of the gravitational field is given by the equation

$$E_G = E_{QV} - 0.5\gamma|g|^2 \left[\mathrm{J/m^3}\right] \text{or} \left[\mathrm{N/m^2}\right], \tag{5}$$

where E_{QV} is the extremely high energy density of the Quantum Vacuum (QV) (see Appendix C). Note that gravity lowers the energy density of QV differently to the energy of electrostatic field and that energy densities and pressure have the same dimensions.

2. ANISOTROPY OF ENERGY DISTRIBUTION AROUND A MASS INDUCED BY AN EXTERNAL G-FIELD

Consider a spherical body with the g-field defined by (B3) in the Appendix B. The self g-field is isotropic, i.e., the magnitude is equal for all directions. In the presence of an external g-field $\boldsymbol{g}_{ext} = g_{ext} \times \overline{z}$ and writing the isotropic surface g-field in the form $\boldsymbol{g}_{surf} = g_r \times \overline{r}$, the resulting surface g-field is given by the formula

$$\boldsymbol{g} = g_{ext} \times \overline{z} - g_{surf} \times \overline{r}. \tag{6}$$

The modulus of this vector is

$$|\boldsymbol{g}|^2 = g_{surf}^2 + g_{ext}^2 - 2g_{surf}g_{ext}\cos(\overline{r} \times \overline{z}). \tag{7}$$

Evidently, the energy density at the surface of the sphere is anisotropic. For example, if $\cos(\bar{r}\times\bar{z})=+1$, we have

$$E=E_{QV}-0.5\gamma\left(g_{ext}-g_{self}\right)^2 \tag{8}$$

and if $\cos(\bar{r}\times\bar{z})=-1$,

$$E=E_{QV}-0.5\gamma\left(g_{ext}+g_{self}\right)^2. \tag{9}$$

This anisotropy is responsible for the existence of a recoil force described in next chapter (See Figure 1).

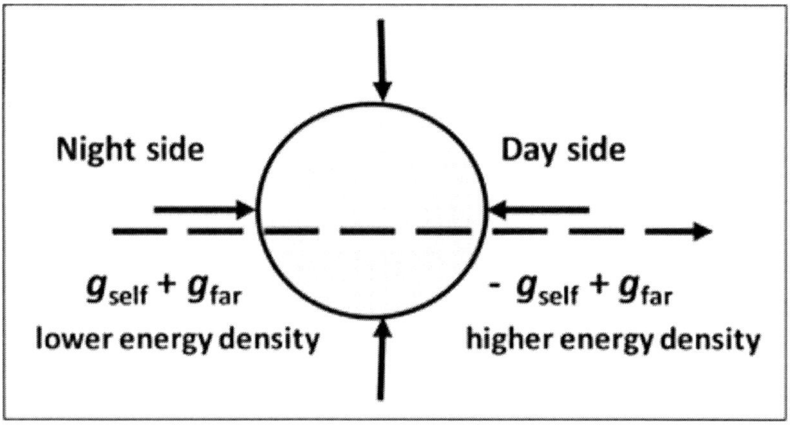

Figure 1. Cross-section of the spherical body (yellow). The self g-field (solid arrows) is spherically symmetric. The g-field of a far body (dotted line) penetrates the body with no change of sign. Therefore at the night side we have summation and at the day side subtraction of the fields. The gravitational energy density is lower at the night side w.r.t. the day side.

3. THE SCHWARTZSCHILD RADIUS AND THE HIGHEST VALUE OF A G-FIELD IN NATURE

The Schwartzschild radius R_{sch} is defined as the radius of a sphere such that, if all of the mass of a body is compressed within that sphere, the escape speed from the surface of the sphere would equal the speed of light. In this paper we try to apply this notion to calculate the maximum value of the modulus of any g-field. Schwartzschild, using equations of general relativity derived the following form

$$R_{sch} = \frac{2GM}{c^2} \ [\mathrm{m}] \quad . \tag{10}$$

For a sphere of radius R_{sch} the equivalent surface mass density $\rho_M = M/\left(4\pi R_{sch}^2\right)$.

Therefore, the surface g-field is

$$|\boldsymbol{g}| = 4\pi G \rho_M = \frac{c^4}{4GM} \ \left[\mathrm{m/s^2}\right] \quad . \tag{11}$$

The gravitational part of the energy density at the surface (see (5) and (11)) is

$$E_{sch} = \frac{1}{8\pi G}|\boldsymbol{g}|^2 = \frac{c^8}{128\pi G^3 M^2}\left[\mathrm{J/m^3}\right] \quad . \tag{12}$$

For comparison let us calculate the Einstein's energy density

$$E_{ein} = \frac{Mc^2}{(4/3)\pi R_{sch}^3} = \frac{3c^8}{32\pi G^3 M^2}\left[\mathrm{J/m^3}\right] \quad . \tag{13}$$

Note that the ratio (13)/(12) equals 12. Let us quote the neutron star PSRJ614-2230 with a radius 19,300 [m] and a mass M = 3. 978 × 10^{30} [kg] (two solar masses). The surface g-field of this star equals 7.13 × 10^{11} [m/s^2]. Its Schwartzschild radius 5911 [m] is only 3.27 times smaller w.r.t. the physical value 19300 [m]. A neutron star with the diameter 5911 [m] and the mass equal to twice the mass of the Sun would have a g-field at the surface given by (11):

$$|g|_{\max} = 7.59\times10^{12}\left[\mathrm{m/s^2}\right] \tag{14}$$

In this paper we apply this value as the largest possible value of any g-field. Note that the highest value of the g-field is defined macroscopically at the surface of a neutron star. Differently, the highest value of electrostatic field is defined microscopically at the surface of the electron (see [1]).

4. DERIVATION OF THE FORMULA FOR CALCULATION OF THE POWER OF THE ENERGY EXCHANGE BETWEEN A MASS M AND THE QUANTUM VACUUM

We formulate a hypothesis that baryonic matter continuously exchange energy with the Quantum Vacuum with zero balance. Let us calculate the power of the emission. In absence of external fields we postulate an isotropic absorption and emission pattern as illustrated in Figure 2. There is no recoil force of radiation. The external g-field induces anisotropy of radiation resulting in a recoil force. Our goal is the calculation of the power of the energy exchange. Our choice is the model of the radiation pattern defined by an ellipsoid

$$\sigma_\Omega = \sigma_{\max}\frac{1-\varepsilon^2}{1+\varepsilon\cos(\varphi)}\ \left[\mathrm{W/Steradian}\right], \tag{15}$$

where e is the eccentricity of the ellipse. This formula uses the polar coordinates centered in the focus of the ellipsoid. The recoil force is given by the integral

$$|\boldsymbol{F}| = \frac{v}{c^2}\int_{4\pi}\sigma_\Omega \boldsymbol{n}_0 \mathrm{d}\Omega, \tag{16}$$

where v is the velocity of radiation, c the velocity of light in free space and n_0 a unit vector directed along the longer axis of the ellipse. The derivation of Appendix A yields for a small value of ε the following formula (v = c)

$$|F|_{recoil} \cong \frac{\varepsilon P}{3c} [\mathrm{N}] \quad (17)$$

This recoil force should be equal to the gravitation force

$$|F|_{grav} = |g| M \; [\mathrm{N}] \quad (18)$$

Equating the above formulae yields the following expression for the power P

$$P = \frac{3c|g|M}{\varepsilon} [\mathrm{W}] \quad (19)$$

Evidently, the calculation of the value of the power P requires the knowledge of the value of the eccentricity e. Following the procedure of defining the eccentricity for electrostatic fields [2] let us define

$$\varepsilon = \frac{|g|}{|g|_{\max}} \quad (20)$$

where $|g|_{\max}$ is defined by (14). We get the following simple formula

$$P = 3c|g|_{\max} M \simeq 6.82 \times 10^{21} M \; [\mathrm{W}] \quad (21)$$

For the model of two parallel planes covered with a mass density $r_M/2$ [kg/m^2]t (21) takes the form

$$P_{\mathrm{Pointing}} = 3c|g|_{\max} \frac{\rho_{\mathrm{M}}}{2} = 3.41 \times 10^{21} \rho_{\mathrm{M}} \; \left[\mathrm{W/m^2}\right] \quad (22)$$

having the dimensions of the Pointing vector of electromagnetic theory. The Table 1 presents the value of P for selected bodies.

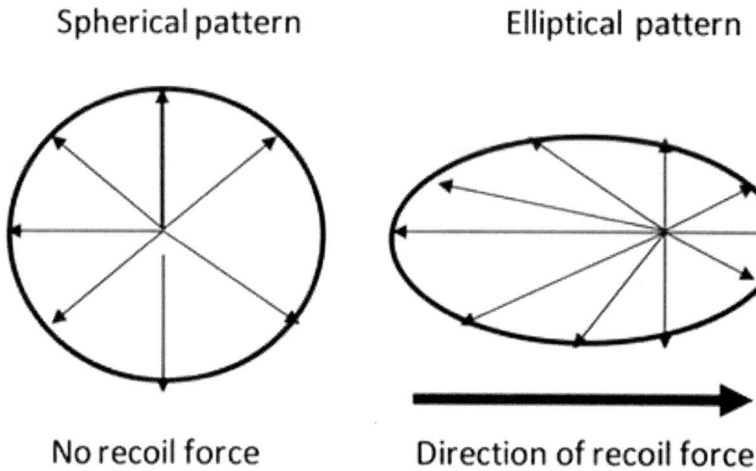

Figure 2. Left. The cross-section of a circular radiation pattern. Right. The cross-section of an elliptical radiation pattern. In examples the eccentricity is extremely small.

Table 1. A list of the values calculated using (21).

Name	Mass [kg]	Power [W]
Elektron	9.109×10^{-31}	6.21×10^{-9}
Neutron	1.675×10^{-27}	1.142×10^{-5}
Autor	88	6.00×10^{23}
Earth	5.973×10^{24}	4.073×10^{46}
Moon	7.347×10^{22}	5.07×10^{44}
Sun	1.989×10^{30}	1.356×10^{52}
Planck mass	2.176×10^{-8}	1.479×10^{14}
Neutron star	3.978×19^{30}	2.713×10^{52}

5. ILLUSTRATION OF THE IDEA OF THE RECOIL NATURE OF GRAVITATIONAL FORCES BY A MODEL OF A BINARY STAR

The Nobel prize in physics in the year 1993 has been awarded to R.A. Hulse and J.H. Taylor for the discovery of a binary neutron PSR B1923+16 and precise measurements of the elongation of the orbital period giving the evidence of radiation gravitational waves [3] - [6] . In this chapter we indicate that possibly gravitational radiation carries negative energy. In the binary PSR B1923+16 the two neutron stars of nearly equal masses are orbiting each along a separate elliptical orbit around a common center of mass (Figure 3) The laureates measured the rate of decrease of the orbital period equal 76.5 [ms/year] and calculated the power of the emission of gravitational waves $P = 7.35\times10^{24}\ [\mathrm{W}]$. Our goal can be achieved presenting calculations for a simplified system of two neutron stars with equal masses and a common circular orbit (Figure 4).

6. KINETIC ENERGY OF THE NEUTRON STARS ON A CIRCULAR ORBIT

We investigate a model of a binary neutron star of equal masses $M_1 = M_2 = M$ orbiting on the initial orbit of radius R. The presented theory is well known. We present a convenient version. Equating the gravitational force with the centripetal force

$$G\frac{M^2}{R^2} = M\omega^2 R \tag{23}$$

yields the following radius of the circular orbit

$$R_{orb} = \sqrt[3]{\frac{GM}{\omega^2}}, \tag{24}$$

where G is the gravitational constant, $\omega = 2\pi/T$ is the initial angular velocity and T the initial orbital period. Note that if we assume a loss of energy due to the emission of gravitational waves, w and T are functions of time. However, during one orbital period the change is negligible. The orbital (tangential) velocity is $v = \omega R$.

The initial kinetic energy of both stars is

$$E_k = 2\frac{Mv^2}{2} = M\omega^2 R^2;\; R = \sqrt[3]{\frac{GM}{\omega^2}};\; \omega = \frac{2\pi}{T}. \tag{25}$$

The insertion yields

$$E_k = G^{\frac{2}{3}} M^{\frac{5}{3}} \frac{(2\pi)^{\frac{2}{3}}}{T^{\frac{2}{3}}}. \tag{26}$$

Note that multiplication of both sides of (23) by R yields the equality of the kinetic and potential energies The potential energy is $E_p = -G\frac{M^2}{R}$ and the kinetic energy is $E_k = M\omega^2 R^2$, i.e., $E_k + E_p = 0$. Both energies differ only by sign. The observations of PSR B1923+16 have shown a decrease of the orbital periods by $\Delta T = 7.65\times 10^{-5}$ [s] per year [3] . Assuming that a similar decrease occurs also in our circular model we get an increase of the kinetic energy

$$E_k' = G^{\frac{2}{3}} M^{\frac{5}{3}} \frac{(2\pi)^{\frac{2}{3}}}{(T-\Delta T)^{\frac{2}{3}}}. \tag{27}$$

The increase of the kinetic energy per year is

$$\Delta E_k = G^{\frac{2}{3}} M^{\frac{5}{3}} (2\pi)^{\frac{2}{3}} \left[\frac{1}{(T-\Delta T)^{\frac{2}{3}}} - \frac{1}{T^{\frac{2}{3}}}\right] = Pt_{\text{year}}, \tag{28}$$

where P is the radiated power and $t_{\text{year}} = 3.15360\times 10^{7}$ [s] is the duration of a year in seconds.

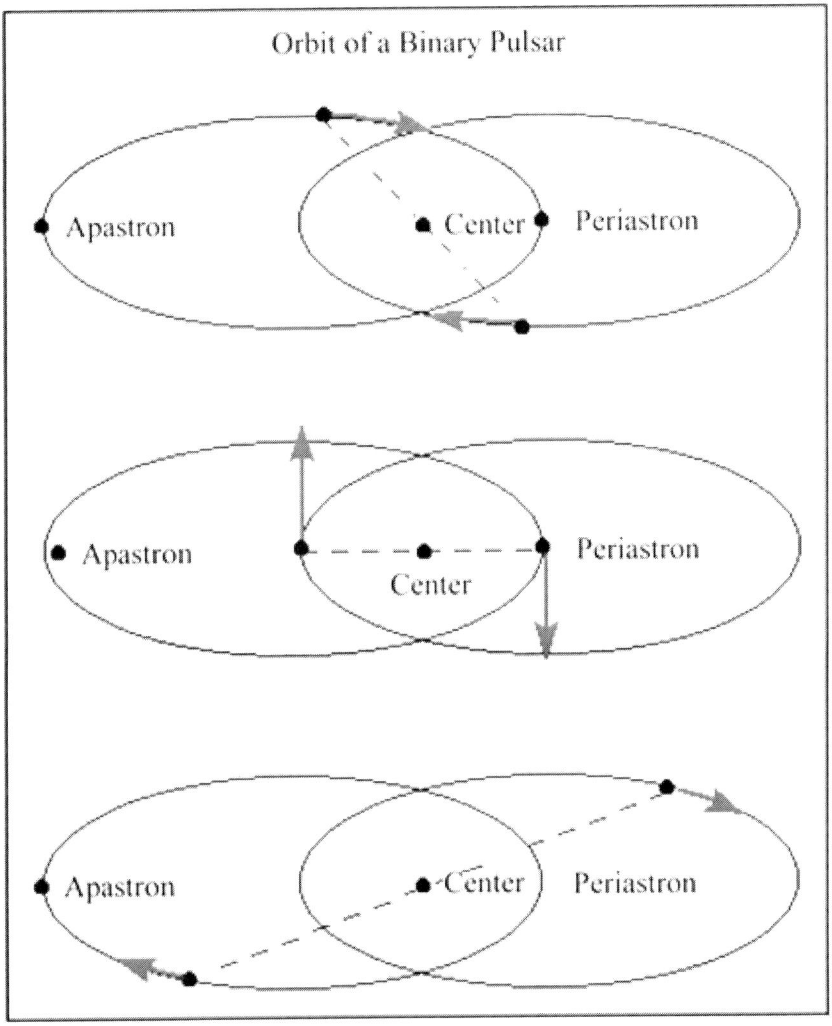

Figure 3. Two elliptical orbits of binary neutron stars with a common center of mass. Adapted from [5] .

Remark: The kinetic energy and T are functions of time. However, we have no need to apply differential equations since the initial value of ΔT is very small. Note that for a given ΔT this equation uniquely defines the value of the power of the emitted gravitational waves. We observe that the emission of gravitational energy causes the increase of the kinetic energy. In consequence, if we assume that the lost of the kinetic energy is caused by the emission of gravitational waves the emitted energy should be classified as negative. This should be understood as periodic lowering of the positive energy of the Quantum Vacuum propagating with the speed of light. In the case of the system reported by Taylor and Hulse the kinetic orbital energy is a periodic function of time. Therefore, (27) and (28) should be replaced by the mean values.

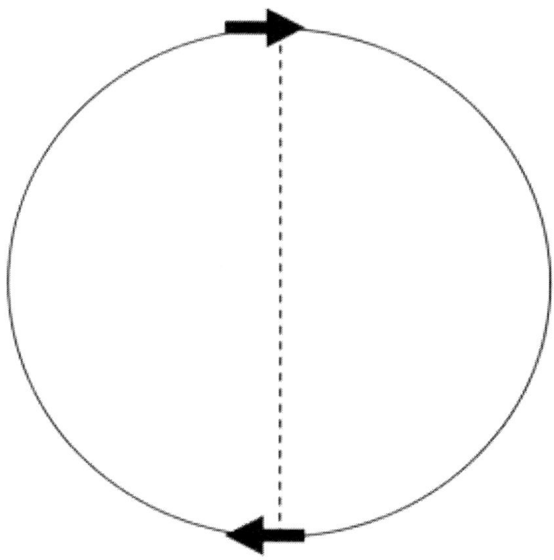

Figure 4. A circular orbit of two neutron stars.

7. SELECTED DATA FOR THE MODEL OF A BINARY STAR WITH A CIRCULAR ORBIT

The following data have been selected from data of the PSR B1913+16. The measured initial orbital period T = 27906.97959 [s] and the calculated masses are $M_1 = 2.8764 \times 10^{30}$ [kg], $M_2 = 2.72050 \times 10^{30}$ [kg]. We applied for our model with the circular orbit the following data:

Equal masses (mean value)

$$M = 2.74084 \times 10^{30} \, [\mathrm{kg}] \tag{29}$$

The radius of each star

$$R_N = 20000\,[\mathrm{m}] \tag{30}$$

The initial angular velocity

$$\omega = \frac{2\pi}{T} = \frac{2\pi}{27807.19557} = 2.255498196\times10^{-4}\,[1/\mathrm{s}] \tag{31}$$

The initial radius of the circular orbit

$$R_{orb} = \sqrt[3]{\frac{GM}{\omega^2}} = 1.4716\times10^{9}\,[\mathrm{m}] = 1471.6\,[\mathrm{km}] \tag{32}$$

The initial tangential orbital velocity is

$$v = \omega R_{orb} = 331.818\,[\mathrm{km/s}] \tag{33}$$

The volume mass density of each star is

$$\rho_V = \frac{3M}{4\pi R_N^3} = 8.3687\times10^{16}\,\left[\mathrm{kg/m^3}\right] \tag{34}$$

corresponding to the equivalent surface mass density

$$\rho_S = \frac{M}{4\pi R_N^2} = 5.579\times10^{20}\,\left[\mathrm{kg/m^2}\right] \tag{35}$$

The intensity of the self gravitational field at the surface is

$$|\mathbf{g}|_S = g_S = \frac{\rho_S}{\gamma} = 4\pi G\rho_S = 4.57898\times10^{11}\,\left[\mathrm{m/s^2}\right] \tag{36}$$

The intensity of the g-field at the center of a single star induced by its companion

$$|\mathbf{g}|_2 = g_2 = \frac{\rho_S}{\gamma}\frac{1}{\left(2R_{orb}\right)^2} = 19.4394\,\left[\mathrm{m/s^2}\right] \tag{37}$$

(about twice of the Earth field). The Schwartzschild radius is

$$R_{sch} = \frac{2GM}{c^2} = 4072.949\,[\mathrm{m}] \tag{38}$$

yielding the value

$$g_{\max} = 5.51329 \times 10^{12}\,[\mathrm{m/s^2}] \tag{39}$$

The difference in comparison to (14) is small. The Einstein's energy density of a single star is

$$u_{\mathrm{Einstein}} = \rho_V c^2 = 9.7964 \times 10^{33}\,[\mathrm{J/m^3}] \tag{40}$$

The energy density at the surface of the star defined by the power $P = 3cg_{maz}M$ is

$$u_P = \frac{P}{4\pi R_N^2 c} = 9.01875 \times 10^{33}\,[\mathrm{J/m^3}] \tag{41}$$

The energy density of the surface self g-field is

$$u_s = 0.5\gamma g_s^2 = 1.2467 \times 10^{32}\,[\mathrm{J/m^3}] \tag{42}$$

8. ARGUMENTS IN FAVOR OF THE PRESENTED HYPOTHESIS ABOUT THE PHYSICAL ORIGIN OF GRAVITATION

1) The existence of the Quantum Vacuum as a medium with extremely high energy density is confirmed by many experiments and is not questioned. Let us mention the experimental confirmation of the predicted Casimir force.
2) Recently researchers have created electrons from "nothing" This nothing is the energy of QV.

3) Recently the scientists from Berkeley University [7] have measured the difference of the Compton frequencies for two groups of cesium atoms, a local group and a second group after a small journey. They confirmed experimentally the Einstein's twilling effect (time dilation). Certainly, this gives the evidence, that cesium atoms emit energy at Compton frequencies.

4) The gravitation attraction force is not the result of the pressure of the gravitation field on the baryonic matter. For example, in the model of two parallel planes covered with a uniform mass density (see Figure 9 in the Appendix B) the pressure inside the plates is higher in comparison to the outside Since the two plates attract this cannot be the result of radiation pressure of the gravitational field.

9. CONCLUSIONS

Many experiments and phenomena confirm that the Quantum Vacuum is a medium with extremely high energy density. Let us mention the Casimir effect and the Lamb shift. A good confirmation gives the electron on the Bohr orbit. Due to the rules of electromagnetism it radiates energy. Without absorption of the energy from the QV it should decay in a short time. This makes the hypothesis that particles absorb and reemit energy from the QV highly probable. Using this hypothesis we derived a formula enabling the calculation of the power of the energy exchange. The value of this power is formidable. However, let us recall the formidable energy density of the QV. Our numerical results depend on the maximum possible value of the intensity of any gravitational field. The eventual application of another value of this constant will change only numerical results. Let us recall that the g-field lowers the energy density of the QV. Our results are valid for the Newton's law of gravity. In frame of this law, the QV is a linear medium. However, we know that for high density g-fields nonlinear effects occur. For example, the speed of light in vacuum is lowered by gravitation. Gravitational deflection of light beams is well known. Assuming the validity of the statement that gravitational waves carry negative energy, any device constructed to detect gravitational waves should be able to measure periodic variations of the energy density of the Quantum Vacuum, for example, looking for periodic variations of the speed of light. Note the extremely small values of the eccentricity of the radiation pattern. In the worst case of the binary neutron stars, the eccentricity is $\varepsilon = (37)/(36) \approx 3.52 \times 10^{-11}$. The small g-field of the companion star controls the emission diagram of the large power defined by (21). We have a kind of amplification.

This paper differs from the reference [8] . We presented the derivation of the formula for the power emitted by the mass M (see (21)) and have shown that possibly the gravitational waves carry negative energy in the sense of (5). The examples with neutron stars are new.

REFERENCES

1. Hahn, S. (1976) Kleinheubacher Berichte, 20, 145-149.
2. Hahn, S.L. (2007) Bulletin of the Polish Academy of Sciences, Technical Sciences, 35, 405-410.
3. Taylor, J.H. and Weisberg, J.M. (1962) Astrophysivcal Journal, 253, 908-920.
4. Press Release, 13 October 1993, about the Nobel Price in Physics 1993, The Royal Swedish Academy of Sciences.
5. The binary Pulsar PSR1913+16. www.astrocornell.edu/academics/PSR1913+16.htm
6. Paoletti, M.S. (1994) Binary Pulsars and Evidence for Gravitational Radiation. University of Maryland, College Park. *) The references to are selected arbitrary from many papers describing the achievement of Taylor and Hulse.
7. A Time to Kilo: The Atomic Interferometer, Berkeley Science Review. (a report of the achievement of Prof Holger Müller and his team).
8. Information about Experiments of Creation of Matter Click "Creation of Matter from Nothing".

CHAPTER 16

Measuring Mass and Spin of Dark Matter Particles with the Aid Energy Spectra of Single Lepton and Dijet at the e^+e^- Linear Collider

I. F. Ginzburg[1,2]

[1] *Sobolev Institute of Mathematics, Novosibirsk, Russia*

[2] *Novosibirsk State University, Novosibirsk, Russia*

ABSTRACT

In many models stability of Dark Matter particles D is ensured by conservation of a new quantum number referred to as D-parity. Our models also contain charged D-odd particles $D^{\pm}$ with the same spin as D. Here we propose a method to precisely measure the masses and spins of D-particles in the process $e^+e^- \to D^+D^- \to DDW^+W^- \to DDW(q\bar{q})(\ell\nu)$ with a signature dijet $+\,\mu(\text{or e})\,+$ nothing. It is shown that the energy distribution of the lepton has singular points (upper edge and kinks or a peak) whose positions are kinematically determined. For precise determination of D and $D^{\pm}$ masses, it is sufficient to measure these singular points and upper edge of dijet energy spectrum. After this procedure, even an approximate measurement of the corresponding cross section allows a determination of the spin of D particles to be performed. New points of this work are: 1) We propose to use only the well measurable energy spectra of individual leptons and the upper edge of the dijet energy spectrum. 2) We propose to identify the spin of D-particles via value of the cross section for the discussed process.

Keywords: Dark Matter, W-Boson, Lepton, Linear Collider

1. INTRODUCTION

1.1. Models

In the broad class of models Dark Matter (DM) consists of particles D similar to those in SM, with the following properties:

- The neutral DM particle D with mass M_D and spin $s_D=0$ or $1/2$ has a new conserved quantum number, which we call the D-parity. All known particles are D-even, while the D is D-odd.
- In addition to D, other D-odd particles exist: a charged $D^{\pm}$ and (sometimes) a neutral D^A, with the same spin s_D and with masses $M_+, M_A > M_D$. The D-parity conservation ensures stability of the lightest D-odd particle D.
- These D-particles interact with the SM particles via the Higgs boson DDh, D^+D^-h, D^AD^Ah and via the covariant derivative in the kinetic term of the Lagrangian. These are the gauge interactions $D^+D^-\gamma$, D^+D^-Z, D^+DW^-, $D^+D^AW^-$, D^ADZ with the standard electroweak couplings g, g' and e (coupling to Z can be added by a mixing factor $\mu_M \le 1$, deviation from 1 appears due to possible mixing of D with other D-odd neutrals).
- The first example of such model provides well known MSSM (see e.g. [1] -[6]) for specific set of parameters. Here our term D-parity means R-parity. For the considered set of parameters, D is the lightest neutralino χ_1^0, the heavier neutralino χ_2^0 can play role D^A and the next in mass D-odd particle is the lightest chargino, spin of these D-particles $s_D = 1/2$. The other D-odd particles (in particular, sleptons and squarks) are supposed to be heavier than the ILC beam energy E.

- The second example of such models provides the Inert Doublet Model (IDM) (see e.g. [7] -[13] and Appendix A). That is the Z_2 symmetric Two Higgs Doublet Model, containing two scalar doublets ϕ_S and ϕ_D. The "standard" scalar (Higgs) doublet ϕ_S is responsible for electroweak symmetry breaking and the masses of fermions and gauge bosons just as in the Standard Model (SM). The second scalar doublet ϕ_D doesn't receive vacuum expectation value and doesn't couple to fermions. In this model the D-parity conservation is ensured by a Z_2 symmetry, four degrees of freedom of the Higgs doublet ϕ_S are the same as in the SM: three Goldstone modes and the standard Higgs boson h. All the components of the scalar doublet ϕ_D are realized as the massive D-particles: two charged $D^{\pm}$ and two neutral ones D, D^A with masses M_+, M_D, M_A respectively with $M_+, M_A > M_D$. IDM contains no other D-odd particles. All D-particles have spin $s_D = 0$.

A possible value of mass M_D is limited by stability of D during the Universe existence [14] -[20] . The non-observation of the processes $e^+e^- \to D^+D^-$ and $e^+e^- \to DD^A$ at LEP gives $M_+ > 90$ GeV and $M_A > 100$ GeV (at $M_A - M_D > 10$ GeV) [11] -[13] . Limitations for masses of neutralino and chargino can be found in [20] . For IDM, limitations for parameters of Z-peak, S and T results in ([11] -[13] [20])

$$|\Delta T| = 2.16\left(\frac{M_+ - M_D}{\mathrm{v}}\right)\cdot\left|\frac{M_+ - M_A}{\mathrm{v}}\right| < 0.15. \tag{1}$$

with $\mathrm{v} = 246$ GeV expectation value of Higgs field. Further, we will have in mind $M_D \lesssim 80$ GeV and assume $M_+ - M_D, M_A - M_D > 20$ GeV.

1.2. The Problem

The neutral and stable D can be produced and detected via production $D^{\pm}$ or D^A and subsequent decay $D^{\pm} \to DW^{\pm}$, $D^A \to DZ$ with either on shell (real) or off shell $W^{\pm}$ or Z. The off shell W emerges as a $q\bar{q}$ pair (dijet[1]) or $\ell\nu$, having the same quantum numbers as W but with an effective mass $M^* < M_W$. From now on, W or Z refers to any of these two cases.

To discover the DM particle, one needs to specify such processes with a clear signature. The e^+e^- Collider ILC/CLIC provides an excellent opportunity for this task (see, e.g., [21] [22]) in the process $e^+e^- \to D^+D^-$ with a clear signature, see Equations (6) and (7) below. The cross section of this process is a large fraction of the total cross section of e^+e^- annihilation, Section 3.3.

The masses M_+ and M_D could be found via the edges of the energy distribution of dijets, originating from W from decay $D^\pm \to DW^\pm$, section 2.2, 2.4 (see [5] [6] for MSSM and [11] -[13] for IDM). However, this method cannot provide a good accuracy in measuring the mass. Indeed, the individual jet energy measurement suffers from a sizable uncertainty. In particular, this uncertainty smoothes the lower edge in the dijet energy spectrum.

On the contrary, the lepton energy can be measured much more precisely. In this paper we show, first, that the energy distribution of leptons has singular points whose positions are kinematically determined. Measuring positions of these singularities will allow, in principle, to determine masses M_D and M_+ with good precision (Sections 2.3 and 2.4). In contrast to [5] - [6] [11] -[13] , our description is suitable for different models.

Moreover, we present a simple method for measuring spin of DM particles in these very experiments.

The discussed problem differs strongly from that for the case when the lightest charged D-odd particle is slepton (another set of parameters of MSSM). In the latter case DM particles are produced via slepton pair $e^+e^- \to \tilde{\ell}^+\tilde{\ell}^- \to \ell^+\ell^-\chi_0\chi_0$. First of all, signature of this process is quite different from that one in our problem (6), (7). Second, the energy of observable lepton decay product of slepton is measurable well in each individual event, in difference with our case, when similar product of decay, W , is seen as dijet or lepton plus neutrino with badly measurable energy in each individual event. Therefore, the approach used in the analysis of slepton production (cf. [23] -[25]) cannot be applied directly to our problem.

The overall picture is summarized in Section 3. Short conclusion is given in Section 4.

In the Appendix B we discuss the potential of the process $e^+e^- \to DD^A \to DDZ$ for similar problems, for completeness. In contrast with previous studies, we find that this potential is not too high.

In the Appendix C we consider possible background processes and show that the most of them can be neglected at the analysis.

1.3. Scale of Cross Sections

We express discussed cross sections via

$$\sigma_0 \equiv \sigma\left(e^+e^- \to \gamma \to \mu^+\mu^-\right) = 4\pi\alpha^2/3s. \tag{2}$$

The total cross section of e^+e^- annihilation at ILC for $\sqrt{s} \equiv 2E > 200$ GeV is $\sim 10\,\sigma_0$. The annual integrated luminosity L for the ILC project [22] gives

$$L\sigma_0 \sim 3\times 10^5. \tag{3}$$

The process $e^+e^- \to D^+D^-$ represents a significant fraction of all e^+e^- annihilation events-see (19), (20), **Figure 2** and **Table2** With the luminosity (3), the annual number of events of discussed type will be $(0.6 \div 3)\times 10^5$, depending on M_+/E and s_D, and about $1/3$ of them (in the mode with e or μ plus dijet) are suitable for our analysis.

2. THE PROCESS $e^+e^- \to D^+D^-$

Note before all that the energies, γ -factors and velocities of $D^\pm$ are

$$E_\pm = E = \sqrt{s}/2,\ \ \gamma_\pm = E/M_+,\ \ \beta_\pm = \sqrt{1 - M_+^2/E^2}. \tag{4}$$

2.1. The Signature

- If $M_A > M_+$ or D^A is absent, once produced, particles $D^\pm$ decay fast (with a unit probability) to $DW^\pm$,

$$e^+e^- \to D^+D^- \to DDW^+W^-. \tag{5}$$

The observable states are decay products of W with a large missing transverse energy $\not{E}_T$ carried away by the invisible D-particle, and the missing mass of particles escaping observation $M(\not{E}_T)$ is large. In contrast to the LHC, where a large flux of low $p_\perp$ particles demands an additional $p_\perp$ cut off, at e^+e^- LC such particles are absent.

Therefore, the signatures of the process in the modes suitable for observation are

$$\boxed{\begin{array}{l} e^+e^- \to DD(W \to q\bar{q})(W \to q\bar{q}): \text{ Two dijets + nothing,} \\ \text{with energy of each dijet } < E, \text{with large } \not{E}_T \text{ and large } M(\not{E}_T). \end{array}} \tag{6a}$$

$$\boxed{\begin{array}{l} e^+e^- \to DD(W \to \ell\nu)(W \to q\bar{q}): \text{ One dijet + e or } \mu + \text{nothing,} \\ \text{with energy of each dijet or lepton } < E, \text{with large } \not{E}_T \text{ and large } M(\not{E}_T). \end{array}} \tag{6b}$$

At $M^* > 5$ GeV, the branching ratios for different channels of W decay are roughly identical for on-shell W [20] and off-shell W. In particular, the fraction of events with signature (6a) is $0.676^2 \approx 0.45$. The fraction of events with signature (6b) is $2 \times 0.676 \times 2 \times (1 + 0.17) \times 0.108 \approx 0.33$ (here 0.17 is a fraction of μ or e from the decay of τ). At $M^* < 5$ GeV, $BR(e\nu)$ and $BR(\mu\nu)$ increase, while the dijet becomes a set of a few hadrons.

If $M_+ > M_A$, when analysing the main process $e^+e^- \to D^+D^-$, one more decay channel is added, $D^\pm \to D^A W^\pm \to DZW^\pm$. Its branching ratio $B = BR(D^+ \to D^A W^+)$ is typically less than 0.5 (see discussion in section 2.4). Particle D^A decays fast to DZ, creating new cascades $e^+e^- \to D^+D^- \to DW^+D^AW^- \to DDW^+W^-Z$, $e^+e^- \to D^+D^- \to D^AW^+D^AW^- \to DDW^+W^-ZZ$.
As a result, the signature of the processes $e^+e^- \to D^+D^-$ in the modes suitable for observation contains both (6) and processes with decay W's or Z's in the mentioned cascades:

$$\boxed{4 \div 1 \text{ dijets and } 0 \div 5 \text{ leptons with large } \not{E}_T \text{ and large } M(\not{E}_T) + \text{nothing.}} \tag{7}$$

Note that the processes with invisible decay $Z \to \nu\bar{\nu}$ (we denote these states as Z_n, their $BR = 20\%$) have signature (6).

2.2. W Energy Distribution in $e^+e^- \to D^+D^- \to DDW^+W^-$

Here we consider the energy distribution of W with an effective mass M^*. At each value of M^*, we have in the rest frame of $D^\pm$ a two-particle decay $D^\pm \to DW^\pm$ with[2]

$$E_W^r\left(M^*\right)=\frac{M_+^2+M^{*2}-M_D^2}{2M_+},\quad p_W^r\left(M^*\right)=\frac{\Delta(M_+^2,M^{*2},M_D^2)}{2M_+},$$
$$\Delta\left(s_1,s_2,s_3\right)^2=s_1^2+s_2^2+s_3^2-2s_1s_2-2s_1s_3-2s_2s_3. \tag{8}$$

Denoting by θ the W^+ escape angle in the D^+ rest frame with respect to the direction of D^+ motion in the laboratory frame and using $c\equiv\cos\theta$, we find the energy of W^+ in the laboratory frame as $E_W^L=\gamma_\pm\left(E_W^r+c\beta_\pm p_W^r\right)$. Therefore, at given M^*, the energy E_W^L of W lies within the interval $\gamma_\pm\left(E_W^r\pm\beta_\pm p_W^r\right)$.

In particular, at $M_+-M_D>M_W$ we have $M^*=M_W$, and the kinematical edges of the W energy distribution are

$$E_{W,on}^{L,\pm}=\gamma_\pm\left(E_W^r\left(M_W\right)\pm\beta_\pm p_W^r\left(M_W\right)\right). \tag{9}$$

At $M_+-M_D<M_W$ we have $0\le M^*\le M_+-M_D$, and obtain similar edges, which are different for each value of M^*. The absolute upper and lower bounds on the energy distribution of W are attained at $M^*=0$, they are equal to

$$E_{W,off}^{L,\pm}=E(1\pm\beta_\pm)\left(1-M_D^2/M_+^2\right)/2. \tag{10}$$

At the highest value $M^*=M_+-M_D$ we have $p_W^r=0$, and an interval (9) is reduced to a point, where the entire W energy distribution has a maximum (peak) of

$$E_{W,p}^L\equiv E_W^{L,\pm}\Big|_{(M^*=M_+-M_D)}=E\left(1-M_D/M_+\right). \tag{11}$$

2.3. Single Lepton Energy Distribution in $e^+e^- \to D^+D^- \to DDW^+W^- \to DDq\bar{q}\ell\nu\,(6b)$

The fraction of such events for each separate lepton, e^+, e^-, μ^+ or μ^-, is about 0.08, their sum is about 0.33 of the total cross section of the process. We will speak, for definiteness, $\ell = \mu^-$ and neglect the muon mass.

Note that in the laboratory frame, for a W with some energy E_W^L, its γ-factor and the velocity are $\gamma_{WL} = E_W^L / M^*$ and $\beta_{WL} \equiv \sqrt{1-\gamma_{WL}^{-2}}$.

We study the distribution[3] of muons over its energy ε, $d\sigma^\mu(\varepsilon | M_+, M_D)/d\varepsilon$. We show that this distribution has singular points, whose positions are kinematically determined, i.e. model independent.

- If $M_+ - M_D > M_W$ we have $M^* = M_W$, and the muon energy and momentum in the rest frame of W are $M_W/2$. Just as above, denoting by θ_1 the escape angle of μ relative to the direction of the W in the laboratory frame and using $c_1 = \cos\theta_1$, we find that the muon energy in the laboratory frame is $\varepsilon = \gamma_{WL}(1 + c_1\beta_{WL})(M_W/2)$. Therefore, for these muons $\varepsilon^+(E_W^L) \geq \varepsilon \geq \varepsilon^-(E_W^L)$ where

$$\varepsilon^+(E_W^L) = E_W^L(1+\beta_{WL})/2 = E_W^L + \sqrt{(E_W^L)^2 - M_W^2}\,/2$$

and $\varepsilon^-(E_W^L) = M_W^2 / (4\varepsilon^+(E_W^L))$.

It is easy to check that the interval corresponding to energy $E_{lW}^L < E_W^L$ is located entirely within the interval, correspondent to energy E_W^L. Therefore, all muon energies lie within the interval determined by the highest value of W energy:

$$\varepsilon^+ \geq \varepsilon \geq \varepsilon^- \equiv \frac{M_W^2}{4\varepsilon^+}, \text{where } \varepsilon^+ \equiv \varepsilon^+\left(E_{W,on}^{L,+}\right) = \frac{E_{W,on}^{L,+} + \sqrt{\left(E_{W,on}^{L,+}\right)^2 - M_W^2}}{2}. \quad (12)$$

(Note that $E_{W,on}^{L,+} = \varepsilon^+ + M_W^2/(4\varepsilon^+)$.)

With a shift of ε from these boundaries inwards, the density of states in the ε distribution grows monotonically due to contributions of smaller values E_W^L up to values $\varepsilon_k^{\pm}$, corresponding to the lowest value of W energy $E_{W,on}^{L,-}$:

$$\varepsilon_k^- \equiv \varepsilon^-\left(E_{W,on}^{L,-}\right) = \frac{E_{W,on}^{L,-} - \sqrt{\left(E_{W,on}^{L,-}\right)^2 - M_W^2}}{2}, \quad \varepsilon_k^+ \equiv \varepsilon^+\left(E_{W,on}^{L,-}\right) = \frac{M_W^2}{4\varepsilon_k^-}. \tag{13}$$

In these points the energy distributions of muons has kinks. Between these kinks, the ε-distribution is approximately flat.

Figure 1, the left plot, shows the energy distribution of muons for the case of the matrix element independent of θ_1. Since positions of kinks are kinematically determined, it is not surprising that calculations for distinct models (containing different angular dependence) demonstrate variations in shapes but do not perturb the position of kinks.

- If $M_+ - M_D < M_W$, the $D^\pm$ decays to $DW^{*\pm}$, where $W^{*\pm}$ is an off-shell W with an effective mass $M^* \le M_+ - M_D$. The calculations for each M^* similar to shown above demonstrate that the muon energies are within the interval appearing at $M^* = 0$:

$$\left(\varepsilon^- = 0; \varepsilon^+ = E_{W,off}^{L,+}\right). \tag{14}$$

Similarly to the preceding discussion, the increase of M^* shifts the interval boundaries inwards. Therefore, the muon energy distribution increases monotonously from the outer bounds up to the maximum (peak) at $M^* = M_+ - M_D$ (cf. (11)):

$$\varepsilon_p = E\left(1+\beta_\pm\right)\left(1 - M_D/M_+\right)/2. \tag{15}$$

To get an idea about the shape of the peak, we use the distribution of W^*'s (dijets or $\ell\nu$ pairs) over the effective masses M^* which is given by the spin-dependent factor $R_{s_D} p^* dM^{*2}$:

$$R_0 = \frac{p^{*2}}{\left(M_W^2 - M^{*2}\right)^2},\quad R_{1/2} = \frac{\left(M_+^2 + M_D^2 - M^{*2}\right)\left(2M_W^2 + M_+^2 + M_D^2\right) - 4M_+^2 M_D^2}{\left(M_W^2 - M^{*2}\right)^2 M_W^2}. \qquad (16)$$

The density of muon states in energy is calculated by convolution of kinematically determined distribution with distribution (16). Neglecting the angular dependence of the matrix element, we obtain the result in form of **Figure 1**, right plot. One can see that the discussed peak is sharp enough for both values of spin $s_D = 0$ and $1/2$.

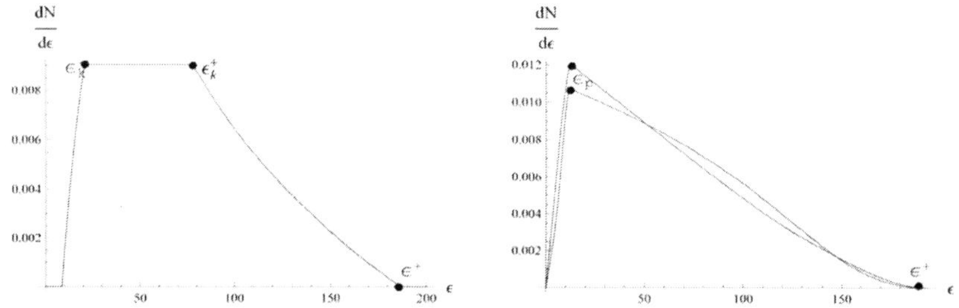

Figure 1. Distributions $dN/d\varepsilon = (1/\sigma)d\sigma/d\varepsilon$ at $E = 250$ GeV, $M_D = 50$ GeV for $M_+ = 150$ GeV—the case with $M_+ - M_D > M_W$ (the right plot) and for $M_+ = 120$ GeV—the case with $M_+ - M_D < M_W$ (the left plot). In the latter case, the higher and lower peaks are for $s_D = 0$ and $s_D = 1/2$, respectively.

Characteristic values for singular points in the energy distributions of muons (kink and peak) together with similar points for the energy distributions of W (dijets) are given in Table1

The cascade $D^- \to DW^- \to D\tau^-\nu \to D\mu^-\nu\nu\nu$ modifies the spectra just discussed. The energy distribution of τ produced in the decay $W \to \tau\nu$ is the same as that for μ or e, discussed above (within the accuracy of $\sim M_\tau/M^*$). Once produced, τ decays to $\mu\nu\nu$ in 17% of cases (the same for decay to $e\nu\nu$). These muons are added to those discussed above.

In the τ rest frame, the energy of muon is $E_\mu^\tau = yM_\tau/2$ with $y \le 1$. The energy spectrum of muons is $dN/dy = 2(3-2y)y^2$ (see textbooks). This spectrum and the distributions obtained above are converted into the energy distribution of muons in the Lab frame. It is clear that this contribution is strongly shifted towards the soft end of the entire muon energy spectrum.

The resulting distribution retains the upper boundary of the energy distribution of muons ε^{+} (12), (14). Numerical examples show that here the upper kink is smeared, while lower kink ε_{k}^{-} become even more sharp without shift from position (13) in wide region of masses M_{+} and M_{D}. The position of peak (15) is also shifted weakly.

2.4. Additional Decay Channels at $M_{+} > M_{A}$

At $M_{+} > M_{A}$ the decay $D^{\pm} \to D^{A}W^{\pm} \to DZW^{\pm}$ become possible and the processes $e^{+}e^{-} \to D^{+}D^{-} \to W^{+}W^{-}D^{A}D \to DDW^{+}W^{-}Z$, etc. with signature (7) should be taken into account.

The total probability of D^{+} decay to DW^{+} and $D^{A}W^{+}$ equals 1. The decay $D^{\pm} \to D^{A}W^{\pm}$ is described by the same equation as $D^{\pm} \to DW^{\pm}$, but with other kinematical factors since $M_{A} \neq M_{D}$. In the IDM the probability of this new decay is lower than that without D^{A} due to smaller final phase space, i.e. $B = BR\left(D^{+} \to D^{A}W^{+}\right) < 0.5$. In the MSSM value of B depends additionally on the mixing angles. We assume that in general case $B \lesssim 0.5$.

Below we limit ourself by the study of processes with signature (6b), (7a). Unfortunately, some of new processes with intermediate D^{A} look as those with signature (6) since large fraction (20%) of decays of Z is invisible ($\nu\bar{\nu}$ final states). We denote these states of Z as Z_{n}.

Let us consider in more detail production of an observed state with signature (6b), (7a) $\mu^{-} + (1-2)$ dijets $+$ nothing. This state can be obtained from two different cascades.

1) The cascade $D^{-} \to DW^{-} \to D\mu^{-}\nu$. The energy distribution of μ^{-} here reproduces $d\sigma^{\mu}\left(\varepsilon \middle| M_{+}, M_{D}\right)/d\varepsilon$, discussed in Section 2.3 with an additional factor $(1-B)$.

2) Cascade $D^{-} \to D^{A}W^{-} \to DZ_{n}\mu^{-}\nu$. Since couplings $D^{-}DW^{-}$ and $D^{-}D^{A}W^{-}$ differ by a phase factor only (and perhaps mixing angle factors), the energy distribution of μ^{-} in this case is described by the same dependence $d\sigma$ but with the change $M_{D} \to M_{A}$, the corresponding contribution to the entire energy distribution is $0.2B d\sigma^{\mu}\left(\varepsilon \middle| M_{+}, M_{A}\right)/d\varepsilon$. For brevity we will write $d\sigma^{\mu}\left(\varepsilon \middle| M_{+}, M_{D}\right) \to d\sigma_{W}^{\mu}$ and

$d\sigma^{\mu}(\varepsilon|M_{+},M_{A}) \to d\sigma^{\mu}_{WZn}$. The resulting energy distribution is

$$d\sigma^{\mu}_{tot}/d\varepsilon = (1-B)d\sigma^{\mu}_{W}/d\varepsilon + 0.2Bd\sigma^{\mu}_{WZn}/d\varepsilon. \quad (17)$$

The shape of the distribution $d\sigma^{\mu}_{WZn}/d\varepsilon$ is similar to that for $d\sigma^{\mu}_{W}/d\varepsilon$, but with different positions of kinks and (or) peak. As $M_A > M_D$, these new kinks and (or) peak are situated below similar points for $d\sigma^{\mu}_{W}/d\varepsilon$. Since this contribution is much smaller than the main contribution $d\sigma^{\mu}_{W}/d\varepsilon$ (with the overall ratio $0.2B/(1-B)$ at $B < 0.5$), it only results in a weak reshaping of the full energy distribution as compared with distributions $d\sigma^{\mu}_{W}/d\varepsilon$.

Table 1 . The singular point energies of lepton and dijet in $e^{+}e^{-} \to D^{+}D^{-} \to DDq\bar{q}\ell\nu$ (in GeV) at $M_D = 50$ GeV.

E	M_{+}	ε^{+}	ε_{1}^{-}	ε_{2}^{-}	ε_{p}	E^{L}_{Wp}	$E^{L,+}_{W}$
250	150	186.3	20.8	77.8	-	-	195.4
250	200	184.9	34.9	46.3	-	-	193.6
250	80	148.3	-	-	91.3	93.8	148.3
100	80	78	-	-	30	37.5	78

Note that in the case $M_A \approx M_D$ the distributions $d\sigma^{\mu}_{WZn}/d\varepsilon$ and $d\sigma^{\mu}_{W}/d\varepsilon$ are close to each other, and $d\sigma^{\mu}_{tot}/d\varepsilon \propto d\sigma^{\mu}_{W}/d\varepsilon$. In the opposite degenerate case $M_{+} \approx M_{A}$, the quantity $B\square\ 1$ and the influence of the intermediate D^{A} state on the result is negligible. (Such very cases are widely discussed in context of MSSM).

3. THE OVERALL PICTURE

Observation of events with signature (6), (7) will be a clear signal for DM particle candidates. The nonobservation of such events will allow to find lower limits for masses M_{+}, like [11] -[13] . One can hope that these limits will be close to the beam energy E.

At $M_{+} < E$, the cross section $e^{+}e^{-} \to D^{+}D^{-}$ is a large fraction of the total cross section of $e^{+}e^{-}$ annihilation, and it makes this observation a very realistic task.

3.1. Distortion of the Obtained Results

A more detailed analysis reveals two sources of distortion of the obtained results (we neglect them in our preliminary analysis).

1) The final width of W and $D^{\pm}$ (Z and D^{A}) leads to a blurring singularities derived. This effect increases with the growth of $M_{+} - M_{D}$.

2) The energy spectra under discussion will be smoothed due to QED initial state radiation (ISR), final state radiation (FSR) and beamsstrahlung (BS). The ISR and FSR spectra are machine independent, while BS spectrum is specific for each machine (but well known during operations). This smoothing decreases accuracy in measuring of masses. However, the precise knowledge of mentioned spectra allows to solve the problem about restoration original accuracy by means methods of deconvolution in so called "incorrect inverse problem". This work and the estimates of the range where masses and spins can be determined with reasonable accuracy will be the subject of the forthcoming paper.

3.2. Masses

Masses M_{D} and M_{+}. In a well known approach, one measures edges in the energy distributions of dijets, representing W in the decay $D^{\pm} \to DW^{\pm}$ [5] [6] . However, the individual jet energies and consequently, effective masses of dijets cannot be measured with a high precision. The observed lower edge of the W energy distribution in the dijet mode and the position of a peak in this distribution (11) are smeared by this uncertainty. One can only hope for a sufficiently accurate measurement of the upper edge of the W energy distribution, $E_{W}^{L,+}$ (9), (10).

We suggest to extract the second quantity for description of masses from the lepton energy spectra. The lepton energy is measurable with a high accuracy. We found above that the singular points of the energy distribution of the leptons in the final state with signature (6a) are kinematically determined, and therefore can be used for a mass measurement.

M1) If a D^A particle is absent or $M_A > M_+$, the results (12)-(15) describe the energy distributions completely. The shape of the energy distribution of leptons (with one peak or two kinks) allows to determine which case is realized, $M_+ - M_D > M_W$ or $M_+ - M_D < M_W$.

At $M_+ - M_D > M_W$, the positions of upper edge in the dijet energy distribution $E^{L,+}_{W,on}$ (9) and the lower kink in the muon energy distribution ε_k^+ (13) give us two equation necessary for determination of M_D and M_+. We reproduce these equations for clarity

$$E^{L,\pm}_{W,on} = \frac{E}{M_+}\left(E_W^r \pm \beta_+ p_W^r\right),\ \varepsilon_k^- = \frac{E^{L,-}_{W,on} - \sqrt{\left(E^{L,-}_{W,on}\right)^2 - M_W^2}}{2}, \text{where}$$

$$E_W^r = \frac{M_+^2 + M_W^2 - M_D^2}{2M_+},\ p_W^r = \frac{\Delta\left(M_+^2, M_W^2, M_D^2\right)}{2M_+},\ \beta_+ = \sqrt{1 - \frac{M_+^2}{E^2}}. \qquad (18)$$

At $M_+ - M_D < M_W$, two similar equations are provided by the position of the upper edge in the dijet energy distribution $E^{L,+}_{W,off}$ (10) and the peak in muon energy distribution ε_p (15).

In both cases the position of the upper edge in the dijet energy distribution $E^{L,+}_{W,on}$ or $E^{L,+}_{W,off}$ should be extracted from all events with signature (6), (7), the position of the lower kink in the muon energy distribution ε_k^+ or peak ε_p can be extracted from events with signature (6b) only.

M2) The signal of realization of the inequality $M_A < M_+$ will be observation of the process $e^+e^- \to DD^A$, having signature (26). In this case the position of the upper edge in the dijet energy distribution is the same as in previous case. The position of lower edge in the dijet energy distribution is either shifted or smeared, in this case the method of [5] [6] becomes completely inapplicable. The entire energy distribution of muons in the observed state $\mu + 1$ or 2 dijet $+$ nothing was described in the Section 2.4. It was shown there that taking into account a new decay channel $D^- \to D^A W^- \to DZ_n \mu^- \nu$ changes the position of the main singularities in the muon energy spectrum very weakly. Therefore the above mentioned procedure for finding M_{+} and M_{D} can be used in this case as well.

The opportunity to extract new singularities from the data, related to $d\sigma^{\mu}_{WZn}/d\varepsilon$ (and giving additionally M_{A}), requires a separate study (see also analysis in Appendix B).

3.3. Spin of D -Particles s_{D}

The amplitude of the process $e^{+}e^{-} \to D^{+}D^{-}$ is the sum of model-independent QED diagram (the photon annihilation), the Z annihilation diagram and in some models t -channel exchange by other D-odd particles. We start with the description of cross section in the minimal approximation, taking into account only photon and Z annihilation diagrams. Neglecting terms $\propto \left(1/4-\sin^{2}\theta_{W}\right)$ (described $\gamma - Z$ interference) we have:

$$\sigma_{\min}(s_{D}) = \sigma_{0}\begin{cases} \beta_{\pm}\left[1+\dfrac{2M_{+}^{2}}{s}+r_{Z}\beta_{\pm}^{2}\right] & \left(s_{D}=\dfrac{1}{2}\right), \\ \beta_{\pm}^{3}\left[\dfrac{1}{4}+r_{Z}\cos^{2}(2\theta_{W})\right] & (s_{D}=0), \end{cases} \tag{19}$$

where $r_{Z} = \dfrac{\mu_{M}}{\left(2\sin(2\theta_{W})\right)^{4}\left(1-M_{Z}^{2}/s\right)^{2}} = \dfrac{0.124\mu_{M}}{\left(1-M_{Z}^{2}/s\right)^{2}}$, factor $\mu_{M} \le 1$ is expressed via parameters of possible mixing, etc. Figure 2 and Table 2 represent dependence of $\sigma_{\min} e^{+}e^{-} \to D^{+}D^{-}/\sigma_{0}$ (19) on beam energy $E=\sqrt{s}/2$ for $\mu_{M}=1$.

The cross section of the process is reduced by contribution of the diagram with t -channel exchange by other D-odd particle D_{F}. This decrease is not so strong if mass of D_{F} is high enough. For example, if mass of selectron is more than 250 GeV (condition 2 in Section 1 and [20]), the cross section for $s_{D}=1/2$ is reduced by a factor ≥ 0.6, $\sigma(s_{D}=1/2) \ge 0.6\sigma_{\min}(s_{D}=1/2)$. Combining with numbers from Figure 2 and Table 2 we obtain (for identical masses M_{+} at a given beam energy E):

$$\sigma(s_{D}=1/2) \ge 0.6\sigma_{\min}(s_{D}=0) \tag{20}$$

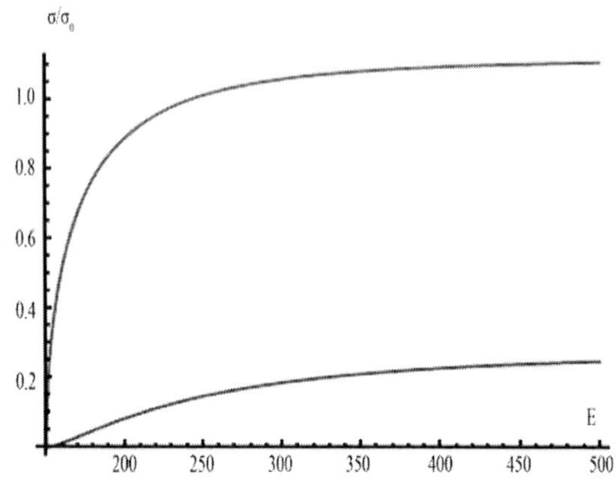

Figure 2. The upper curve for $s_D = 1/2$, the lower for $s_D = 0$; $M_+ = 150$ GeV.

Table 2. Some values of $\sigma_{min}\left(e^+e^- \to D^+D^-\right)/\sigma_0$.

E, GeV	100	250	250	250
M_+, GeV	80	80	150	200
$s_D = 0$: σ/σ_0	0.066	0.245	0.162	0.062
$s_D = 1/2$: σ/σ_0	0.84	1.107	1.02	0.82

The experimental value of the $e^+e^- \to D^+D^-$ cross section is obtained by summing over all processes with signature (6), (7) (that is about 3/4 of the total cross section). By taking into account the known BR's for W decay the accuracy of this restoration of $\sigma\left(e^+e^- \to D^+D^-\right)$ can be improved.

When masses M_+ become known, the cross section $\sigma_{min}\left(e^+e^- \to D^+D^-\right)$ is calculated with reasonable precision with Equation (19). The strong inequality (20) allows to determine spin s_D from the obtained values of cross sections even with a handful of well-reconstructed events.

4. CONCLUSIONS

We consider models in which stability of dark matter particles D is ensured by conservation of new quantum number referred to as D-parity. Besides these models contain charged particles $D^\pm$ with the same D-parity. (Examples Inert Doublet Model with scalar D-particles and MSSM with D-

particle of spin 1/2 and D- parity equal R-parity). In these models we have studied the energy distribution of single lepton in the process like $e^+e^- \to D^+D^- \to DDW^\pm(\to q\bar{q})W^\mp(\to \ell\nu)$, having high enough cross section. Simple analysis allows us to establish that this distribution has singular points, kinks, peaks and end points, which are driven by kinematics only, and therefore are model-independent. Based on this analysis, we propose to use the mentioned distribution at future linear e^+e^- collider ILC, CLIC, etc. for precise measuring of masses of dark matter particles and charged particles $D^\pm$.

This method is in several aspects superior to the standard approaches discussed elsewhere.

1) It uses leptons which are copious and can be accurately measured in contrast with jets that individual energy can be measured only with lower precision.

2) These singularities are robust and survive even when superimposed on top of any smooth background.

In addition, even a rough measurement of cross sections with a very clean signature allows us to determine spin of DM particles based on the results of mentioned kinematical measurements.

ACKNOWLEDGMENTS

I. F. Ginzburg, This work was supported in part by grants RFBR and NSh-3802.2012.2, Program of Dept. of Phys. Sc. RAS and SB RAS "Studies of Higgs boson and exotic particles at LHC" and Polish Ministry of Science and Higher Education Grant N202 230337. I am thankful to A. Bondar, E. Boos, A. Gladyshev, A. Grozin, S. Eidelman, I. Ivanov, D. Ivanov, D. Kazakov, J. Kalinowski, K. Kanishev, P. Krachkov and V. Serbo for discussions.

REFERENCES

1. Hooper, D. (2009) TASI 2008 Lectures on Dark Matter. arXiv:0901.4090 [hep-ph]
2. Maniatis, M. (2009) The Next-to-Minimal Supersymmetric extension of the Standard Mod. arXiv:0906.0777 [hep-ph]
3. Kazakov, D.I. (2010) Supersymmetry on the Run: LHC and Dark Matter. arXiv:1010.5419 [hep-ph]
4. Ellis, J. (2010) New Light on Dark Matter from the LHC. arXiv:1011.0077 [hep-ph]
5. Li, Y. and Domerotski, A. (2010) Chargino and Neutralino Masses at ILC. arXiv: 1007.0698

6. Asano, M., Fujii, K., Hundi, R.S., Itoh, H., Matsumoto, S., Okada, N., Saito, T., Suehara, T., Takubo, Y. and Yamamoto, H. (2011) Identification of new physics and general WIMP search at the ILC. arXiv:1007.2636 [hep-ph]
7. Deshpande, N.G. and Ma, E. (1978) Physical Review D, 18, 2574-2576.
8. Barbieri, R., Hall, L.J. and Rychkov, V.S. (2006) Improved Naturalness with a Heavy Higgs: An Alternative Road to LHC Physics. arXiv:hep-ph/0603188.
9. Ginzburg, I.F., Kanishev, K.A., Krawczyk, M. and Sokolowska, D. (2010) Evolution of Universe to the present inert phase. arXiv:1009.4593 [hep-ph]
10. Gustafsson, M., Rydbeck, S., Lopez-Honorez, L. and Lundstrom, E. (2012) Physical Review D, 86, Article ID: 075019.
11. Espirito Santo, M., Hultqvist, K., Johansson, P. and Lipniacka, A. (2003) DELPHI 2003-002 PHYS 928. http://delphiwww.cern.ch/pubxx/delnote/dn2003.html
12. Lundstrom, E., Gustafsson, M. and Edsjo, J. (2009) Inert Doublet Model and LEP II Limits. arXiv:0810.3924 [hep-ph]
13. Aoki, M., Kanemura, S. and Yokoya, H. (2013) Reconstruction of Inert Doublet Scalars at the International Linear Collider. arXiv:1303.6191 [hep-ph]
14. Dolle, E.M. and Su, S. (2009) Physical Review D, 80, Article ID: 055012.
15. Dolle E., Miao X., Su S., Thomas B. (2010) Physical Review D, 81, Article ID: 035003.
16. Lopez Honorez, L., Nezri, E., Oliver, J.F. and Tytgat, M.H.G. (2007) The Inert Doublet Model: An Archetype for Dark Matter. arXiv:0612275 [hep-ph]
17. Nezri, E., Tytgat, M.Y.G. and Vertongen, G. (2009) Positrons and antiprotons from inert doublet model dark matter. arXiv:0901.2556 [hep-ph]
18. Andreas, S., Hambye, T. and Tytgat, M.H.G. (2008) WIMP Dark Matter, Higgs Exchange and DAMA. arXiv: 0808.0255 [hep-ph]
19. Honorez, L.L. and Yaguna, C.E. (2010) The Inert Doublet Model of Dark Matter Revisited. arXiv:1003.3125 [hep-ph]
20. Particle Data Group (2012) Physical Review D, 86, Article ID: 010001.
21. Heuer, R.D., et al. (2001) TESLA Technical Design Report, DESY 2001-011, TESLA Report 2001-23, TESLA FEL 2001-05.
22. International Linear Collider TDR (2013) The International Linear Collider Technical Design Report—Volume 2: Physics. arXiv:1306.6352 [hep-ph]
23. Moortgat-Pick, G. (2008) Journal of Physics: Conference Series, 110, Article ID: 072027.
24. Freitas, A., Martyn, H.U., Nauenberg, U. and Zerwas, P.M. (2004) Sleptons: Masses, Mixings, Couplings. arXiv: hepph/0409129
25. Conley, J.A., Dreiner, H.K. and Wienemann, P. (2012) Measuring a Light Neutralino Mass at the ILC: Testing the MSSM Neutralino Cold Dark Matter Model. arXiv:1012.1035 [hep-ph]
26. Ginzburg, I.F (2010) Simple and robust method for search Dark Matter particles and measuring their properties at ILC in various models of DM. arXiv:1010.5579 [hep-ph]

CHAPTER 17

The Relativistic Geometry of the Complex Matter Space

Reza Ahangar

Department of Mathematics, Texas A&M University, Kingsville, TX, USA

ABSTRACT

Duality behavior of photons in wave-particle property has posed challenges and opportunities to discover other frontiers of fundamental particles leading to the relativistic and quantum description of matter. The speed of particles faster than the speed of light could not be recognized, and matter was always described as a real number. A new fundamental view on matter as a complex value has been introduced by many authors who present a paradigm that is shifted from real or pure imaginary particles to Complex Matter Space. A new assumption will be imposed that matter has two intrinsic components: i) mass, and ii) charge. The mass will be measured by real number systems and charged by an imaginary unit. The relativistic concept of Complex Matter Space on energy and momentum is investigated and we can conclude that the new Complex Matter Space (CMS) theory will help get one step closer to a better understanding toward: 1) Un-Euclidean description of Minkowski Geometry in the context of the Complex Matter Space, 2) transformation from Euclidean to Minkowski space and its relativistic interpretation. Finally, geometrical foundations are essential to have a real picture of space, matter, and the universe.

Keywords: Relativistic Geometry, Mass and Charge, Complex Matter Space, Un-Euclidean, Minkowski's Geometry

1. INTRODUCTION

Several thousand years of human experiences in geometry, which was structured by Pythagoras and Euclid, has undergone challenges during the past few centuries. Among all were the attempts to prove Euclid's fifth postulate independently from other postulates and distinguish all undefined terms like point, line, plane, from terms that can be redefined like distances and parallel lines.

In seeking the truth behind the parallel postulate in 19th century, some alternatives and geometric models to Euclid's parallel postulate were presented.

In 1898, one of the great mathematicians of all time, David Hilbert presented a rigorous axiomatic approach to Euclidean Geometry (see Hilbert, 1902 [1]). Many others like Riemann, Bolyai, and Lobachevski created Un-Euclidean Geometry.

Thus, Hyperbolic Geometry was born and many geometrical models were developed (see Milnor, 1996 [2]).

Many were skeptical of the validity of this bizarre geometry which was axiomatically consistent and agreed with the models developed. But all these theories were not experimented physically to be consistent with the natural universe.

Certainly, in a small velocity for a shorter distance, Euclidean Geometry is a simplest model for our physical universe. But it is not the proper model for the entire universe particularly when the velocity of the object is near or approaches to the speed of light.

Einstein's General Relativity predicted that gravity alters the velocity of light and this causes the light ray to bend. Or, it can be interpreted that light travels in a geodesic line. This is a revolutionary conclusion that the physical space with the existing mass and gravity is not Euclidean (see Taylor et al., 2004 [3] , and Feynman, 1963 [4]).

NASA Experiment, May 4, 2011: Einstein was right again. There is a space-time vortex around Earth, and its shape precisely matches the predictions of Einstein's theory of gravity. Researchers confirmed these points at a press conference today at NASA headquarters where they announced the long-awaited results of Gravity Probe B (GP-B). "The space-time around Earth appears to be distorted just as general relativity predicts", says Stanford University physicist Francis Everitt, principal investigator of the Gravity Probe B mission (NASA, 2011 [5]).

A shortest path in traveling between two points can be defined by an arc-length or geodesic line. It may not be considered as straight lines because acceleration creates distortion. Due to high speed of the particle or a constant curvature there may two possibilities. The yard stick or rulers can get longer or shrink as they move away from the stationary point (Lavenda, 2009 [6]).

A massive object in space curves the space and bends the light, when a photon moves past a star; it's the star's gravity that causes it to deflect, regardless of whether we consider photon a particle or wave.

All evidence indicating the existence of dark matter and dark energy suggests that we may need to redefine matter. As a result we present the following two postulates:

Complex Matter Space (CMS) (see Ahangar, 2014 [7] [8] [9])

- Postulate (I): mass and charge are two intrinsic components of matter and they can be measured and determined in a complex plane. The real part is equal to the mass and the imaginary part is the charge of the particle. That is $M = m + iq$.
- Postulate (II): In a one dimensional space, position and velocity are assumed to be real numbers and the rate of change of the position of a particle can be assumed to be faster than the speed of light.

According to de Broglie's duality of particles used to describe matter—wave characteristic of photons (see Mayer, 2007 [10] and Gauthier, 2009 [11]). It is well known that the complex plane can help us to explain the duality of waves and explain a wave with two components in the complex plane with real and imaginary components:

$$\psi(\boldsymbol{r},t) = \psi_{\text{real}} + \mathrm{i}\,\psi_{\text{imag}} \tag{1}$$

We will refer the value $|\psi|^2$ as the intensity of the wave function where the magnitude of the complex number is denoted by $|\psi| = \sqrt{|\psi_{\text{real}}|^2 + |\psi_{\text{imag}}|^2}$ [3] .

Traveling Beyond the Speed of Light:

According to the "Special Theory of Relativity", it is not allowed to cross the boundary of the speed of light c, in the real universe. But if we can pretend from this assumption that we exist in the imaginary world then it may be allowed to go faster than light, since our conjugate space will be the real world.

In section two we present an introduction to the geometric structure of Complex Matter Space. We will show its relation with Minkowski Geometry and pseudo-Euclidean space.

In section three, we use a bilinear transformation in hyperbolic geometry. In section four, we investigate the hyperbolic rotation which is equivalent to Lorentz transformation. In section five, we will show that the relativistic view of geometry is consistent with hyperbolic geometry. As a conclusion, the law of cosines is redeveloped in a hyperbolic triangle.

2. GEOMETRICAL STRUCTURE OF THE CMS

Assume V is a vector space and T represents a bilinear transformation from $V \times V$ into the R. Let us select vectors $u = \langle u^0, u^1, u^2, \cdots u^n \rangle$ and $v = \langle v^0, v^1, v^2, \cdots, v^n \rangle$ in the vector space V, where all of the first $(n+1)$ components are real $u \text{ and } v \in V$ define the transformation T from $V \times V \to R$ by

$$T(u,v) = \sum_{k=0}^{k=n} u^k v^k = u^0 v^0 + u^1 v^1 + u^2 v^2 + \cdots + u^n v^n \tag{2}$$

In the following theorem, we will show that the transformation T defined in (2) is a bilinear transformation.

Theorem (1): The operator $T: V \times V \to R$ defined by (2) is a bilinear operator meaning that for any real numbers a and b

$$T(au_1 + bu_2, v) = aT(u_1, v) + bT(u_2, v) \tag{3}$$

Proof: Notice that the superscripts are not exponents, thus

$$\begin{aligned} T(au_1 + bu_2, v) &= \sum_{k=0}^{k=n} \left[a(u_1^k) + b(u_2^k) \right] v^k \\ &= a\sum_{k=0}^{k=n} (u_1^k) v^k + b\sum_{k=0}^{k=n} (u_2^k) v^k \\ &= aT(u_1, v) + bT(u_2, v) \end{aligned}$$

which proves that T is a bilinear operator that can be used to define the Mikowski norm.

Also notice that the operator T mapping a vector u to itself will produce a value in R:

$$T(u,u)=\sum_{k=1}^{k=n}\left(u^k\right)^2=\left(u^0\right)^2+\left(u^1\right)^2+\left(u^2\right)^2+\cdots+\left(u^n\right)^2 \tag{4}$$

If the vector space V is selected from real valued vectors then

$$\forall k\in N, u^k\in R, \left(u^k\right)^2\geq 0 \tag{5}$$

implies that the relation (4), is a positive definite inner product, that is for nonzero vector u, the transformation $T(u,u)>0$.

But when the vectors are selected from the complex field then the relation (4) can be one of the following cases:

1) $T(u,u)>0$, the transformation T is positive definite, when the vector u is in the region which is called space-like.
2) $T(u,u)<0$ the transformation T is negative definite, when u is in the time-like, otherwise
3) $T(u,u)=0$ and the transformation is called indefinite, when the vector u belongs in the light-like.

Many questions exist to justify the operator T for objects in complex space as well as explanation for the properties of the transformation T. The most important one is when T is positive or negative definite and the operator T from $R^{n+1}\times R^{n+1} --> R$ satisfies the properties of a metric space (see Naber, 1992 [12]).

To answer this question we need the definition of Metric Space first.

Metric Space: A set X, whose elements we shall call points, is said to be a Metric Space if with any two points P and Q of X there is associated a real number d(P,Q), called distance from P to Q such that

a) $\mathrm{d}(P,Q)>0$ if $P\neq Q, \mathrm{d}(P,P)=0$,

b) $\mathrm{d}(P,Q)=\mathrm{d}(Q,P)$, and c) $\mathrm{d}(P,Q)\leq \mathrm{d}(P,R)+\mathrm{d}(R,Q)$ for any R in X

One can verify that the transformation T defined in (2), (3), and (4) fails the postulate (a) of metric space, due to the case where $T(u,v)$ can be negative definite.

Thus as a conclusion T is not a Metric Space.

II) Special Complex Geometrical Structure: The standard Euclidean inner product of two vectors in (2) and (3) can be defined by the following matrix form

$$\boldsymbol{u} \cdot \boldsymbol{v} = \boldsymbol{u}^t I \boldsymbol{v} \tag{6}$$

where, I is the identity matrix, $\boldsymbol{u}^t = \left\langle u^0, u^1, u^2, \cdots u^n \right\rangle, \boldsymbol{v}^t = \left\langle v^0, v^1, v^2, \cdots, v^n \right\rangle$, and superscript t represents the transpose of these vectors.

To develop Minkowski space in (2) for k = 0, consider special coordinate for

$u^0 = v^0 = \text{i}ct$,

where the symbol i is the imaginary number $\text{i} = \sqrt{-1}$. The value c is assumed to be a constant real number representing the speed of light.

Thus the operator $T: V \times V \to R$ defined by the following

$$T(u,v) = \sum_{k=1}^{k=n} u^k v^k + (\text{i}ct)(\text{i}ct) = \sum_{k=1}^{k=n} u^k v^k - \left(c^2 t^2\right) \tag{7}$$

is also a bilinear operator. The relation (7) for n = 3 is a four dimensional Minkowski space for events u and v and the transformation can be described by

$$T(u,v) = \sum_{k=1}^{k=3} u^k v^k + (\text{i}ct)(\text{i}ct) = u^1 v^1 + u^2 v^2 + u^3 v^3 - c^2 t^2 \tag{8}$$

3. PROPERTIES OF THE BILINEAR OPERATOR

The transformation $T: V \times V \to R$ generates Minkowski inner product space and

$$\|\boldsymbol{u}\|^2 = T(u,u) = \sum_{k=1}^{k=n} \left(u^k\right)^2 - (ct)^2 \tag{9}$$

Proof: The transformation is indefinite if it is neither positive nor negative definite.

Notice that the relations (2), (3), and (4) will demonstrate the four dimensional Minkowski space.

We can use the relation (8) to establish Minkowski Norm as follows:

$$\|u\|^2 = T(u,u) = \sum_{k=1}^{k=3}\left(u^k\right)^2 - (ct)^2 = \left(u^1\right)^2 + \left(u^2\right)^2 + \left(u^3\right)^2 - (ct)^2 \tag{10}$$

where the Euclidean norm in 3D real event: $r^2 = \left(u^1\right)^2 + \left(u^2\right)^2 + \left(u^2\right)^2$ can change (10) into the following form of

$$T(u,u) = r^2 - (ct)^2 = \left\langle ict, u^1, u^2, u^3 \right\rangle I_{4\times4} \left\langle ict, u^1, u^2, u^3 \right\rangle \tag{11}$$

where I represents the identity matrix, as the identity matrix used in (6)

$$I_{4\times4} = \begin{bmatrix} 1 & 0 & 0 & 0 \\ 0 & 1 & 0 & 0 \\ 0 & 0 & 1 & 0 \\ 0 & 0 & 0 & 1 \end{bmatrix}.$$

The Minkowski inner product by the transformation $T(u,u)$ in (11) applied for event vector u may be positive definite, negative definite, or indefinite.

Orthogonal Subspace:

When the transformation $T(u,v) = 0$, then two vectors u and v are said to be orthogonal.

Assume a subspace $S \subset V$, such that

$$S^{\perp} = \left\{ v\varepsilon V : T(v,w) = 0 \text{ for all } w\varepsilon S \right\} \tag{12}$$

It can be concluded that the set $S^{\perp} \subset S$ as a subset of S is also a subspace of S?

In the next step we will explore the consistency of a two dimensional analogy of Minkowski-Hyperbolic system with Complex Matter space.

Proposition (1): Let $M(x,y)$ be a point on a hyperbola

$$x^2 - y^2 = 1 \tag{13}$$

For positive values of x there exists a unique value of u such that

$$x = \cosh(u) \text{ and } y = \sinh(u) \tag{14}$$

Proof: Let us take the transformation, $u = \sinh^{-1}(y)$. The result can be demonstrated by the geometric property of the hyperbolic identity

$$x^2 - y^2 = \cosh^2 u - \sinh^2(u) = 1 \tag{15}$$

Notice that the relation (13) for (x, y) is a two dimensional example in a complex plane when the transformation y = ct is described by

$$x^2 + (ict)^2 = 1 \Leftrightarrow \quad x^2 - (c^2t^2) = 1.$$

Pseudo-Euclidean Space: In our introduction it has been verified that the CMS transformation is not a metric space. We will show that this is a special pseudo Euclidean Space.

We define a general form of pseudo-Euclidean space as a finite dimensional n-space together with a quadratic form q. The quadratic form defined for any vector with a suitable choice of basis $(e_1, e_2, \cdots, e_n)$ can be applied to a vector u giving (with $1 \le k < n$)

$$u = \sum_{k=1}^{n} u^k e_k = u^1 e_1 + u^2 e_2 + \cdots + u^n e_n \tag{16}$$

for a quadratic form

$$q(u) = (u^1)^2 + (u^2)^2 + \cdots + (u^k)^2 - \left[(u^{k+1})^2 + \cdots + (u^n)^2\right] \tag{17}$$

is called the magnitude of the vector u. When $k = 1 \cdots n$, then it is Euclidean Space. Notice that the superscripts k, for convenience, selected to be from 1 to n. This is slightly different from the transformation (2), (3), and (4). We will have the general Minkowski Space when k = 0.

In a pseudo-Euclidean space, unlike in a Euclidean space, there exist vectors with negative magnitude.

$$q(u) = \|u\|^2 = (u^1)^2 + (u^2)^2 + \cdots + (u^n)^2 - (u^0)^2 \tag{18}$$

for all positive integer k where, $0<k\leq n.$

A very important pseudo-Euclidean space is Minkowski's Space which is the mathematical setting in which Albert Einstein's theory of special relativity is conveniently formulated. The following is another example of pseudo-Euclidean space using Minkowski space, n = 4 and k = 3 so that

$$q(u)=u_1^2+u_2^2+u_3^2+u_4^2 \tag{19}$$

where assuming $\|u\|=\sqrt{q(u)}$ and $u_4=\text{i}ct$ the above relation will be

$$\|u\|=\sqrt{u_1^2+u_2^2+u_3^2+(\text{i}ct)^2}=\sqrt{u_1^2+u_2^2+u_3^2-(ct)^2} \tag{20}$$

The geometry associated with this pseudo-metric was investigated by Poincare and the rotation group is called Lorentz group.

This magnitude is used in Lorentzian manifold which represents the flat space-time in R^4 uses the coordinates (x, y, z, t) and the metric

$$ds^2=\begin{bmatrix}\text{i}cdt & dx & dx & dy\end{bmatrix}\begin{bmatrix}1&0&0&0\\0&1&0&0\\0&0&1&0\\0&0&0&1\end{bmatrix}\begin{bmatrix}\text{i}cdt\\dx\\dy\\dz\end{bmatrix}$$

$$ds^2=dx^{\mu}1_{4\times4}dx_{\nu}$$

$$ds^2=(\text{i}cdt)^2+dx^2+dy^2+dz^2$$

4. FROM EUCLIDEAN ROTATION TO LORENTZ TRANSFORMATION

When we replace x^2 symbolically for the sum of the squares of all real components and $u^0=ct$ in the relation (18) we will obtain the quadratic form

$$q(u) = x^2 - (ct)^2$$

Definition of Lorentz transformation Matrix: We also use the relations (14) to define the following transformation:

$$T(u) = \begin{bmatrix} \cosh(u) & \sinh(u) \\ \sinh(u) & \cosh(u) \end{bmatrix} \tag{21}$$

where $x = \cosh(u)$ and $y = \sinh(u)$ imply that b $\tanh(u) = \frac{y}{x}$.

The transformation T in (21) acting on an event vector $\begin{bmatrix} \zeta \\ \iota \end{bmatrix}$ on a hyperbola $\zeta^2 - \iota^2 = k$ such that

$$\begin{bmatrix} t \\ z \end{bmatrix} = \begin{bmatrix} \cosh(u) & \sinh(u) \\ \sinh(u) & \cosh(u) \end{bmatrix} \begin{bmatrix} \tau \\ \zeta \end{bmatrix}$$

will be converted into a new hyperbola $z^2 - t^2 = k$ and preserves Minkowski's norm.

The concept of space and time has been evolved and became paradoxical from the time of Descarts to Minkowskiin twentieth century.

We are introducing the space-time coordinate (x, y, z, t) such that it will transform into a new position point (x', y', z', t').

The units on all components x, y, z can be considered real numbers but we assume that the unit for time as an imaginary unit "ic" such that $i = \sqrt{-1}$ and c the speed of light c. Thus the component for t can be demonstrated symbolically by

$$\tau = ict \tag{22}$$

To keep the concepts simple, assume y and z stay fixed and the coordinate $(x \& \tau)$ under a rotation angle theta $= \theta$, then the new coordinate will be in the following form:

$$x = x'\cos(\theta) - \tau'\sin(\theta)$$
$$\tau = \tau'\cos(\theta) + x'\sin(\theta)$$

or in the matrix form

$$\begin{bmatrix} x \\ \tau \end{bmatrix} = \begin{bmatrix} \cos(\theta) & -\sin(\theta) \\ \sin(\theta) & \cos(\theta) \end{bmatrix} \begin{bmatrix} x' \\ \tau' \end{bmatrix} \tag{23}$$

Let us call the initial reference frame $S = \{(\tau, x) \in C \times R\}$ and $S' = \{(\tau', x') \in C \times R\}$ after rotation with angle theta. In this reference angle, the evolution on time axis is very interesting. The tau axis in the S will be evolved into $(tau)'$ in the S' system. Any point on the tau axis will be transformed into a new point $(tau)'$ in the equation when the angle theta is 90°.

In other words let us replace $x' = 0$ in the system (23) then

$$\begin{aligned} x &= -\tau' \sin(\theta) \\ \tau &= \tau' \cos(\theta) \end{aligned} \tag{24}$$

This is showing the transform of all of the points on the time axis. By dividing these two relations it will give us

$$\tan(\theta) = -\frac{x}{\tau} \tag{25}$$

Replace tau from the relation (4.1) we will obtain

$$\tan(\theta) = -\frac{x}{ict} = \frac{iv}{c} \tag{26}$$

According to the relativity postulates, the time evolution is a function of the velocity. Thus we can find sine and cosine values of the angle theta using (26):

$$\cos(\theta) = \frac{1}{\sqrt{1 + \tan^2(\theta)}} = \frac{1}{\sqrt{1 + \left(\frac{iv}{c}\right)^2}} = \frac{1}{\sqrt{1 - \left(\frac{v}{c}\right)^2}}$$

$$\sin(\theta) = \cos(\theta)\tan(\theta) = \frac{\frac{iv}{c}}{\sqrt{1-\left(\frac{v}{c}\right)^2}}$$

Substitute in the system (23)

$$\begin{bmatrix} x \\ \tau \end{bmatrix} = \gamma \begin{bmatrix} 1 & -\frac{iv}{c} \\ \frac{iv}{c} & 1 \end{bmatrix} \begin{bmatrix} x' \\ \tau' \end{bmatrix} \tag{27}$$

where gamma is equal to: $\gamma = \frac{1}{\sqrt{1-\left(\frac{v}{c}\right)^2}}$. The relation (27) is the Lorentz transformation. In addition to these sine and cosine functional transformations, one can develop the additively property of rotations and velocity:

$$\tan(\theta+\phi) = \frac{\tan(\theta)+\tan(\phi)}{1-\tan(\theta)\cdot\tan(\phi)} = i\frac{v+w}{1-v\cdot w}$$

Minkowski investigated the Lorentz transformation as an Euclidean rotation. This can be achieved by finding (x',τ') from the system (23) and replacing (22) for tau.

$$\begin{bmatrix} x' \\ ict' \end{bmatrix} = \begin{bmatrix} \cos(\theta) & ict\cdot\sin(\theta) \\ -\sin(\theta) & ict\cdot\cos(\theta) \end{bmatrix} \begin{bmatrix} x \\ ict \end{bmatrix} \tag{28}$$

We also replace the angle theta by pure imaginary angle (iw) and using the following identities

$$\cos(\theta) = \cos(iw) = \cosh(w) \text{ and } \sin(\theta) = \sin(iw) = i\sinh(w) \tag{29}$$

Using these identities the hyperbolic version of the relation (28) will be

$$\begin{cases} ct' = ct\cdot\cosh(w) - x\sinh(w) \\ x' = -ct\cdot\sinh(w) + x\cosh(w) \end{cases}$$

and the inverse

$$\begin{cases} ct = ct' \cdot \cosh(w) + x' \cdot \sinh(w) \\ x = ct' \cdot \sinh(w) + x' \cdot \cosh(w) \end{cases} \tag{30}$$

These relations can be described by the following matrix form:

$$\begin{bmatrix} x' \\ ct' \end{bmatrix} = \begin{bmatrix} \cosh(w) & -\sinh(w) \\ -\sinh(w) & \cosh(w) \end{bmatrix} \begin{bmatrix} x \\ ct \end{bmatrix}$$

and

$$\begin{bmatrix} x \\ ct \end{bmatrix} = \begin{bmatrix} \cosh(w) & \sinh(w) \\ \sinh(w) & \cosh(w) \end{bmatrix} \begin{bmatrix} x' \\ ct' \end{bmatrix}$$

The hyperbolic systems in (30) explains the CMS with hyperbolic angle of rotation w and rapidity

$$\tanh(w) = \frac{v}{c} \tag{31}$$

Comparing the two tangent relationships of (26) and (31) we can discover many interesting connections between two spaces i) Minkowski Space and ii) CMS.

One is in the real world and another in the imaginary space.

In the first we are dealing with Euclidean trigonometry and in the second with the hyperbolic trigonometry.

It can be concluded that

$$\cosh(w) = \frac{1}{\sqrt{1-\left(\tanh(w)^2\right)}} = \frac{1}{\sqrt{1-\left(\frac{v}{c}\right)^2}} = \gamma$$

$$\sinh(w) = \tanh(w) \cdot \cosh(w) = \frac{\frac{v}{c}}{\sqrt{1-\left(\frac{v}{c}\right)^2}}$$

Geometry of Complex Matter in Complex Plane:

Assume a complex point z represents a particle $M = m + iq$ such that

$$|z| = m = \mathrm{Re}(M) \quad \text{and} \quad \arg(z) = \frac{q}{m} = \frac{\mathrm{Im}(z)}{|z|} = \arg(z) = \frac{\mathrm{Im}(z)}{\mathrm{Re}(z)} \tag{32}$$

In this case we need to consider two different cases in which that can happen:

When $|x| > |y|$ and $|x| < |y|$

Case (I): If $|x| > |y|$ then

$$r = |z| = \pm\sqrt{x^2 - y^2}$$

As a result:

$$\left(\frac{x}{r}\right)^2 - \left(\frac{y}{r}\right)^2 = \frac{x^2 - y^2}{x^2 - y^2} = 1$$

This statement describes the existence of an angle in Minkowskian Geometry such that

$$\cosh(\varnothing) = \frac{x}{r} = \frac{\mathrm{Re}(z)}{|z|} = \frac{m}{\sqrt{m^2 + q^2}}$$

$$\sinh(\varnothing) = \frac{y}{r} = \frac{\mathrm{Im}(z)}{|z|} = \frac{q}{\sqrt{m^2 + q^2}}$$

$$\tanh(\varnothing) = \frac{\sinh(\varnothing)}{\cosh(\varnothing)} = \frac{y}{x} = \frac{\mathrm{Im}(z)}{\mathrm{Re}(z)} = \frac{q}{m}$$

case (II): If $|x| < |y|$ then

$$r = |z| = \pm\sqrt{y^2 - x^2}$$

$$\left(\frac{y}{r}\right)^2 - \left(\frac{x}{r}\right)^2 = \frac{y^2 - x^2}{y^2 - x^2} = 1$$

Consequently, there exists a unique number phi such that:

$$\sinh(\varnothing) = \frac{x}{r} = \frac{\mathrm{Re}(z)}{|z|} = \frac{m}{\sqrt{m^2 + q^2}}$$

$$\cosh(\varnothing) = \frac{y}{r} = \frac{\mathrm{Im}(z)}{|z|} = \frac{q}{\sqrt{m^2 + q^2}}$$

$$\tanh(\varnothing) = \frac{\sinh(\varnothing)}{\cosh(\varnothing)} = \frac{x}{y} = \frac{\mathrm{Re}(z)}{\mathrm{Im}(z)}$$

As a result of this demonstration, we conclude that the complex number z in CMS can be described symbolically by

$$\begin{cases} z = r\left(\cosh(\varnothing) + \mathrm{i}\sinh(\varnothing)\right) = r * \mathrm{CISH}(\varnothing) & \text{if } |x| > |y| \\ z = r\left(\sinh(\varnothing) + \mathrm{i}\cosh(\varnothing)\right) = r * \mathrm{SICH}(\varnothing) & \text{if } |x| < |y| \end{cases} \tag{33}$$

5. RELATIVISTIC AND MINKOWSKI GEOMETRY IN CMS

The relations between Relativistic and Minkowski Geometry have been investigated extensively by many in the twentieth century. In this paper we would like to clarify the geometric concept and consistency of the Complex Matter Space with these two mathematical landmarks of the century.

By the relation (9) the Minkowski norm can be used to define an event vector

$$\boldsymbol{p} = [p_1, p_2, \cdots, p_n]$$

in the space-time $R^n \times C$ which is defined by

$$E(\boldsymbol{p}, t) = [p_1, p_2, \cdots, p_n, \mathrm{i}t] \tag{34}$$

where R^n is real Euclidean space and C is complex space, for $i=\sqrt{-1}$.

Definition of CMS inner-product: For two event vectors, $E_1 = E(\boldsymbol{p}) = (\boldsymbol{p}, t_1)$ and $E_2 = E(\boldsymbol{q}) = (\boldsymbol{q}, t_2)$ we define the CMS inner product by:

$$E_1 \cdot E_2 = \sum_{k=1}^{n} p_k q_k + (\mathrm{i}t_1)(\mathrm{i}t_2) \tag{35}$$

As we defined in (18) and (19), the CNS pseudo-norm can be rearranged by

$$\|E\|^2 = (p^1)^2 + (p^2)^2 + \cdots + (p^n)^2 - (t)^2 = \|p\|^2 - t^2 \tag{36}$$

This value can be positive, negative, or indefinite. Consequently, the vector event will be inside the time-like, light-like, or space-like.

Lemma 5.1: Assume two vector events E_1 and E_2, the inner product for CMS

$$E_1 \cdot E_2 = (E_1)^t JE_2 \tag{37}$$

where J is the $(n+1)\times(n+1)$ identity matrix.

Proof: By the definition:

$$E_1 = T(\boldsymbol{p}, t_1) = \langle \boldsymbol{p}, \mathrm{i}t_1 \rangle$$

$$E_2 = E(\boldsymbol{q}, t_2) = \langle \boldsymbol{q}, \mathrm{i}t_2 \rangle$$

Using matrix multiplication, we can define

$$E_1 \cdot E_2 = (E_1)^t JE_2 = [p_1, p_2, \cdots, p_n, \mathrm{i}t_1] \begin{pmatrix} 1 & \cdots & 0 \\ \vdots & \ddots & \vdots \\ 0 & \cdots & 1 \end{pmatrix} \begin{bmatrix} q_1 \\ q_2 \\ \vdots \\ q_n \\ \mathrm{i}t_2 \end{bmatrix} = (\boldsymbol{p}, \mathrm{i}t_1)^t J(\boldsymbol{q}, \mathrm{i}t_2) \tag{38}$$

where the superscript t represents the transpose of the matrix.

In the next step, we will use the development that used for the future events described in "The Geometry of Spacetime ..." by James Callahan (2000) [13] .

We are targeting to apply for complex vector events which we need in the Complex Matter Space theory.

Event Transformation in CMS: Affine transformation includes translation and rotation that have been studied both in Euclidean and Minkowski geometries.

Let us denote the transformation of the, $\text{vector event} = \boldsymbol{r}$ by

$$T_\theta(\boldsymbol{r}) = A\boldsymbol{r}$$

where A is a matrix:

$$A = \begin{bmatrix} \cosh(\theta) & \sinh(\theta) \\ \sinh(\theta) & \cosh(\theta) \end{bmatrix} \tag{39}$$

With a little algebra one can verify the following inverse transform:

$$A^{-1} = \begin{bmatrix} \cosh(\theta) & -\sinh(\theta) \\ -\sinh(\theta) & \cosh(\theta) \end{bmatrix} = (T_\theta(\boldsymbol{r}))^{-1} \tag{40}$$

The angle θ of hyperbolic rotation in this transformation of the future vector $\boldsymbol{r}$ can be measured by hyperbolic radians.

Lemma 5.1: The hyperbolic transformation T has additive property on the rotation angle. That is

$$T_\theta(T_\alpha) \equiv T_{\theta+\alpha} \tag{41}$$

Proof: Assume a point P is the initial position on a unit hyperbola representing the event $E = \boldsymbol{OP} = \begin{bmatrix} \cosh(\alpha) \\ \sinh(\alpha) \end{bmatrix}$, such that the vector $P = (\cosh(\alpha), \sinh(\alpha))$

Thus using the definition,

$$T_\theta(E) = T_\theta \begin{bmatrix} \cosh(\alpha) \\ \sinh(\alpha) \end{bmatrix} = \begin{bmatrix} \cosh(\theta) & \sinh(\theta) \\ \sinh(\theta) & \cosh(\theta) \end{bmatrix} \begin{bmatrix} \cosh(\alpha) \\ \sinh(\alpha) \end{bmatrix}$$

By using algebra of multiplication of two matrices, we conclude that

$$T_\theta(E) = \begin{bmatrix} \cosh(\theta+\alpha) \\ \sinh(\theta+\alpha) \end{bmatrix} \tag{42}$$

The new position is a new point E' on the hyperbola with the angle rotation alpha and the initial hyperbolic angle theta. That is $T_\theta(E) = E'$.

Proposition 5.1: The Hyperbolic transformation in (39) preserves the CMS inner-product (40).

Proof: Assume two vector events $E_1 = [P_1, t_1]'$ and $E_2 = [P_2, t_2]'$. Using definition (37)

$$\begin{aligned} &T_\theta(E_1)\cdot T_\theta(E_2) = [T_\theta(E_1)]' JT_\theta(E_2) \\ &= [E_1' \cdot T_\theta'] JT_\theta(E_2) \\ &= E_1' \cdot \begin{bmatrix} \cosh(\theta) & \sinh(\theta) \\ \sinh(\theta) & \cosh(\theta) \end{bmatrix} \begin{bmatrix} 1 & 0 \\ 0 & -1 \end{bmatrix} \begin{bmatrix} \cosh(\theta) & \sinh(\theta) \\ \sinh(\theta) & \cosh(\theta) \end{bmatrix} \cdot E_2 \\ &= E_1' JE_2 = E_1 \cdot E_2 \end{aligned}$$

Proposition 5.2: Assume two vector events $E_1' = E(p,t_1) = [p_1, p_2, \cdots, p_n, \mathrm{i}\cdot t_1]$, and $E_2' = E(q,t_2) = [q_1, q_2, \cdots, q_n, \mathrm{i}\cdot t_2]$ with a hyperbolic angle between the two event vectors $< E_1 E_2 = \beta$ then

$$\boldsymbol{E}_1 \cdot \boldsymbol{E}_2 = \|\boldsymbol{E}_1\| \|\boldsymbol{E}_2\| \cosh(\beta) \tag{43}$$

or equivalently; the angle between two hyperbolic events can be described by:

$$\beta = \cosh^{-1}\left(\frac{\boldsymbol{E}_1 \cdot \boldsymbol{E}_2}{\|\boldsymbol{E}_1\| \|\boldsymbol{E}_2\|} \right) \tag{44}$$

Proof: Let us assume the following two unit vectors for only two events:

$$\boldsymbol{U}_i = \frac{\boldsymbol{E}_i}{\|\boldsymbol{E}_i\|} \quad \text{for} \quad i = 1, 2$$

where the angle between the two unit vectors have the same direction as the event vectors

$$\angle \boldsymbol{U}_1\boldsymbol{U}_2 = \angle E_1E_2 = \beta$$

Select the transformation T with initial hyperbolic angle theta such that

$$T_\theta(\boldsymbol{U}_1) = \begin{bmatrix} 1 \\ 0 \end{bmatrix}$$

to calculate the transformation of the second unit vector using the conclusion of (42):

$$T_\theta(\boldsymbol{U}_2) = \begin{bmatrix} \cosh(\theta+\beta) \\ \sinh(\theta+\beta) \end{bmatrix} = \begin{bmatrix} \cosh(\beta) \\ \sinh(\beta) \end{bmatrix}$$

Since the unit vector is in the direction of the horizontal axis when theta is zero we will find the second unit vector such that the hyperbolic rotation preserves the Minkowski inner product. That is

$$T_\theta(\boldsymbol{U}_1)\cdot T_\theta(\boldsymbol{U}_2) = \boldsymbol{U}_1 \cdot \boldsymbol{U}_2$$

$$\begin{bmatrix} 1 \\ 0 \end{bmatrix}^t \begin{bmatrix} \cosh(\beta) \\ \sinh(\beta) \end{bmatrix} = \boldsymbol{U}_1 \cdot \boldsymbol{U}_2$$

$$\cosh(\beta) = \boldsymbol{U}_1 \cdot \boldsymbol{U}_2 \Rightarrow \cosh(\beta) = \frac{\boldsymbol{E}_1}{\|\boldsymbol{E}_1\|}\cdot\frac{\boldsymbol{E}_2}{\|\boldsymbol{E}_2\|}$$

Consequently we can have the final result:

$$\boldsymbol{E}_1 \cdot \boldsymbol{E}_2 = \|\boldsymbol{E}_1\|\|\boldsymbol{E}_2\|\cosh(\beta) \qquad (45)$$

Proposition 5.3: Hyperbolic Triangular Inequality:

In every triangle OE_1E_2 when E_1 and E_2 are two event vectors with the angle between two vectors is $\angle E_1E_2 = \beta$ then

$$\|\boldsymbol{E}_1 - \boldsymbol{E}_2\|^2 = \|\boldsymbol{E}_1\|^2 + \|\boldsymbol{E}_2\|^2 - 2\|\boldsymbol{E}_1\|\|\boldsymbol{E}_2\|\cosh(\beta) \qquad (46)$$

The arc length connecting two points E_1 and E_2:

$$\|E_1E_2\|^2 = \|E_1\|^2 + \|E_2\|^2 - 2\|E_1\|\|E_2\|\cosh(\beta) \tag{47}$$

Proof: (i) To prove the first relation (i), it would be helpful to see event vectors $E_1 - E_2$ as a complex valued and their complex conjugates $E_1 - E_2$ or $(E_1 - E_2)^*$

$$\begin{aligned}\|E_1 - E_2\|^2 &= (E_1 - E_2)\cdot(E_1 - E_2)^* \\ &= (E_1 - E_2)\cdot(E_1^* - E_2^*) \\ &= E_1E_1^* - E_1\cdot E_2^* - E_2\cdot E_1^* + E_2E_2^* \\ &= \|E_1\|^2 + \|E_2\|^2 - 2E_1E_2^*\end{aligned}$$

Substitute the conclusion in (45)

$$\|E_1\|^2 + \|E_2\|^2 - 2E_1\cdot E_2 = \|E_1\|^2 + \|E_2\|^2 - 2\|E_1\|\|E_2\|\cosh(\beta).$$

For the proof of the second part, consider $\|E_1E_2\|$ is the hyperbolic distance between two event vectors E_1 and E_2. The relation (47) is a direct conclusion of the

$$\|E_1 - E_2\| = \|E_1E_2\|.$$

6. CONCLUDING REMARKS

The Beltrami-Klein Model: It is helpful, first to try redefining the primitive terms like "point" "line", "lies on", and "between".

The Beltrami-Klein space is the set of all interior points of a circle which is called point in hyperbolic plane. Two points A and B inside a circle and a chord passing through these two points are imagined. The line is defined by an open chord joining these two points, which is called a hyperbolic line.

There exists an isomorphism between the Klein and Poincare models. That means there is a one-to-one correspondence can be set up between points and lines in one model and those in the others which preserve incidence, congruence, and betweenness.

By accepting the fifth postulate in Euclidean geometry, we can prove existence of a rectangle (quadrilateral with four right angles). Denying or having an alternative to the parallel postulate, particularly in hyperbolic geometry, one can prove that a triangle with angle sum equivalent to 2π radians or quadrilateral with four right angles does not exist (see Figures 1-3).

Thus a quadrilateral with three right angles is called a Lambert Quadrilateral (see Figure 4).

Using the definition of hyperbolic tangent, it can be verified that the mapping $u \dashrightarrow \tanh(u)$ is a one-to-one mapping from the real line into the interval (−1, 1).

Notice that a free online software Geogebra used to draw the following graphs [14] .

With a new advanced proposed postulate for a better unified approach in matter, energy, and momentum, we are challenged to connect the Complex Matter Theory to the two historical approaches of Relativity and Hyperbolic Geometry.

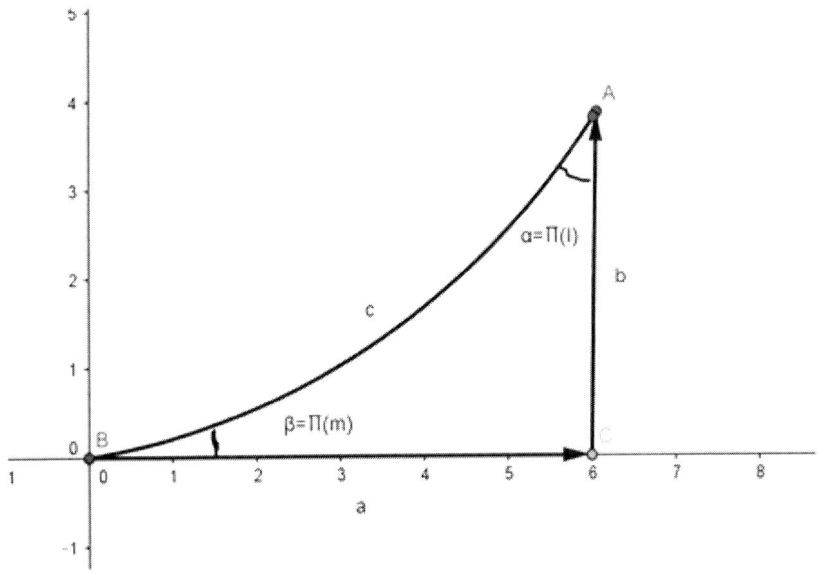

Figure 1. Hyperbolic right triangle with limiting angle of parallelism.

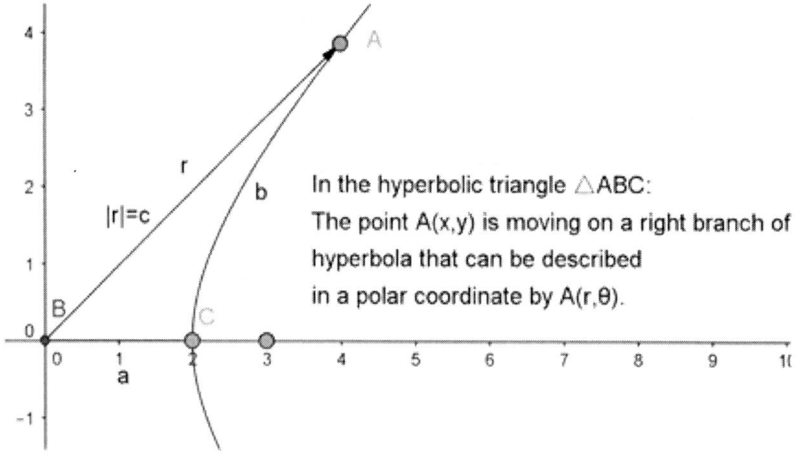

Figure 2. A point $A(x,y)$ is moving on a right branch of a unit hyperbola $x^2 - y^2 = 1$, with a hyperbolic triangle ABC.

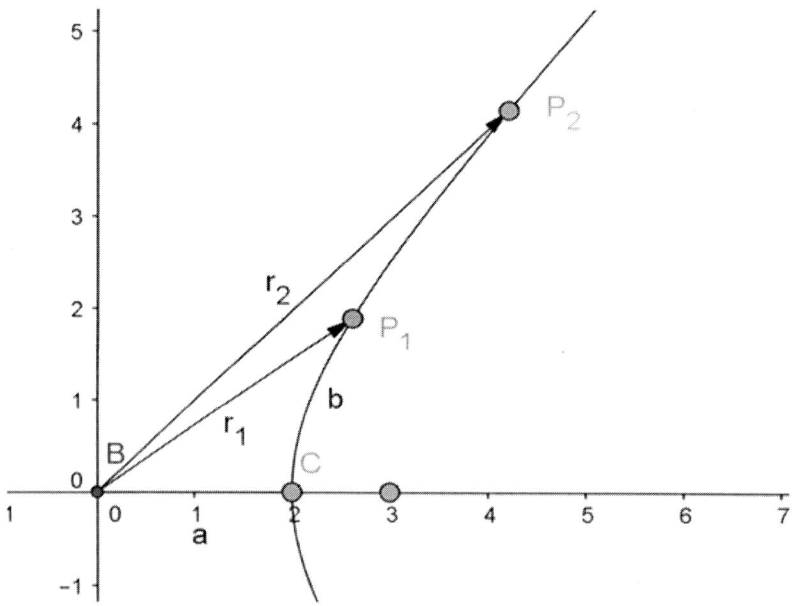

Figure 3. Two points P_1 and P_2 on the hyperbolic plane.

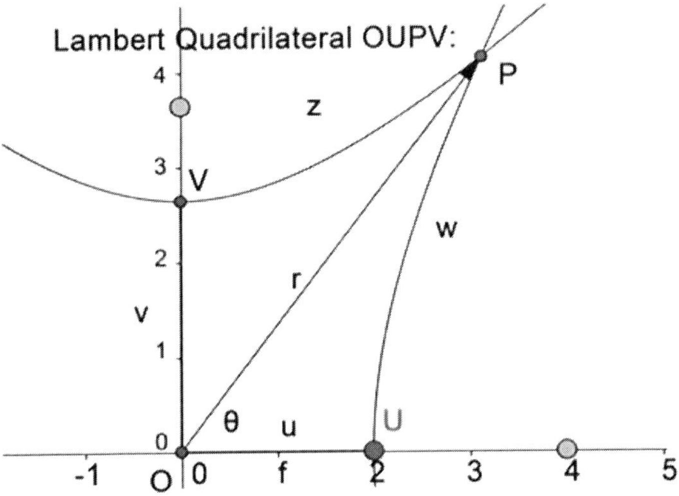

Figure 4. Lambert quadrilateral.

In this work, we conclude that in the new complex matter approach, not only it is a new theory consistent with others, but it describes all shadows and unclear relations by a very beautiful method.

- Matter is redefined with two components of mass and charge which are the intrinsic part of the matter.
- Complex Matter Space Theory: all three space-time regions are space-like, time-like, and light-like. A particle like moving faster than the speed of light can be described by an imaginary mass and a particle-like with imaginary mass can travel faster than the speed of light.
- The geometry of the Complex Matter Space is consistent with the Lorentz transformation, Minkowski, and hyperbolic geometry.
- Given two vector events $\boldsymbol{E}_1$ and $\boldsymbol{E}_2$, the angle between two events can be
- described by a known formula $\cosh(\beta) = \frac{\boldsymbol{E}_1 \cdot \boldsymbol{E}_2}{\|\boldsymbol{E}_1\|\|\boldsymbol{E}_2\|}$ in hyperbolic geometry.
- In every triangle OE_1E_2 when $\boldsymbol{E}_1$ and $\boldsymbol{E}_2$ are two event vectors with the angle that between two vectors is $\angle E_1E_2 = \beta$ then

$$\|\boldsymbol{E}_1 - \boldsymbol{E}_2\|^2 = \|\boldsymbol{E}_1\|^2 + \|\boldsymbol{E}_2\|^2 - 2\|\boldsymbol{E}_1\|\|\boldsymbol{E}_2\|\cosh(\beta)$$

To compare these results with Euclidean space, one can see Greenberg (1980) [15] and Kay (2001) [16] .

REFERENCES

1. Hilbert, D. (1902) The foundations of Geometry. The Open Court Publishing Company, Chicago, IL.
2. Milnor, J. (1996) Hyperbolic Geometry: The First 150 Years. The Mathematics Heritage of Henry Poincare, Proceedings of Symposia in Pure Mathematics of the American Mathematical Society, 39, 25-40.
3. Taylor, J., Zafiratos, C. and Dubson, M.A. (2004) Modern Physics for Scientists and Engineers. 2nd Edition, Pearson Prentice Hall, New Jersey.
4. Richard, F. (1963) The Feynman Lectures on Physics. Addison Wesley, Reading, MA, Volume III, 1-6, 7-5.
5. NASA (2011) Gravity Probe-B. https://science.nasa.gov/science-news/science-at-nasa/2011/04may_epic
6. Lavenda, B.H. (2009) The Relativity of Hyperbolic Space. Universitadegli Studi, Camerino, Italy.
7. Ahangar, R.R. (2014) Foundation of Complex Matter Space and Special Theory of Relativity, a Unifying Approach. Journal of Nuclear and Particle Physics, 4, 147-153.
8. Ahangar, R.R. (2014) Quantum Complex Matter Space. International Journal of Theoretical and Mathematical Physics, 4, 159-163.
9. Ahangar, R.R. (2014) Complex Matter Space and Relativistic Quantum Mechanics. Applied Mathematics, 5, No. 21.
10. Mayer, A.F. (2007) Wave Energy in Quantum Mechanics. Journal of Physics, Conference Series, 70, 012013.
11. Gauthier, R. (2009) The Dirac Equation and the Superluminal Electron Model. http://www.superluminalquantum.org
12. Naber, G.L. (1992) The Geometry of Minkowski Spacetime: An Introduction to the Mathematics of Special Theory of Relativity. Dover, New York.
13. Callahan, J. (2000) The Geometry of Spacetime: An Introduction to Special and General Relativity. Springer, New York.
14. Geogebra Software. For Hyperbolic Geometry See. http://www.geogebra.org http://www.geogebra.org/material/simple/id/7005
15. Greenberg, M.J. (1980) Euclidean and Non-Euclidean Geometries, Development and History. 2nd Edition, WH Freeman, New York.
16. Kay, D.C. (2001) College Geometry a Discovery Approach. 2nd Edition, Addison Wesley Longman, Boston.

CHAPTER 18

Dark Energy Calculations Using the Paraquantum Gamma Factor ($\gamma_{P\psi}$) on the Relativistic Energy Equation

João Inácio Da Silva Filho[1,2]

[1] *Group of Applied Paraconsistent Logic, Santa Cecília University-UNISANTA, Santos, Brazil*

[2] *Institute for Advanced Studies of the University of São Paulo-IEAUSP, São Paulo, Brazil*

ABSTRACT

A Paraconsistent Logic (PL) is a non-classical logic which revokes the principle of non-Contradiction and admits the treatment of contradictory information in its theoretical structure. Paraquantum Logic (PQL) is based on a type of PL denominated Paraconsistent Annotated Logic with annotation of two values (PAL2v). The PAL2v have a representative Lattice of four vertices (Lattice FOUR) where are made interpretations with construction of Paraquantum Logical Model and equations capable computation values extract of Observable Variable measurements. The studies of the PQL are based on propagation of Paraquantum logical states ψ in a Paraquantum Universe represented by PQL-Lattice of four vertices. These studies of PQL are based in two Paraquantum factors: the Paraquantum Gamma Factor ($\gamma_{P\psi}$) that has his action in the measurements of Observable Variables in the Physical world and the Paraquantum Factor of quantization h_{ψ}, which has his action in the Paraquantum Universe. In this paper we analyze the application of Paraquantum Gamma Factor $\gamma_{P\psi}$ and its intrinsic characteristics that add important information into the equation of Einstein's relativistic Energy ($E = MC^2$). In this article were made several calculations to demonstrate the effects of applying the Paraquantum Gamma Factor ($\gamma_{P\psi}$) in relativistic energy equation. It is found that the factors of using the Paraquantum Logical Model make an adjustment in the equation of Einstein's relativistic Energy and identify related values with recent results found for the Dark Energy and dark matter. In the Paraquantum/Relativistic Energy equation the $\gamma_{P\psi}$appears as an important factor of transition between the relativistic universe and the Newtonian Universe. The results suggest that its use would be very important in the interpretation of the behavior of other astronomical factors as the cosmological constant and gravitation.

Keywords: Paraconsistent Logic, Paraquantum Logic, Dark Energy, Dark Matter, Relativity Theory

1. INTRODUCTION

Models capable of simulating with precision our reality need to work with uncertainties, ambiguities and inconsistencies when it is extracting information in measurements of real world. For calculations based on Classic Logic (binary logic), the uncertain values cause great difficulties for the creation of efficient models to simulate phenomena of real physical systems. Nowadays many researches are being developed to find models based on non-Classic Logics [1] . In this work we use the fundamentals of a non-classical Logic called Paraconsistent Logic (PL) [2] . Paraconsistent logic, in its extended form named Paraquantum Logic (PQL), was used in Physics with good results, where the calculations are based on a factor related to measurements of Observable Variables in the physical world. This factor, called Paraquantum Gamma Factor $(\gamma_{P\psi})$ was originated in the interpretations of the special relativity theory. In this paper we will explore the characteristics of the Paraquantum Gamma Factor to provide a mathematical modelling that can help to clarify the values relating to dark energy and dark matter found in recent cosmological measurements. Below we present the main concepts of Paraconsistent Logic.

1.1. Paraconsistent Annotated Logic

Paraconsistent Logic is a non-classical logic which revokes the principle of non-Contradiction and admits the treatment of contradictory information in its theoretical structure [1] -[4] . The Paraconsistent Annotated Logic with annotation of two values (PAL2v) [5] is a class of Paraconsistent Logics particularly represented through a Lattice FOUR (Hasse Diagram) [2] . As seen in [5] [6] , with PAL2v we can get equations that allow quantization calculations of values extracted from measurements on Observable Variables. For calculations the measurement values are expressed as evidence Degrees μ and λ, where: $\mu, \lambda \in \Re$ and $1 \leq \mu \leq 0$ and $1 \leq \lambda \leq 0$. By geometric transformations the representative values of uncertainties can be considered in two axes of the PAL2v-Lattice [5] : First is through the certainty Degree on the horizontal axis:

$$D_{C(\mu,\lambda)} = \mu - \lambda \tag{1}$$

Second is through the Contradiction Degree on the vertical axis:

$$D_{ct(\mu,\lambda)} = \mu + \lambda - 1 \tag{2}$$

These two values will compose a Paraconsistent Logical State (e_τ), located within of the PAL2v-Lattice.

$$\varepsilon_{\tau(\mu,\lambda)} = (D_C, D_{ct}) \tag{3}$$

where: e_τ is the Paraconsistent Logical State.

D_C is the Certainty Degree obtained from the evidence Degrees μ and λ.

D_{ct} is the Contradiction Degree obtained from the evidence Degrees μ and λ.

Some features found in the analyses of the PAL2v allow us to consider it as a quantum logic called Paraquantum Logic (PQL) [6] [7] . That way, a Paraquantum function $y_{(Py)}$ is defined as the Paraquantum logical State y:

$$\psi_{(PQ)} = \left(D_{C(\mu,\lambda)}, D_{ct(\mu,\lambda)}\right) \tag{4}$$

For each measurement performed in the physical world of μ and λ, we obtain a unique duple $\left(D_{C(\mu,\lambda)}, D_{ct(\mu,\lambda)}\right)$

which represents a unique Paraquantum logical state ψ which is a point of the PQL-Lattice [6] -[8] . On the vertical axis of contradictory degrees, the two extreme real Paraquantum logical states are:

a) The contradictory extreme Paraquantum logical state which represents Inconsistency T:

$$\psi_{T} = \left(D_{C(1,1)}, D_{ct(1,1)}\right) = (0,1)$$

b) The contradictory extreme Paraquantum logical state which represents Indeterminate ^:

$$\psi_{\perp} = \left(D_{C(0,0)}, D_{ct(0,0)}\right) = (0,-1)$$

On the horizontal axis of certainty degrees, the two extreme real Paraquantum logical states are:

c) The real extreme Paraquantum logical state which represents Veracity t:

$$\psi_{t} = \left(D_{C(1,0)}, D_{ct(1,0)}\right) = (1,0)$$

d) The real extreme Paraquantum logical state which represents Falsity F:

$$\psi_{F} = \left(D_{C(1,0)}, D_{ct(1,0)}\right) = (-1,0)$$

The Vector of State P(ψ) will always be the vector addition of its two component vectors:

X_C Vector with same direction as the axis of the certainty degrees (horizontal) whose module is the complement of the intensity of the certainty degree: $X_C = 1 - |D_C|$

Y_{ct} Vector with same direction as the axis of the contradiction degrees (vertical) whose module is the contradiction degree: $Y_{ct} = D_{ct}$. Figure 1 represents a Paraquantum Logical State with all these features.

1.2. The Paraquantum Factor of Quantization h_ψ

As it was seen in [6] [7] the continuous variation of measurements in the Observable Variables in the physical world results in Paraquantum Logical State (y) propagation in the PQL-Lattice (Paraquantum Universe). When the superposed Paraquantum logical state (y_{sup}) propagates on the lattice of the PQL a value of quantization for each equilibrium point is established [6] . This point is the value of the contradiction degree of the Paraquantum logical State of quantization y_{hy}, such that:

$$h_\psi = \sqrt{2} - 1 \tag{5}$$

where: h_y is the Paraquantum Factor of quantization.

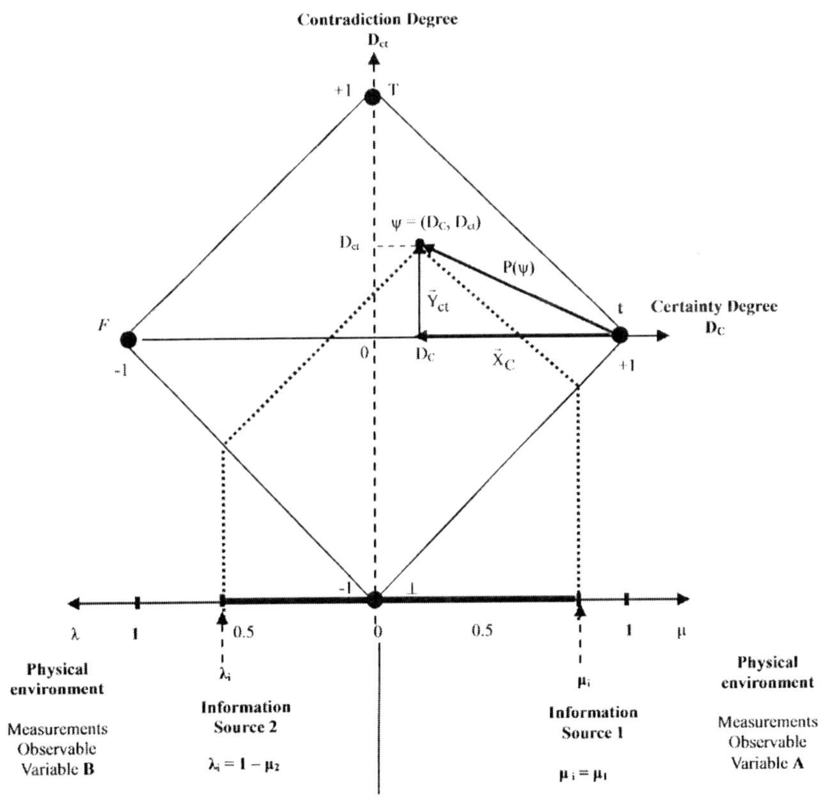

Figure 1. Paraquantum Logical State (ψ) into the Paraquantum Lattice on the point ($D_{C(\mu,\lambda)}$, $D_{ct(\mu,\lambda)}$).

The factor h_y quantifies the levels of energy through the equilibrium points where the Paraquantum Logical State of quantization y_{hy} is located [7] . In order to completely express it, we have to take into account the factor related to the Paraquantum Leaps which will be added to or subtracted from the Paraquantum Factor of quantization [6] such that:

$$h_{\psi t} = h_{\psi} \pm h_{\psi leap} \quad (6)$$

where, we can make the calculations:

$$h_{\psi Leap} = \left(\sqrt{1+h_{\psi}^{2}} - 1\right) \quad (7)$$

So, the Paraquantum Factor of quantization in its complete or total format, which acts on the quantities, is:

$$h_{\psi t} = h_{\psi} \pm \left(\sqrt{1+h_{\psi}^{2}} - 1\right) \quad (8)$$

1.3. Newton Gamma Factor

In the International System of units (SI), to express the value of force (F) in a unit of Newton (N) an adjustment on the value of mass is necessary [9] . Doing such comparisons and analogies between the International unit Systems (SI) and the British System we obtain for the proportionality factor k in the equation which expresses Newton's second law (see [9] [10]).

$k_{br} = 1.38254952 \approx \sqrt{2}$ in the British System.

$k_{SI} = 0.7233013951 \approx \frac{1}{\sqrt{2}}$ in the International System of units (SI).

Given the importance of the Factor k_{br}, which will be largely used in the equations of the PQL, its value is called Newton Gamma Factor whose symbol is γ_N [11] [12] . Therefore, in order to apply classical logics in the Paraquantum Logical model, the Newton Gamma Factor is $\gamma_N = \sqrt{2}$.

1.4. Paraquantum Gamma Factor $\gamma_{P\psi}$

The Lorentz Factor γ can be identified in the infinite Power Series of the binomial expansion [13] [14] related to the series obtained from consecutively applying the Newton Gamma Factor γ_N (see [11]). In the Paraquantum analysis [8] we define a correlation value called Paraquantum Gamma Factor $\gamma_{P\psi}$ such that:

$$\gamma_{P\psi} = \gamma + \frac{\gamma}{\gamma_N} - 1 \tag{9}$$

where: γ_N is the Newton Gamma Factor: $\gamma_N = \sqrt{2}$

γ is the Lorentz factor which is: $\gamma = \dfrac{1}{\sqrt{1-\left(\frac{v}{c}\right)^2}}$

Using the Paraquantum Gamma Factor $(\gamma_{P\psi})$ allows the computations, which correlate values of Observable Variables on the Physical World to the values related to quantization through the Paraquantum Factor of quantization (h_ψ) in the PQL-Lattice (Paraquantum Universe) [6] . The quantitative analysis on the PQL-Lattice defines a quantitative value (Q_{Value}) of any physical quantity, which can be represented on the PQL-Lattice horizontal axis of the certainty Degrees and on the vertical axis of the contradiction degrees. Since the maximum value is normalized on the PQL Fundamental Lattice [6] [8] [11] [12] , considering the Paraquantum factor of quantization only, we can write:

$$1 = h_\psi + (1 - h_\psi) \tag{10}$$

In the PQL Fundamental Lattice the unitary value of the quantization is equivalent to a paraquantum quantization represented in the Paraquantum Logical state ($\psi_{h\psi}$) added to the value of its complement. We have:

$$Q_{Value\max Fund} = h_\psi Q_{Value\max Fund} + (1 - h_\psi) Q_{Value\max Fund} \tag{11}$$

where: $Q_{Value\max Fund}$ is the value of the total amount represented on the unitary axis of the PQL Fundamental Lattice.

Equation (11) shows that the maximum amount of any quantity in the physical environment is composed by two quantized fractions where: one is determined on the Paraquantum Logical state of Quantization $\psi_{h\psi}$ by the Paraquantum Factor of quantization h_ψ and the other is determined by its complement $(1 - h_\psi)$. When the Paraquantum Gamma Factor $\gamma_{P\psi}$ is applied on the paraquantum quantities, besides correlating the paraquantum values to the physical environment, it also works as a factor of expansion or contraction of the PQL-Lattice [11] [12] . Figure 2represents the PQL-Lattice with a Paraquantum logical State in equilibrium point by these two Factors.

2. THE PARAQUANTUM ANALYSIS IN NEWTONIAN UNIVERSE

As it was seen in [11] [12] , on the Paraquantum analysis of the Physical Systems in the Newtonian universe the space as Variable Observable, and the time as Variable Observable, are considered separately. In this way, the action of the Paraquantum Gamma Factor $(\gamma_{P\psi})$ is only in the Variable Observable of the time [11] -[13] . In this case, information Source 1 presents the variation of the space is: $Q\Delta s_{measuredS}$.

And information Source 2 presents the variation of the time is:

$$Q\Delta t_{measuredS}\gamma_{P\psi}.$$

This manner, in the Newtonian universe the favorable Evidence Degree of Observable Variable velocity is extracted of two source of information.

For the unitary value of the Discourse Universe (or Interest Interval), in the Newtonian universe the Paraquantum Gamma Factor acting in the time correlates in the equilibrium point the three greatness physics [8] [13] , such that:

$$v_{(\psi)}\frac{1}{\gamma_{P\psi}} = \frac{(\Delta s_{measured})}{(\Delta t_{measured})\gamma_{P\psi}} \qquad (12)$$

where: $v_{(\psi)}$ = paraquantum velocity.

$\Delta t_{measured}$ = measured variation of time.

$\Delta s_{measured}$ = measured variation of space.

$\gamma_{P\psi}$ = Paraquantum Gamma Factor by (9), were in Newtonian universe, is: $\gamma_{P\psi} = \dfrac{1}{\gamma_N} = \dfrac{1}{\sqrt{2}}$

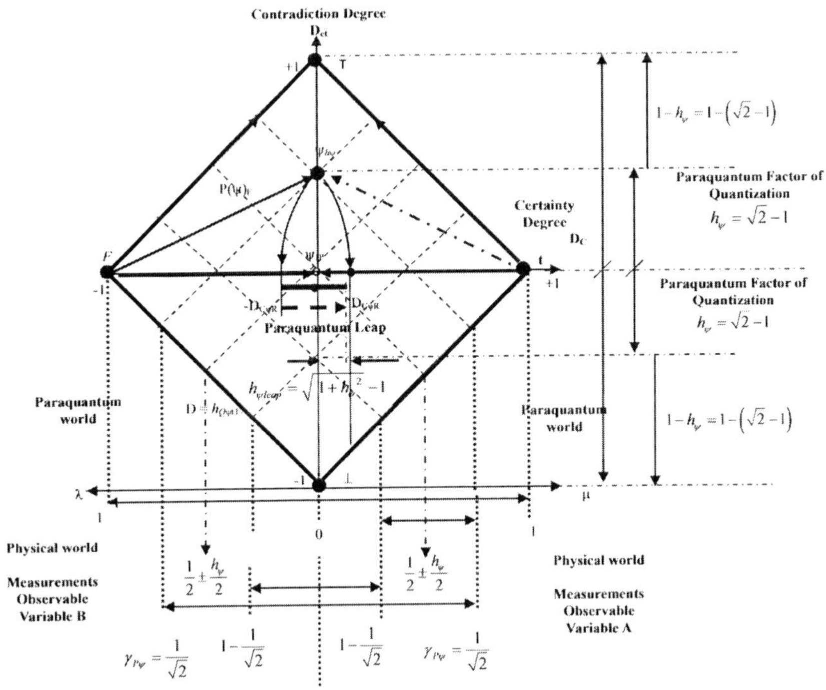

Figure 2. The Paraquantum Factor of quantization h_ψ related to the evidence Degrees with value $\gamma_{P\psi}$ in the equilibrium point.

The Paraquantum velocity is the representation of the velocity of the body which is considered as a physical state of motion in the PQL because it is represented by a Paraquantum Logical State (ψ). For a type of static analysis, therefore without propagation of the Paraquantum logical State (ψ) [6] [8] [13] , the quantized Paraquantum velocity in the equilibrium point is:

$$v_{(\psi)h\psi}\frac{1}{\gamma_{P\psi}}=\frac{(\Delta s_{measured})}{(\Delta t_{measured})\gamma_{P\psi}}\times h_{\psi} \tag{13}$$

where: $v_{(\psi)h\psi}$ = Paraquantum quantized velocity.

According to the foundations of the PQL, the quantized paraquantum velocity in the equilibrium point is represented in the vertical y-axis of the contradiction Degrees of PQL-Lattice. When the analysis is made for a dynamical process Equation (13) of quantization velocity considers the Paraquantum Leap and becomes [8] [13] :

$$v_{(\psi)h\psi}\frac{1}{\gamma_{P\psi}}=\frac{(\Delta s_{measured})}{(\Delta t_{measured})\gamma_{P\psi}}\times h_{\psi t} \tag{14}$$

From Equation (8): $h_{\psi t}=h_{\psi}+\left(\sqrt{1+h_{\psi}^{2}}-1\right)$ then:

$$v_{(\psi)h\psi t}\frac{1}{\gamma_{P\psi}}=\frac{(\Delta s_{measured})}{(\Delta t_{measured})\gamma_{P\psi}}\left(h_{\psi}+\left(\sqrt{1+h_{\psi}^{2}}-1\right)\right) \tag{15}$$

We can rewrite (14) as follows:

$$v_{(\psi)}\frac{1}{\gamma_{P\psi}}=(V_{2}-V_{1})\frac{1}{\gamma_{P\psi}}\left(h_{\psi}+\left(\sqrt{1+h_{\psi}^{2}}-1\right)\right)$$

According to the physical laws [9] [10] [14] -[16] , and using the previous equation, we can obtain the paraquantum acceleration in the Newtonian universe [8] [13] :

$$a_{(\psi)}=\frac{(V_{2}-V_{1})}{(\Delta t_{measured})(\gamma_{P\psi})^{2}}\left(h_{\psi}+\left(\sqrt{1+h_{\psi}^{2}}-1\right)\right) \tag{16}$$

where: V_1 = value of velocity measured at the start.

$\Delta t_{measured}$ = measured time variation.

$a_{(\psi)}$ = quantized value of the state of acceleration of the system.

V_2 = value of velocity measured at the end.

$\gamma_{P\psi}$ = Paraquantum Gamma Factor (Equation (14)).

Δs = measured of traveled space.

h_ψ = Paraquantum Factor of quantization.

Considering the equation that expressed Newton's second law [9] [10] [14] -[16] , we can isolate the Force (F) of the Paraquantum analysis, such that: $(F)\gamma_{P\psi} = (m)a_\psi = F_\psi$. Being Force obtained by: $(F) = (m)\frac{1}{\gamma_{P\psi}}a_\psi$

With this equation, the expression of the Paraquantum acceleration (16), we have:

$$(F) = (m)\frac{(V_2 - V_1)}{(\Delta t_{measured})(\gamma_{P\psi})^3}\left(h_\psi + \left(\sqrt{1+h_\psi^2} - 1\right)\right) \tag{17}$$

This last Equation (17) can be rewritten in such a way the one of the Observable Variables may be represented with quantized values by the action of the Paraquantum Gamma Factor, such that:

$$(F) = \left(m\frac{1}{\gamma_{P\psi}}\right)\frac{(V_2 - V_1)\frac{1}{\gamma_{P\psi}}}{(\Delta t_{measured})\gamma_{P\psi}}\left(h_\psi + \left(\sqrt{1+h_\psi^2} - 1\right)\right) \tag{18}$$

In the equilibrium point from Equation (16) we can isolate V2 obtaining:

$$V_2 = V_1 + \frac{a_{(\psi)}(\Delta t_{measured})}{2\left(h_\psi + \left(\sqrt{1+h_\psi^2} - 1\right)\right)}.$$

The average velocity in the Paraquantum Logical Model has a value computed according to the laws of physics [11] :

$$V_{average} = \frac{V_1 + V_2}{2} \rightarrow V_{average} = (V_1 + V_2)\frac{1}{2}.$$

However, in the Newtonian universe, inverse value of the Newton Factor is considered, so the equation of average velocity can be expressed in a paraquantum form as follows:

$$V_{average} = (V_1 + V_2)\left(\frac{1}{\gamma_N}\right)^2 \tag{19}$$

The equation of traveled space can be written in a paraquantum fashion as follows:

$$\Delta s_{\psi} = (V_1 + V_2)\left(\frac{1}{\gamma_N}\right)^2 \times \Delta t_{measured}\gamma_{P\psi} \tag{20}$$

Replacing this value of V_2 of traveled space Δs, we results in:

$$\Delta s = \left(V_1 \Delta t_{measured} + \frac{a_{(\psi)}(\Delta t_{measured})^2}{2} \times \frac{1}{2\left(h_{\psi} + \left(\sqrt{1 + h_{\psi}^2} - 1\right)\right)}\right)\frac{1}{\sqrt{2}} \tag{21}$$

Going back to the value of the Paraquantum Gamma Factor, the equation of space which due to the use of paraquantum largenesses expresses a paraquantum value producing it by:

$$\Delta s_{\psi} = \left[V_1 \Delta t_{measured} + \frac{1}{2}a_{(\psi)}(\Delta t_{measured})^2 \frac{1}{2\left(h_{\psi} + \left(\sqrt{1 + h_{\psi}^2} - 1\right)\right)}\right]\gamma_{P\psi} \tag{22}$$

where: Δs_{ψ} = variation of space traveled obtained from paraquantum values.

V_1 = measured value of initial velocity.

$\gamma_{P\psi}$ = Paraquantum Gamma Factor.

$\Delta t_{measured}$ = measured variation of time.

a_{ψ} = value of acceleration of the body in study related to the paraquantum world.

h_{ψ} = Paraquantum Factor of quantization.

Isolating $(\Delta t_{measured})$ in Equation (16), we have:

$$\frac{(V_2 - V_1)}{a_{(\psi)}} 2\left(h_{\psi} + \left(\sqrt{1+h_{\psi}^2} - 1\right)\right) = (\Delta t_{measured})$$

Which replaced in (22) produces:

$$\Delta s_{\psi} = \frac{(V_2)^2 - (V_1)^2}{a_{(\psi)}} \frac{1}{\sqrt{2}}\left(h_{\psi} + \left(\sqrt{1+h_{\psi}^2} - 1\right)\right)$$

Going back to the value corresponding to the Paraquantum Gamma Factor in the Newtonian universe, we have:

$$\Delta s_{\psi} = \frac{\left((V_2)^2 - (V_1)^2\right)\gamma_{P\psi}}{a_{(\psi)}}\left(h_{\psi} + \left(\sqrt{1+h_{\psi}^2} - 1\right)\right) \tag{23}$$

When the Observable Variables that produce the Evidence Degrees are values of space traveled with square velocity, we can obtain the equation that determines the Paraquantum acceleration:

$$a_{(\psi)} = \frac{\left((V_2)^2 - (V_1)^2\right)}{\Delta s}\gamma_{P\psi}\left(h_{\psi} + \left(\sqrt{1+h_{\psi}^2} - 1\right)\right) \tag{24}$$

Using Equation (24) of acceleration and equation which expresses Newton's second law mathematically [10] [11] , Force can be computed by multiplying mass by the Paraquantum acceleration, such that:

$$F = \frac{m\left((V_2)^2 - (V_1)^2\right)}{\Delta s}\gamma_{P\psi}\left(h_\psi + \left(\sqrt{1+h_\psi^2} - 1\right)\right) \tag{25}$$

From the laws of classical physics [11] we obtain Work multiplying the value of Force by the displacement, then:

$$W_\psi = m\left((V_2)^2 - (V_1)^2\right)\gamma_{P\psi}\left(h_\psi + \left(\sqrt{1+h_\psi^2} - 1\right)\right) \tag{26}$$

Paraquantum Kinect Energy

A full Paraquantum Work (W_ψ) is identified with the total kinetic Energy at the equilibrium point where the Paraquantum Logical State is located: $E_{k\psi} = W_\psi$. Therefore, the total kinetic Energy is expressed by:

$$E_{k\psi} = \gamma_{P\psi} m(V_2)^2 - \gamma_{P\psi} m(V_1)^2\left(h_\psi + \left(\sqrt{1+h_\psi^2} - 1\right)\right) \tag{27}$$

And the equation of the Paraquantum kinetic energy, represented at the equilibrium point of the PQL-Lattice, is expressed by:

$$E_{k\psi}\frac{1}{\gamma_{P\psi}} = \left(m(V_2)^2 - m(V_1)^2\right)\left(h_\psi + \left(\sqrt{1+h_\psi^2} - 1\right)\right) \tag{28}$$

where: $E_{k\psi}\frac{1}{\gamma_{P\psi}}$ = is the kinetic energy quantized with respect to the physical world.

m = Mass of the body being considered.

The Work performed in the physical universe is identified with the quantity of paraquantum kinetic energy quantized with respect to the physical environment. Being the kinetic Energy the system's energy variation [9] [10] [14] -[16] , then, from the previous equation we can obtain the paraquantum final energy such that:

$$E_{initialk\psi} = \gamma_{P\psi}\left(m(V_1)^2\right)\left(h_\psi + \left(\sqrt{1+h_\psi^2} - 1\right)\right) \tag{29}$$

where: $E_{Initial\psi}$ = Quantized initial kinetic energy.

V_1= value of the velocity measured at the start.

And the paraquantum final energy such that:

$$E_{Final\psi} = \gamma_{P\psi} m (V_2)^2 \left(h_\psi + \left(\sqrt{1 + h_\psi^2} - 1 \right) \right) \tag{30}$$

where: $E_{Final\psi}$ = Quantized final kinetic energy.

V_2= value of the velocity measured at the end.

The total amount of energy involved in the Paraquantum Logical Model [13] in a static state is that one computed at the value of the measured final velocity such that:

$$E_{Total\psi} = \gamma_{P\psi} m (V_2)^2 \tag{31}$$

when related to the physical environment, the Paraquantum total energy in the static state is expressed by:

$$E = E_{Total\psi} \frac{1}{\gamma_{P\psi}} = m (V_2)^2 \tag{32}$$

where:

E = Total energy of the system involved by the Paraquantum Logical Model in the static state.

$E_{Total\psi}$ = Total quantized Energy of the system involved by the Paraquantum Logical Model.

V_2= Final value of measured velocity.

As the analysis is done in the Newtonian Universe the velocity in these equations is not related to the speed of light in a vacuum c, but obtained by dividing the space and time, where only time suffers the action of Paraquantum Gamma Factor [8] [13] . As, for these equations the velocity that is related to the speed of light in vacuum is equal to zero, then the Lorentz factor is unitary$(\gamma = 1)$. This causes the value of the Paraquantum

Gamma Factor [13] , which acts in these equations extracted in the Newtonian universe, is always the inverse of the factor of Newton: $\gamma_{P\psi} = \frac{1}{\gamma_N} = \frac{1}{\sqrt{2}}$. Figure 3 shows the representation of time (t), space (s) and of the mass (m) that is multiplying the velocity squared (V^2).

3. THE PARAQUANTUM ANALYSIS IN RELATIVISTIC UNIVERSE

Is verified that for Newtonian universe the mass is constant and the total Paraquantum energy $E_{t(\psi)}$ is represented only in the y-axis of the PQL-Lattice. Comparing the amounts of total energy with the quantized pure final kinetic energy at the equilibrium point, therefore, comparing the Equation (11) with Equation (31), we have:

$$Q_{Value\max Fund} h_{\psi} = \gamma_{P\psi} m(V_2)^2 h_{\psi}$$

Being the quantized pure final kinetic energy obtained at the equilibrium point represented by:

$$E_{Final\psi\, pure} \frac{1}{\gamma_{P\psi}} = m(V_2)^2 h_{\psi} \tag{33}$$

Since the Paraquantum Logical Model is normalized, there is a complemented quantized pure kinetic energy, therefore:

$$E_{Final\psi\, pureC} \frac{1}{\gamma_{P\psi}} = 1 - m(V_2)^2 h_{\psi} \tag{34}$$

So, the total amount of total kinetic Energy, without adding the effect of the Paraquantum Leaps to it, appears as the unit on the axis of the contradiction degrees of the PQL-Lattice, so that the normalized value is computed by:

$$E_{\psi} \frac{1}{\gamma_{P\psi}} = \left(m(V_2)^2 h_{\psi}\right) + \left(1 - m(V_2)^2 h_{\psi}\right) \tag{35}$$

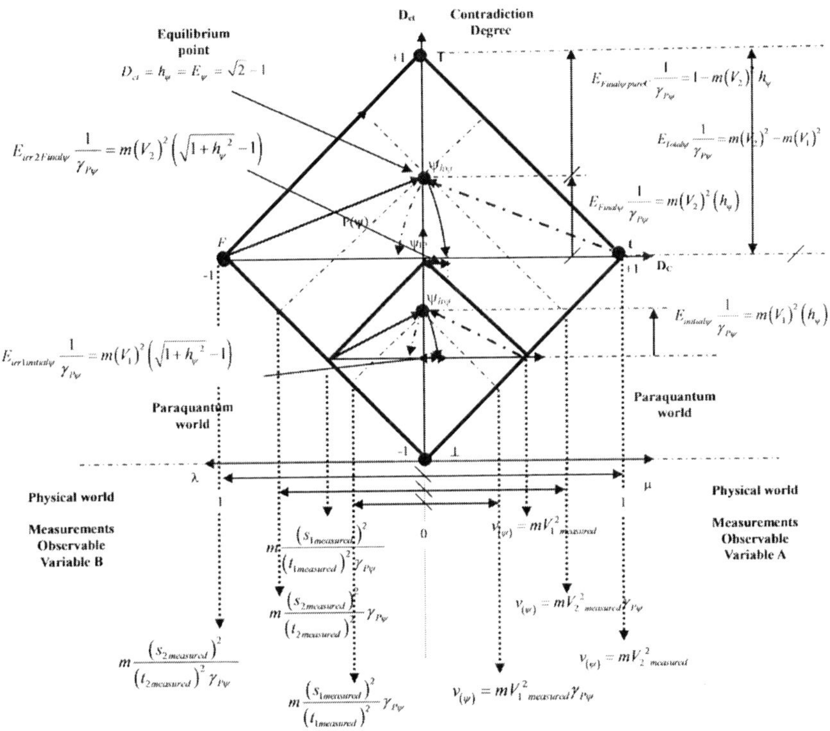

Figure 3. Representation of time (t), space (s) and of the mass (m) that is multiplying the velocity squared (V^2)—with the Paraquantum kinetic energy $E_{(\psi)}$ represented in they-axis of the PQL-Lattice.

Equation (35) is for generalized values, however, for the relativity theory the pure complemented final kinetic energy can be written taking the unitary value in which it is bounded by the speed of the light c in the vacuum [14] [16] . That is, for the speed V listed as a fraction of the speed of light in vacuum: $1 = m(c)^2$.

$$E_{\psi}\frac{1}{\gamma_{P\psi}} = \left(m_{bo}\left(V_2\right)^2 h_{\psi}\right) + \left(m(c)^2 - m_{bo}\left(V_2\right)^2 h_{\psi}\right) \tag{36}$$

where: m = Relativistic mass, that is, the mass produced by Relativistic effects.

m_{bo} = body mass, that is, the mass of the body studied.

The first part of energy Equation (36) is the Kinect Energy of the universe Newtonian [14] :

$$E_{\psi} \frac{1}{\gamma_{P\psi}} = \left(m_{bo} \left(V_2 \right)^2 h_{\psi} \right) \tag{37}$$

which related to the physical environment it is expressed by:

$$E_{\psi} = \gamma_{P\psi} h_{\psi} m_{bo} \left(V_2 \right)^2 \tag{38}$$

The second part of Equation (36) is the energy complemented, is the part corresponding relativistic analysis, therefore:

$$E_{\psi C} \frac{1}{\gamma_{P\psi}} = m \left(c \right)^2 - h_{\psi} m_{bo} \left(V_2 \right)^2 \tag{39}$$

From Equation (39) we can write the Paraquantum Relativistic Energy equation such that:

$$E_{\psi} = \gamma_{P\psi} m \left(c \right)^2 - \gamma_{P\psi} h_{\psi} m_{bo} \left(V_2 \right)^2 \tag{40}$$

where: E_{ψ} = complemented final quantized kinetic Energy or Paraquantum relativistic Energy.

c = Constant value of the velocity of light in the vacuum, imposed as maximum value.

V_2 = measured value of the particle's velocity related c.

m = Relativistic mass, that is, the mass produced by Relativistic effects.

m_{bo} = Mass of the particle.

$\gamma_{P\psi}$ = Paraquantum Gamma Factor by Equation (9).

For this condition the PQL-Lattice is bounded by the velocity c of light in the vacuum [14] [16] , such that now it is possible to obtain the paraquantum pure total energy based on the maximum bound which is the speedof light in the vacuum. Equation (39) represents the value of the kinetic energy from the equilibrium point depending on the speed of the body related to the speed of light in vacuum. For the final value of the amount of energy in a given speed, the equation will be:

$$E_{\psi\, final} = \gamma_{P\psi} m(c)^2 + \gamma_{P\psi} h_\psi m_{bo} (V_2)^2 \tag{41}$$

If we consider in the Equation (41) the condition: $V_2 \ll c$ or small values of mass, then we can make an equation of approximated values:

$$E_{\psi\, final} = \gamma_{P\psi} m(c)^2 \tag{42}$$

where: $E_{\psi\, final}$ = Paraquantum pure total kinetic Energy of the system.

c = Constant value of the light speed in the vacuum, imposed as maximum value.

$\gamma_{P\psi}$ = Paraquantum Gamma Factor.

3.1. Paraquantum/Relativistic Energy Equation

A full Paraquantum Relativistic Energy equation is obtained if we consider the irradiant Energy caused by the Paraquantum leap [6] [8] [11] [12] [17]. In this way Equation (36) will be written as:

$$E_{\psi t} \frac{1}{\gamma_{P\psi}} = \left(m_{bo} (V_2)^2 h_{\psi t}\right) + \left(m(c)^2 - m_{bo} (V_2)^2 h_{\psi t}\right) \tag{43}$$

As from Equation (8), $h_{\psi t} = h_\psi + \left(\sqrt{1 + h_\psi^2} - 1\right)$, then we have the total energy equation:

$$E_{\psi t} \frac{1}{\gamma_{P\psi}} = \left(m_{bo} (V_2)^2 \left(h_\psi + \left(\sqrt{1 + h_\psi^2} - 1\right)\right)\right) + \left(m(c)^2 - m_{bo} (V_2)^2 \left(h_\psi + \left(\sqrt{1 + h_\psi^2} - 1\right)\right)\right) \tag{44}$$

The first part of the Energy Equation (44) is the Kinect Energy of the Newtonian universe:

$$E_{\psi t}\frac{1}{\gamma_{P\psi}} = m_{bo}(V_2)^2\left(h_\psi + \left(\sqrt{1+h_\psi^2}-1\right)\right) \tag{45}$$

which related to the physical environment it is expressed by:

$$E_{\psi t} = \gamma_{P\psi}\left(h_\psi + \left(\sqrt{1+h_\psi^2}-1\right)\right)m_{bo}(V_2)^2 \tag{46}$$

As in the Newtonian universe $\gamma_{P\psi} = \frac{1}{\gamma_N} = \frac{1}{\sqrt{2}}$, then:

$$E_{\psi t} = \frac{1}{\sqrt{2}}\left(h_\psi + \left(\sqrt{1+h_\psi^2}-1\right)\right)m_{bo}(V_2)^2 \tag{47}$$

The second part of Equation (44) is the energy complemented of the part corresponding at relativistic analysis, therefore:

$$E_{\psi t}\frac{1}{\gamma_{P\psi}} = m(c)^2 - m_{bo}(V_2)^2\left(h_\psi + \left(\sqrt{1+h_\psi^2}-1\right)\right) \tag{48}$$

From Equations (47) and (48) we can write the Paraquantum Relativistic Energy equation, such that:

$$E_{\psi t} = \gamma_{P\psi}m(c)^2 - \frac{1}{\sqrt{2}}\left(h_\psi + \left(\sqrt{1+h_\psi^2}-1\right)\right)m_{bo}(V_2)^2 \tag{49}$$

The Equation (49) represents the value of the kinetic energy from the Paraquantum equilibrium point h_ψ [6] depending on the mass and speed of the body related to the speed of light in vacuum.

In the Paraquantum equilibrium point between the Newtonian and Relativistic Universe is possible to match the value of rest energy with the Newtonian kinetic energy value. So, making the equality of relativistic part with Newtonian part of Equation (49), we have:

$$E_{pointP\psi} = m(c)^2 = \left(h_\psi + \left(\sqrt{1+h_\psi^2}-1\right)\right)m_{bo}(V_2)^2$$

For Paraquantum equilibrium point $\gamma_{P\psi} = \frac{1}{\gamma_N} = \frac{1}{\sqrt{2}}$, then the Equation (49) can be write:

$$E_{\psi t} = \left(\gamma_{P\psi} - \frac{1}{\sqrt{2}}\right) m(c)^2 \quad (50)$$

And from Equation (44) we have the final equation of the energy:

$$E_{\psi t} = \left(\gamma_{P\psi} - \frac{1}{\sqrt{2}}\right) m(c)^2 + \gamma_{P\psi}\left(h_\psi + \left(\sqrt{1+h_\psi^2} - 1\right)\right) m_{bo}(V_2)^2 \quad (51)$$

Equation (51) is the Paraquantum/Relativistic Energy Equation.

3.2. Expanded Paraquantum/Relativistic Energy Equation

As $\gamma_{P\psi} = \gamma + \frac{\gamma}{\gamma_N} - 1$ the final Paraquantum/relativistic Energy Equation (51) can be writing:

$$E_{\psi t} = \left(\left(\gamma + \frac{\gamma}{\gamma_N} - 1\right) - \frac{1}{\sqrt{2}}\right) m(c)^2 + \left(\gamma + \frac{\gamma}{\gamma_N} - 1\right)\left(h_\psi + \left(\sqrt{1+h_\psi^2} - 1\right)\right) m_{bo}(V_2)^2 \quad (52)$$

And γ_N is the Newton Factor with value $\sqrt{2}$ then:

$$E_{\psi t} = \left(\gamma + \frac{\gamma}{\sqrt{2}} - \left(1 + \frac{1}{\sqrt{2}}\right)\right) m(c)^2 + \left(\gamma + \frac{\gamma}{\sqrt{2}} - 1\right)\left(h_\psi + \left(\sqrt{1+h_\psi^2} - 1\right)\right) m_{bo}(V_2)^2 \quad (53)$$

The rigid body or referenced object with null speed ($V_2 = 0$) the Lorentz Factor is equal to 1 $(\gamma = 1)$. This results in:

$$E_{\psi t} = \left(1 + \frac{1}{\sqrt{2}} - \left(1 + \frac{1}{\sqrt{2}}\right)\right) m(c)^2 + \left(1 + \frac{1}{\sqrt{2}} - 1\right)\left(h_\psi + \left(\sqrt{1+h_\psi^2} - 1\right)\right) m_{bo}(0)^2 = 0$$

If we consider in Equation (45), the condition: $V_2 \square c$ or small values of mass, then $\gamma \approx 1$, and:

$$\left(\gamma+\frac{\gamma}{\sqrt{2}}-\left(1+\frac{1}{\sqrt{2}}\right)\right)m(c)^2 \approx 0$$

and Kinect energy is calculated by Equation (47):

$$E_{kf}\sqrt{2}=\left(h_\psi+\left(\sqrt{1+h_\psi^2}-1\right)\right)m_{bo}(V_2)^2$$

where: the Paraquantum Factor of Quantization $h_\psi=\sqrt{2}-1\approx 0.414213562\cdots$

And: the Paraquantum Leap value $h_{P\psi leap}=\sqrt{1+h_\psi^2}-1=0.08223922\cdots$

$$h_\psi+\left(\sqrt{1+h_\psi^2}-1\right)=0.496605762\cdots$$

But since the rigid body or referenced object are moving there is a Lorentz factor value, therefore the equation (51) can always be used in its overall shape. This means that the final energy equation can be used in the relativistic Universe, as well as in the Newtonian universe.

3.3. Results and Discussion about Paraquantum/Relativistic Energy Equation

As seen in [9] [11] [14] -[16] when the Lorentz factor tends to 1 the relativistic math expressions tend to classic expressions. Thus, the Lorentz factor can be listed as:

$$\gamma=\frac{E}{Mc^2}=1+\frac{E_K}{Mc^2} \tag{54}$$

From Equation (54) the relativistic Energy (E) by the special relativity theory [14] [16] can be written as:

$$E=\gamma Mc^2 \tag{55}$$

Or:

$$E = E_K + Mc^2 \tag{56}$$

Using Equation (54) the relativistic kinetic energy can be written as:

$$E_K = (\gamma - 1)Mc^2 \tag{57}$$

If $\gamma = 1$, $E_K = 0$, then, from Equation (56) we have:

$$E = Mc^2 \tag{58}$$

For the conventional kinetic energy calculation [14] [16] is used Equation (57) and with the use of the Paraquantum Logical Model is applied to Equation (50). In relation to the results of Equation (57) it is an established fact that the values obtained by Equation (50) always is added $\frac{1}{\sqrt{2}} = 0.707106781\cdots$

In percentage means that the Energy value obtained by Equation (50) has 70.7106781% more than the value obtained by applying Equation (57). This percentage value is equivalent to the dark energy [18] [19] obtained in the latest astronomical observations. This happens because the use of conventional relativistic equation (Equation (50) does not take into account Relativistic effects related to the matter. Using Equation (51) it is an established fact that this effect can only be despised when speed is very low in relation to speed of light, or when the mass of the body is very small. Therefore, compared to conventional energy equation (Equation (57)), Equation (51) adds the value of the relativistic effects and also the value related to speed and with mass of body in the Newtonian universe.

This value added to more amounts to $1 - \frac{1}{\sqrt{2}} = 0.292893218\cdots$ or 29.2893218% of remaining energy, or dark matter [18] [19] . This added value of 29.2893218% can still be distributed for Relativistic effects, which vary by the action of the Lorentz factor, constant mass-related effects and effects caused by Paraquantum Leap, that it is related to acceleration [6] -[8] (Paraquantum Irradiant Energy). This energy distribution is studied by expansion of second part of Equation (51), as is seen below.

From Equation (51):

$$E_{\psi t} = \left(\gamma_{P\psi} - \frac{1}{\sqrt{2}}\right) m(c)^2 + \gamma_{P\psi}\left(h_\psi + \left(\sqrt{1+h_\psi^2} - 1\right)\right) m_{bo}(V_2)^2$$

We can make the distribution of the Paraquantum Gamma Factor for the second part, such that:

$$E_{\psi t} = \left(\gamma_{P\psi} - \frac{1}{\sqrt{2}}\right) m(c)^2 + \left(\gamma + \frac{\gamma}{\gamma_N} - 1\right)\left(h_\psi + \left(\sqrt{1+h_\psi^2} - 1\right)\right) m_{bo}(V_2)^2$$

Therefore, as seen in Equation (51) this second part is about the rest energy, now with the relativistic effects:

$$E_{rest} = \left(\gamma + \frac{\gamma}{\gamma_N} - 1\right)\left(h_\psi + \left(\sqrt{1+h_\psi^2} - 1\right)\right) m_{bo}(V_2)^2$$

Making the distributive to only the Lorentz Factor-related values, we get:

$$\gamma h_\psi m_{bo}(V_2)^2 + \gamma\left(\sqrt{1+h_\psi^2} - 1\right) m_{bo}(V_2)^2$$

Continuing, we have:

$$\gamma h_\psi m_{bo}(V_2)^2 + \gamma\sqrt{1+h_\psi^2}\, m_{bo}(V_2)^2 - \gamma m_{bo}(V_2)^2$$

With the values for the constants:

$h_\psi = \sqrt{2} - 1 \approx 0.414213562\cdots$ and $h_{P\psi leap} = \sqrt{1+h_\psi^2} = 1.08223922\cdots$

$$0.414213562\gamma m_{bo}(V_2)^2 + 1.0823922\gamma m_{bo}(V_2)^2 - \gamma m_{bo}(V_2)^2$$

$$0.414213562\gamma m_{bo}(V_2)^2 + 0.0823922\gamma m_{bo}(V_2)^2$$

So this is the relativistic value of part related to Lorentz factor:

$$\text{Relativistic Value } \gamma = 0.496605762\gamma m_{bo}(V_2)^2$$

Similarly we can calculate the value for the Lorentz factor divided by root of 2, $\gamma_N = \frac{\gamma}{\sqrt{2}}$:

$$\frac{\gamma}{\sqrt{2}}h_\psi m_{bo}(V_2)^2 + \frac{\gamma}{\sqrt{2}}\left(\sqrt{1+h_\psi^2}-1\right)m_{bo}(V_2)^2$$

Continuing, we have:

$$\frac{\gamma}{\sqrt{2}}h_\psi m_{bo}(V_2)^2 + \frac{\gamma}{\sqrt{2}}\sqrt{1+h_\psi^2}\, m_{bo}(V_2)^2 - \frac{\gamma}{\sqrt{2}}m_{bo}(V_2)^2$$

Considering the constant values:

$$\frac{0.414213562}{\sqrt{2}}\gamma m_{bo}(V_2)^2 + \frac{0.0823922\gamma}{\sqrt{2}}m_{bo}(V_2)^2$$

So this is the relativistic value of part related to Lorentz factor divided by root of 2:

$$\text{Relativistic Value } \gamma/\sqrt{2} = 0.351153302\gamma m_{bo}(V_2)^2$$

Similarly we can calculate the value related at constant −1:

$$-h_\psi m_{bo}(V_2)^2 - \left(\sqrt{1+h_\psi^2}+1\right)m_{bo}(V_2)^2$$

Continuing, we have:

$$-h_\psi m_{bo}(V_2)^2 - \sqrt{1+h_\psi^2}\, m_{bo}(V_2)^2 + m_{bo}(V_2)^2$$

$$-0.414213562m_{bo}\left(V_2\right)^2-1.0823922m_{bo}\left(V_2\right)^2+m_{bo}\left(V_2\right)^2$$

$$-0.414213562m_{bo}\left(V_2\right)^2-0.0823922m_{bo}\left(V_2\right)^2$$

So this is the non-relativistic value of part related to constant −1:

$$\text{Constant Value }(-1)=-0.496605762m_{bo}\left(V_2\right)^2$$

Added the relativistic effects, we have:

$$\text{RelativisticValue }\gamma+\text{RelativisticValue }\gamma/\sqrt{2}=\text{Dark Matter}$$

$$0.496605762\gamma m_{bo}\left(V_2\right)^2+0.351153302\gamma m_{bo}\left(V_2\right)^2=0.847758974\gamma m_{bo}\left(V_2\right)^2$$

The percentage in the rest Energy:

$$\frac{0.847758974\times 0.292893218}{100}=24.8302854\%$$

That is 24.8302854% of Relativistic Dark Matter.

Then $0.292893218-0.248302854=0.04590364$

That is 4.590364% of Ordinary Matter.

The total of relativistic effects is: 70.7106781% plus 24.8302854%.

That is 95.5409635%

And the effect of constant −1 is the Kinect Energy value calculates by Newtonian Universe.

$$E_K=-0.496605762m_{bo}\left(V_2\right)^2$$

These values found by Paraquantum Logical Model are compatible with the values found by The Wilkinson Microwave Anisotropy Probe (WMAP) [18] [19] . The WMAP [18] [19] is also known as the Microwave Anisotropy Probe (MAP), and Explorer 80. It is a spacecraft which measures differences in the temperature of the Big Bang's remnant radiant heat – the Cosmic

Microwave Background Radiation—across the full sky. Figure 4 shows the relationship between the values found by WMAP and the values calculated by the equations of the Paraquantum Logical Model.

In the values obtained by WMPA the universe comprises only 4.6% atoms. A much greater fraction, 24% of the universe, is a different kind of matter that has gravity but does not emit any light, called "dark matter". The biggest fraction of the current composition of the universe, 71%, is a source of anti-gravity (sometimes called "dark energy") that is driving an acceleration of the expansion of the universe [18] [19] .

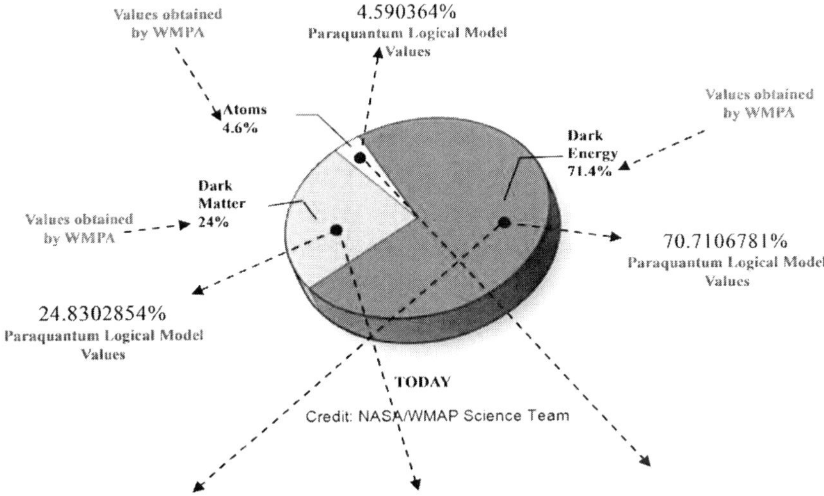

$$E_{\psi t} = \left(\gamma_{P\psi} - \frac{1}{\sqrt{2}}\right) m(c)^2 + \gamma_{P\psi}\left(h_\psi + \left(\sqrt{1+h_\psi^{\,2}} - 1\right)\right) m_{bo}(V_2)^2$$

Paraquantum/Relativistic Energy Equation

$$E_{\psi t} = \left(\left(\gamma + \frac{\gamma}{\gamma_N} - 1\right) - \frac{1}{\sqrt{2}}\right) m(c)^2 + \left(\gamma + \frac{\gamma}{\gamma_N} - 1\right)\left(h_\psi + \left(\sqrt{1+h_\psi^{\,2}} - 1\right)\right) m_{bo}(V_2)^2$$

Figure 4. Relationship of the values of dark energy and dark matter found by calculation of the paraquantum logical model and those observed by WMAP.

4. CONCLUSION

João Inácio Da Silva Filho, In this paper we presented a method calculation based on Paraconsistent Logic that uses structured interpretations in an equilibrium point related to Paraquantum Factor of quantization (h_ψ). This factor is located in the Paraquantum universe represented by a lattice of four vertices. Relating the equilibrium point with values measured in Observable Variables in physical world we can work with another factor called Paraquantum Gamma factor$(\gamma_{P\psi})$. The equations related to the h_{ψ} and $\gamma_{P\psi}$ factors allowed that we could obtain an equation of Paraquantum/Relativistic Energy with adjustments that allowed calculating values similar to those of Dark Energy and Dark Matter observed recently. It turns out that the equation for the relativistic Energy presented in this work is unifying, because it can be used in calculations in the Newtonian universe and also in the Relativistic universe. Results demonstrate that the application of Paraquantum Logic demonstrates a great efficiency to answer questions related to Dark Energy and Dark Matter. The expansion of the Paraquantum/Relativistic Energy Equation presented in this work shows that there are intrinsic parts in future works that may lead to elucidation of other constants and physical quantities, such as gravitation and cosmological constant.

REFERENCES

1. Krause, D. and Bueno, O. (2007) Scientific Theories, Models, and the Semantic Approach. Principia, 11, 187-201.
2. Da Costa, N.C.A. and Marconi, D. (1989) An Overview of Paraconsistent Logic in the 80's. The Journal of Non-Classical Logic, 6, 5-31.
3. Da Costa, N.C.A. (1974) On the Theory of Inconsistent Formal Systems. Notre Dame Journal of Formal Logic, 15, 497-510.
4. Jas'kowski, S. (1969) Propositional Calculus for Contradictory Deductive Systems. Studia Logica, 24, 143-157.
5. Da Silva Filho, J.I., Lambert-Torres, G. and Abe, J.M. (2010) Uncertainty Treatment Using Paraconsistent Logic—Introducing Paraconsistent Artificial Neural Networks. IOS Press, Amsterdam, 328.
6. Da Silva Filho, J.I. (2011) Paraconsistent Annotated Logic in analysis of Physical Systems: Introducing the Paraquantum Factor of quantization h_{ψ}. Journal of Modern Physics, 2, 1397-1409.
7. Da Silva Filho, J.I. (2012) An Introductory Study of the Hydrogen Atom with Paraquantum Logic. Journal of Modern Physics, 3, 312-333.
8. Da Silva Filho, J.I. (2011) Analysis of Physical Systems with Paraconsistent Annotated Logic: Introducing the Paraquantum Gamma Factor γ_{ψ}. Journal of Modern Physics, 2, 1455-1469.
9. Mckelvey, J.P. and Grotch, H. (1978) Physics for Science and Engineering. Harper and Row Publisher, Inc., New York, London.
10. Tipler, P.A. (1976) Physics. Worth Publishers Inc., New York.
11. Da Silva Filho, J.I. (2012) Relativity Theory and Paraquantum Logic—Part I: The Time and Space in the Paraquantum Logical Model. Journal of Modern Physics, 3, 957-971.
12. Da Silva Filho, J.I. (2012) Relativity Theory and Paraquantum Logic—Part II: Fundamentals of a Unified Calculation. Journal of Modern Physics, 3, 972-988.
13. Da Silva Filho, J.I. (2013) Introducing the Paraquantum Equations and Applications. Journal of Modern Physics, 4, 712-733.

14. Tipler, P.A. and Llewellyn, R.A. (2007) Modern Physics. 5th Edition, W. H. Freeman and Company, New York.
15. Tipler, P.A. and Tosca, G.M. (2007) Physics for Scientists. 6th Edition, W. H. Freeman and Company, New York.
16. Einstein, A. (1955) Relativity the Special and the General Theory. Methuen & Co. Ltd., London.
17. Da Silva Filho, J.I. (2012) Analysis of the Spectral Line Emissions of the Hydrogen Atom with Paraquantum. Journal of Modern Physics, 3, 233-254.
18. Spergel, D.N., et al. (2006) WMAP Collaboration. Wilkinson Microwave Anisotropy Probe (WMAP) Three Year Results: Implications for Cosmology.
19. Ade, P.A.R., Aghanim, N., Armitage-Caplan, C., et al. (2013) (Planck Collaboration) Planck 2013 Results. I. Overview of Products and Scientific Results—**Table 9**. Astronomy and Astrophysics (Submitted).

CHAPTER 19

Creation of a Non-Expanding, Non-Singular Universe

Walter Petry

Mathematical Institute, University of Duesseldorf, Duesseldorf, Germany

ABSTRACT

A theory of gravitation in flat space-time is applied to homogeneous, isotropic cosmological models. There are non-singular cosmological models. A natural interpretation is a non-expanding universe. The redshift is an intrinsic effect and not a Doppler effect. The universe contains only energy in the beginning, i.e. no matter exists. In the course of time matter and radiation are created from energy where the whole energy is conserved. Matter increases with time but a certain time after the beginning of the universe the creation of matter is finished and the universe appears like a static one. A modified Hubble law is considered which may explain the high redshifts of objects in the universe without the assumption of dark energy.

Keywords: Universe, Non-Expanding, Non-Singular, Creation of Matter, Conservation of Total Energy, Redshift Is No Doppler Effect, Modified Hubble Law

1. INTRODUCTION

Einstein's general theory of relativity yields a singularity in the beginning of a homogeneous, isotropic universe. Hence, space and time do not exist before and are created in the beginning. The redshift is interpreted as Doppler effect which yields the expansion of space. This is the mainstream interpretation of cosmological models studied by general relativity. It is implied by the singularity of the models.

A theory of gravitation in flat space-time is studied by the author in several articles, e.g. in article [1] and in the book [2] . The theory of gravitation has a flat space-time metric. Gravitation is described by gravitational potentials which form a tensor of rank two. The source of these potentials is the total energy-momentum tensor including that of the gravitational field which is a tensor by this theory. This theory is a field theory in analogy to the well-known Maxwell's electrodynamics. For weak gravitational fields the results of this theory agree with those of general relativity to measurable accuracy. In the beginning of the universe where strong gravitational fields work there are differences between the results of the two theories.

In the beginning of the universe only energy exists, i.e. there is no matter. In the curse of time matter and radiation are created at coasts of energy where the total energy is conserved. Non-singular cosmological models exist, i.e. a big bang doesn't exist. The creation of matter is finished a certain time after the beginning of the universe, i.e. the creation of the universe is finished. It is now like a stationary universe. A natural interpretation is a non-stationary, non-expanding universe, the creation of which is finished. The redshift is an intrinsic gravitational effect and not a Doppler effect. A non-expanding universe is also experimentally indicated by several authors, e.g. the articles [3] - [8] . Theoretical studies of a non-expanding universe are given by several authors (see e.g. [9] - [14]). A fast increase of the inhomogeneity in the matter dominated universe by flat space-time theory of gravitation is given in article [15] . This is important to explain the presently observed inhomogeneity in the universe. General relativity gives only slow increase of the inhomogeneity which does not allow to explain the presently observed universe.

Chapter 2 contains the theory of gravitation in flat space-time which is a field theory. The metric is a flat space-time, e.g. the pseudo-Euclidean metric. The gravitational field satisfies covariant differential equations of order two with the total energy-momentum as source. Conservation of the total energy-momentum and the equations of motion for matter are stated (see e.g. article [1] and the book [2]).

In Chapter 3 a homogeneous, isotropic universe is studied. The pseudo-Euclidean metric is used implying a flat space as indicated by measurements. The matter consists of dust and radiation. The universe is described by two functions. There is no singularity, i.e. no big bang. The absolute time is introduced. There is no age problem. A redshift formula is derived.

A modified Hubble law is described in Chapter 4 by the assumption that everybody is surrounded by a medium similar to that in electrodynamics. Absolute time is used and a modified Hubble law is derived. Quasars and galaxies with nearly the same redshift can have quite different distances as already experimentally indicated by Arp. Redshifts can give smaller distances of the objects than by the standard Hubble law. It may be that dark energy in the universe is not needed because the high redshifts are explained by the modified Hubble law.

The theory of gravitation in flat space-time and the general theory of relativity are compared with one another from the theoretical point of view in article [16] . A modified Hubble law is stated in article [17] which may be important to explain the high redshift of quasars without the assumption of dark energy. The existence of nearby quasars and galaxies with quite different redshifts is already stated by Arp [18] . This is also in agreement with the modified Hubble law in article [17] . Results about a non-singular, non-expanding universe can also be found in the book [2] .

2. GRAVITATION IN FLAT SPACE-TIME

A theory of gravitation in flat space-time is studied in several articles (see e.g. article [1] and the book [2]) gives for small gravitational fields the same results as general relativity to measurable accuracy. In this section the theory is shortly summarized.

Let (x^i) be the coordinates of space-time and (η_{ij}) the symmetric metric tensor of flat space-time with line-element

$$(\mathrm{d}s)^2 = -\eta_{ij}\mathrm{d}x^i\mathrm{d}x^j . \tag{3.1}$$

A special case is the pseudo-Euclidean geometry where (x^1, x^2, x^3) are the Cartesian coordinates, $x^4 = ct$ and $(\eta_{ij}) = (1,1,1,-1)$. The gravitational field is described by a symmetric tensor (g_{ij}). The proper-time τ is defined by

$$(c\,\mathrm{d}\tau)^2 = -g_{ij}\mathrm{d}x^i\mathrm{d}x^j. \tag{3.2}$$

Put

$$\eta = \det(\eta_{ij}), \quad G = \det(g_{ij}).$$

Define the tensor (g^{ij}) by

$$g_{ik}g^{kj}=\delta_i^j .$$

Then, the Lagrangian for the gravitational potentials (g_{ij}) is defined by

$$L(G)=-\left(\frac{-G}{-\eta}\right)^{1/2} g_{ij}g_{kl}g^{mn}\left(g^{ik}_{/m}g^{jl}_{/n}-\frac{1}{2}g^{ij}_{/m}g^{kl}_{/n}\right) \tag{3.3}$$

where the bar/denotes the covariant derivative relative to the metric (3.1). Let k be the gravitational constant and put

$$\kappa=4\pi k/c^4 .$$

Then, the energy-momentum $T^i_j(G)$ of the gravitational field from the Lagrangian is

$$T^i_j(G)=\frac{1}{8\kappa}\left\{\left(\frac{-G}{-\eta}\right)^{1/2} g_{kl}g_{mn}g^{ir}\left(g^{km}_{/j}g^{ln}_{/r}-\frac{1}{2}g^{kl}_{/j}g^{mn}_{/r}\right)+\frac{1}{2}\delta^i_jL(G)\right\}. \tag{3.4}$$

This is a tensor whereas the energy-momentum of general relativity is not a tensor.

The energy-momentum tensor of matter is

$$T^i_j(M)=(\rho+p)g_{jk}u^ku^i+\delta^i_jpc^2. \tag{3.5}$$

Here, ρ, p and (u^i) denote the density, the pressure and the four-velocity $\left(\frac{dx^i}{d\tau}\right)$ of matter. It holds by relation (3.2)

$$c^2=-g_{ij}u^iu^j. \tag{3.6}$$

Define the differential operator of order two

$$D_j^i = \left(\left(\frac{-G}{-\eta} \right)^{1/2} g^{mn} g_{jk} g_{/n}^{ki} \right)_{/m} . \tag{3.7}$$

Then, the field equations for the potentials from the Lagrangian (3.3) are

$$D_j^i - \frac{1}{2}\delta_j^i D_k^k = 4\pi\kappa T_j^i \tag{3.8}$$

where T_j^i is the total energy-momentum tensor of matter and of the gravitational field, i.e.

$$T_j^i = T_j^i(M) + T_j^i(G). \tag{3.9}$$

The equations of motion are

$$\left(T_i^k(M)\right)_{/k} = \frac{1}{2} g_{kl/i} T^{kl}(M) \tag{3.10}$$

where

$$T^{ij}(M) = g^{jk} T_k^i(M). \tag{3.11}$$

is the symmetric energy-momentum tensor of matter. In addition to the field Equations (3.8) we have the conservation of the total energy-momentum tensor, i.e.

$$T_{i/k}^k = 0. \tag{3.12}$$

The field Equations (3.8) and the equations of motion (3.10) imply the conservation law (3.12) of the total energy-momentum. In addition, the field Equations (3.8) and the conservation law (3.12) of the total energy-momentum give the equations of motion (3.10).

The field Equations (3.8) are formally similar to those of general relativity with quite different meanings of D_j^i and T_j^i.

The theory of gravitation in flat space-time and that of general relativity are compared with one another from the theoretical point of view in article [16] .

3. HOMOGENEOUS, ISOTROPIC UNIVERSE

Flat space-time theory of gravitation is applied to a homogeneous, isotropic universe. The results can be found in several articles, especially in [11] and in the book [2] .

The metric is the pseudo-Euclidean geometry, i.e.

$$(\mathrm{d}s)^2 = -(\mathrm{d}x^1)^2 - (\mathrm{d}x^2)^2 - (\mathrm{d}x^3)^2 + (\mathrm{d}ct)^2 . \tag{4.1}$$

The potentials are

$$g_{ij} = \begin{cases} a(t)^2, & (i = j = 1,2,3) \\ 1/h(t), & (i = j = 4) \\ 0, & (i \neq j) \end{cases} \tag{4.2}$$

which implies the proper-time

$$(c\mathrm{d}\tau)^2 = -a(t)^2 \left\{ (\mathrm{d}x^1)^2 + (\mathrm{d}x^2)^2 + (\mathrm{d}x^3)^2 \right\} + \frac{1}{h(t)} (\mathrm{d}ct)^2 . \tag{4.3}$$

Assume that matter is at rest, i.e.

$$u^i = 0 \ (i = 1,2,3)$$

which gives by (3.2) the four-velocity

$$(u^i) = (0,0,0,ch^{1/2}).$$

The matter tensor has the form

$$T_j^i(M)=\begin{cases} p_r c^2 & (i=j=1,2,3) \\ (\rho_m+\rho_r)c^2 & (i=j=4) \\ 0 & (i\neq j) \end{cases} \tag{4.4}$$

Here, ρ_m and ρ_r are the densities of matter and radiation and p_r is the pressure of radiation, i.e.

$$p_r=\rho_r/3. \tag{4.5a}$$

The pressure of matter is that of dust, i.e.

$$p_m=0. \tag{4.5b}$$

The energy-momentum tensor of gravitation is given by

$$T_j^i(G)=\begin{cases} \frac{1}{16\kappa}L(G) & (i=j=1,2,3) \\ -\frac{1}{16\kappa}L(G) & (i=j=4) \\ 0 & (i\neq j) \end{cases} \tag{4.6}$$

with

$$L(G)=\frac{1}{c^2}a^3\sqrt{h}\left(-6\left(\frac{a'}{a}\right)^2+6\frac{a'}{a}\frac{h'}{h}+\frac{1}{2}\left(\frac{h'}{h}\right)^2\right). \tag{4.7}$$

The prime denotes the time-derivative.

The equations for the gravitational field have the form

$$\frac{\mathrm{d}}{\mathrm{d}t}\left(a^3\sqrt{h}\frac{a'}{a}\right)=2\kappa c^4\left(\frac{1}{2}\rho_m+\frac{1}{3}\rho_r\right) \tag{4.8a}$$

$$\frac{\mathrm{d}}{\mathrm{d}t}\left(a^3\sqrt{h}\frac{h'}{h}\right)=4\kappa c^4\left(\frac{1}{2}\rho_m+\rho_r+\frac{1}{8\kappa c^2}L(G)\right). \tag{4.8b}$$

The conservation of the total energy gives

$$\left(\rho_m+\rho_r\right)c^2+\frac{1}{16\kappa}L(G)=\lambda c^2 \tag{4.9}$$

where λ is a constant of integration. The equations of motion imply under the assumption that matter and radiation do not interact with one another:

$$\rho_m=\rho_{m0}/\sqrt{h},\ \rho_r=\rho_{r0}/\left(a\sqrt{h}\right) \tag{4.10}$$

where ρ_{m0} and ρ_{r0} are the present densities at time $t_0=0$.

The initial conditions for the differential Equations (4.8) are the values at present time $t_0=0$:

$$a(t_0)=h(t_0)=1, a'(t_0)=H_0, h'(t_0)=h_0' \tag{4.11}$$

where H_0 Is the Hubble constant and h_0' is a further constant which is zero for general relativity.

Let Ω_r and Ω_m be the well-known density parameters of radiation and matter. Put

$$\varphi_0=3H_0\left(1+\frac{1}{6}\frac{h_0'}{H_0}\right) \tag{4.12a}$$

and

$$\Omega_m K_0=\frac{1}{3}\left(\frac{2\kappa c^4\lambda}{H_0^2}-\left(\frac{1}{2}\frac{\varphi_0}{H_0}\right)^2\right). \tag{4.12b}$$

The it follows from (4.9) for $t = t_0 = 0$

$$\Omega_m K_0 = \Omega_m + \Omega_r - 1. \tag{4.13}$$

Longer calculations imply the relations

$$a^3 \sqrt{h} = 2\kappa c^4 \lambda t^2 + \varphi_0 t + 1, \tag{4.14a}$$

$$\left(\frac{a'}{a}\right)^2 = \frac{H_0^2}{\left(2\kappa c^4 \lambda t^2 + \varphi_0 t + 1\right)^2}\left(-\Omega_m K_0 + \Omega_r a^2 + \Omega_m a^3\right). \tag{4.14b}$$

The condition

$$K_0 > 0$$

gives by the use of (4.12b) that the denominator of (4.14b) is positive for all $t \in \left]-\infty, +\infty\right[$. Hence, non-singular solutions exist. Let us assume

$$0 < K_0 \ll 1 \tag{4.15}$$

then, there exists a time

$$t_1 \ll t_0 = 0 \text{ with } 0 < a_1 = a(t_1) \ll 1,\ \Omega_m a_1^3 + \Omega_r a_1^2 - \Omega_m K_0 = 0 \tag{4.16}$$

and for all $t \in \left]-\infty, +\infty\right[$ holds

$$a(t) \geq a_1$$

We get for $\rho_{r0} = 0$

$$K_0 = a_1^3,\ \frac{h_0'}{H_0} = -\frac{2}{3} K_0 < 0 \tag{4.17a}$$

$$\Omega_m = 1/(1-K_0). \tag{4.17b}$$

Put

$$H_0\tilde{t}_1 = -1\bigg/\left(\frac{1}{2}\frac{\varphi_0}{H_0}\right) \tag{4.18a}$$

and assume

$$H_0\tilde{t}_1 \ll H_0 t \tag{4.18b}$$

then, the solutions are approximated by

$$a^3(t) \approx \frac{4}{9}\frac{1}{\Omega_m}\left(\frac{1}{2}\frac{\varphi_0}{H_0}\right)^4 \left(H_0 t - H_0\tilde{t}_1\right)^2, \tag{4.19a}$$

$$\sqrt{h} \approx 1 - \frac{1}{9}K_0. \tag{4.19b}$$

In the beginning of the universe, i.e. $t \to -\infty$, the functions are

$$a(-\infty) \approx 1.8a_1, \tag{4.20a}$$

$$\sqrt{h(t)} \approx \left(\frac{1}{2}\frac{\varphi_0}{H_0}t + 1\right)^2 \bigg/ a^3(-\infty). \tag{4.20b}$$

This means that the universe starts with a small positive value $a(-\infty)$ at $t = -\infty$ and decreases till t_1 with $a(t_1) = a_1$. Then, it increases to infinity as $t \to \infty$. The function $h(t)$ is infinite at $t = -\infty$ and it is always decreases for all $t \in]-\infty, +\infty[$ to a value a little bit smaller than 1. In the beginning of the universe there is no matter and no radiation. In the course of time matter and radiation are created. A certain time after the beginning of the universe the creation of matter is finished, i.e. the total universe is created. Hence, the universe appears as a static one. A static universe is also claimed by Watts [5] .

In addition to the time t there exist the proper time $\tilde{\tau}$ defined by

$$\mathrm{d}\tilde{\tau} = \frac{1}{\sqrt{h(t)}}\mathrm{d}t \tag{4.21a}$$

and the absolute time t' defined by

$$\mathrm{d}t' = \frac{1}{a(t)\sqrt{h(t)}}\mathrm{d}t = \frac{1}{a(t)}\mathrm{d}\tilde{\tau} \tag{4.21b}$$

The proper time $\tilde{\tau}$ is measured by atomic clocks. The introduction of the absolute time t' gives

$$(c\mathrm{d}\tau)^2 = -a^2\left(|\mathrm{d}x|^2 - (\mathrm{d}ct')^2\right) \tag{4.22}$$

i.e. the light-velocity in the universe is at every time equal to the vacuum light-velocity c.

The universe happens in the pseudo-Euclidean geometry with the time transformation (4.21b). Hence, space is not expanding.

Two light rays are emitted at distance r at time t'_e resp. $t'_e + \mathrm{d}t'_e$ and they are received at $r = 0$ at time 0 and $0 + \mathrm{d}t'_r$. Hence, it follows

$$r = c\int_{t'_e}^{0} \mathrm{d}t = -ct'_e$$

$$r = c\int_{t'_e+\mathrm{d}t'_e}^{\mathrm{d}t'_r} \mathrm{d}t = -c\left(t'_e + \mathrm{d}t'_e - \mathrm{d}t'_r\right).$$

The last two relation yield

$$dt'_e = dt'_r. \tag{4.23}$$

Therefore, the time-intervals are at any time always identical.

The Equation (4.14b) is by the use of (4.14a) and the use of the absolute time t' of (4.21b) rewritten the form

$$\left(\frac{da}{dt'}\right)^2 = \frac{H_0^2}{a^2}\left(-\Omega_m K_0 + \Omega_r a^2 + \Omega_m a^3\right) \tag{4.24}$$

The frequency ν of an atom emitted at time t'_e in the universe is by (4.22)

$$\nu = a\left(t'_e\right)\nu_0 \tag{4.25}$$

where ν_0 is the emitted frequency of the same atom at present time. By the use of (4.22) the change of the photon-energy during the motion of the photon in the universe is given by

$$\frac{d}{dt'}\left(g_{4k}\frac{dx^k}{d\tau}\right) = \frac{1}{2c}\frac{da^2}{dt'}\left(c^2 - \left|\frac{dx}{dt'}\right|^2\right)\frac{dt'}{d\tau} = \frac{da}{dt'}\left(c^2 - \left|\frac{dx}{dt'}\right|^2\right)^{1/2} = 0.$$

Hence the frequency (energy) is not changed when light moves in the universe. Therefore, the measured redshift at the observer at present time is

$$z = 1/a\left(t'_e\right) - 1. \tag{4.26}$$

The redshift formula (4.26) implies by the use of (4.24), differentiation of (4.24) and the constant velocity of light c in the universe

$$z = H_0\frac{r}{c} + \frac{3}{4}\Omega_m\left(H_0\frac{r}{c}\right)^2. \tag{4.27}$$

The age of the universe measured with absolute time t' from time t'_1 till the present time is by the use of (4.24), (4.13), (4.15) and (4.16)

$$t_0' = \int_{t_1'}^{0} dt' = \int_{a_1}^{1} \frac{dt'}{da} da = \frac{1}{H_0} \int_{a_1}^{1} \frac{a da}{\left(-\Omega_m K_0 + \Omega_r a^2 + \Omega_m a^3\right)^{1/2}}$$

$$\geq \frac{1}{H_0} \int_{a_1}^{1} \frac{a da}{\left(-\Omega_m K_0 + \left(\Omega_r + \Omega_m\right) a^2\right)^{1/2}} \approx \frac{1}{H_0}. \tag{4.28a}$$

The age of the universe from the beginning till t_1' is by the use of (4.24), (4.16) and (4.20a)

$$t_b' = \int_{a(-\infty)}^{a_1} \frac{dt'}{da} da = \frac{1}{H_0} \int_{a_1}^{a(-\infty)} \frac{a da}{\left(-\Omega_m K_0 + \Omega_r a^2 + \Omega_m a^3\right)^{1/2}}$$

$$\leq \frac{1}{H_0} \int_{a_1}^{a(-\infty)} \frac{a da}{\left(-\Omega_m K_0 + \Omega_r a^2 + \Omega_m a_1 a^2\right)^{1/2}}$$

$$\leq \frac{1}{H_0} \frac{1}{\sqrt{\Omega_r + \Omega_m a_1}} a(-\infty) \leq \frac{1}{H_0} 1.8 \sqrt{\frac{a_1}{\Omega_m}} \ll \frac{1}{H_0}. \tag{4.28b}$$

Hence, the age of the universe measured with absolute time from the beginning till t_1' is very small compared to the total age of the universe.

There is no age problem for the universe by the use of the absolute time which is more natural than the proper time $\tilde{\tau}$ by virtue of the constancy of the light-velocity c in the universe.

There exist several articles which require a universe with non-expanding space (see e.g. [3] - [10] [12]). Non- singular universes are studied in the articles [11] - [14] .

A theoretical comparison of the theory of gravitation in flat space-time with general relativity is given in the article [16] . The total energy-momentum tensor of matter and gravitation is the source of the gravitational potentials whereas general relativity has only the energy-momentum of matter as source since the energy-mo- mentum of gravitation is not a tensor. Therefore, the Ricci tenor is used to describe the gravitational metric which implies non-Euclidean geometries. Experimental results indicate that the space of our universe is flat in agreement with flat space-time theory of gravitation. A flat space yields difficulties to be explained by general relativity since in the beginning of the universe there is a high matter density which implies high curvature. Therefore, the inflationary universe is introduced.

It is worth to mention that flat space-time theory of gravitation gives in the matter dominated universe a fast increase of the inhomogeneity of matter (see [15]). This is necessary to explain the observed inhomogeneity by galaxies and quasars. General relativity has difficulties to explain this inhomogeneity because only weak increases can arise.

4. MODIFIED HUBBLE LAW

The interpretation of an expanding space gives with recently observed high redshift quasars an acceleration of the expansion of the space. This is partly explained by the introduction of a cosmological constant in the general theory of relativity. This is also possible for flat space-time theory of gravitation.

Here we will consider another possibility. Let us assume that a body is surrounded by a medium analogous to that of electrodynamics with electric permittivity ε and magnetic permeability μ. We put

$$n = (\epsilon\mu)^{1/2}. \tag{5.1}$$

We follow along the lines of article [17] and that of the book [2] . Put

$$F_{ij} = \frac{\partial A_j}{\partial x^i} - \frac{\partial A_i}{\partial x^j}, \quad F^{ij} = g^{ik} g^{jl} F^{kl} \tag{5.2}$$

with

$$g_{ij} = \mu^{1/2} \mathrm{diag}\left(1,1,1,-1/n^2\right). \tag{5.3}$$

Then, the proper-time is given by

$$(c d\tau)^2 = -\mu^{1/2}\left(|dx|^2 - \frac{1}{n^2}(dct)^2\right). \tag{5.4}$$

Let $J = \left(J^1, J^2, J^3, J^4\right)$ be the electric four-current density and consider the covariant differential equations

$$\frac{\partial}{\partial x^k} F^{ki} = \frac{4\pi}{c} J^i \tag{5.5a}$$

$$\frac{\partial}{\partial x^k} F_{ij} + \frac{\partial}{\partial x^i} F_{jk} + \frac{\partial}{\partial x^j} F_{ki} = 0 \tag{5.5b}$$

The electric and the magnetic field are defined by

$$E = \left(F_{41}, F_{42}, F_{43}\right), \quad B = \left(F_{32}, F_{13}, F_{21}\right) \tag{5.6a}$$

and the derived fields are

$$H = \left(F^{32}, F^{13}, F^{21}\right), \quad D = \left(F^{14}, F^{24}, F^{34}\right). \tag{5.6b}$$

Then, it follows from (5.5) with (5.2) and (5.3)

$$D = \varepsilon E, \quad B = \mu H \tag{5.7}$$

and with the abbreviation $\tilde{J} = \left(J^1, J^2, J^3\right)$ we have the well-known Maxwell equations in a medium

$$rot\ H - \frac{1}{c}\frac{\partial D}{\partial t} = \frac{4\pi}{c}\tilde{J}, \quad div\ D = 4\pi J^4 \tag{5.8a}$$

$$rot\ E + \frac{1}{c}\frac{\partial B}{\partial t} = 0, \quad div\ B = 0. \tag{5.8b}$$

Let us now consider a medium in the universe described by the absolute time t_1' and $a(t')$. We define the approximating potentials

$$\left(g'_{ij}\right) = a^2\left(t'\right)\mu^{1/2}\mathrm{diag}\left(1,1,1,-\frac{1}{n^2}\right). \tag{5.9}$$

The metric is given by

$$\mathrm{d}s^2 = -\left(\left|\mathrm{d}x\right|^2 - a^2 h\left(\mathrm{d}ct'\right)^2\right) \tag{5.10a}$$

and the proper-time τ has the form

$$\left(c\mathrm{d}\tau\right)^2 = -a^2\mu^{1/2}\left(\left|\mathrm{d}x\right|^2 - \frac{1}{n^2}\left(\mathrm{d}ct'\right)^2\right). \tag{5.10b}$$

Assume that an atom at rest emits a light ray at time t'_e in the universe. The Equation (5.10b) gives for the proper-time

$$\mathrm{d}\tau_e = a\left(t'_e\right)\mu^{1/4}\frac{1}{n}\mathrm{d}t'. \tag{5.11}$$

Hence, the emitted frequency is

$$\nu_e = a\left(t'_e\right)\frac{\mu^{1/4}}{n}\nu_0. \tag{5.12}$$

Assume that the light-ray moves only a small part of his way to the observer in this medium. In the exterior of the medium relation (4.22) holds, i.e. the frequency is not changed during the motion. Hence, the redshift at the observer is

$$z \approx \frac{\nu_0}{\nu_e} - 1 = \frac{1}{a\left(t'_e\right)}\frac{n}{\mu^{1/4}} - 1 \tag{5.13}$$

which gives analogous to Section 4

$$z = \frac{n}{\mu^{1/4}} - 1 + \frac{n}{\mu^{\frac{1}{4}}}\left(H_0 \frac{r}{c}\right) + \frac{n}{\mu^{\frac{1}{4}}} \frac{3}{4} \Omega_m \left(H_0 \frac{r}{c}\right)^2 . \tag{5.14}$$

Hence, we get an intrinsic redshift by the medium, i.e. the standard Hubble law doesn't give a real distance indicator. We must use the modified Hubble law (5.14).

We can assume

$$\frac{n}{\mu^{1/4}} > 1. \tag{5.15}$$

Then, we conclude:

1) The linear modified Hubble law overestimates the Hubble constant.

2) For a fixed redshift the resulting distance of the object is smaller than by the standard Hubble law.

3) Quasars and galaxies with nearly the same distance can have quite different redshifts depending on the medium in which light is emitted (see e.g. Arp [18]).

4) Quasars may be nearer than by the standard Hubble law. Hence, the measured energy emitted from these quasars is smaller than generally assumed.

5) It is possible that the assumption of dark energy in the universe is not necessary because the higher redshift implying the acceleration of the universe may be explained by the surrounding medium of the objects, as e.g. of galaxies and quasars.

All these results can be found in the article [17] and in the book [2] .

5. CONCLUSION

The theory of gravitation in flat space-time gives a non-expanding, non-singular universe, i.e. there exists no big bang. The redshift is an intrinsic effect and not a Doppler effect. Space is not expanding. In the beginning of the universe only energy exists and no matter. In the course of time matter (dust) and radiation are created whereas the total energy is conserved.

Matter increases with time and a certain time after the beginning the creation of matter is finished, i.e. the total universe appears static. A modified Hubble law is given which may explain the high redshifts of objects in the universe without the assumption of dark energy.

REFERENCES

1. Petry, W. (1981) General Relativity and Gravitation, 13, 865-872.
2. Petry, W. (2014) A Theory of Gravitation in Flat Space-Time. Science Publishing Group.
3. Lerner, L.J. (2005) arXiv: astro-phys. 0509611.
4. Lerner, L.J., Faloma, R. and Scarpa, R. (2014) International Journal of Modern Physics, D23, Article ID: 1450058.
5. Watts, A. (2014) "Settled Science" Paper Claims the Universe Is Static, Not Expanding. http://www.sci-news.com/astronomy/science-universe-is-not-expanding
6. Soares, D.S.L. (2006) arXiv: physics/0605098.
7. Afonso-Faus, A. (2010) arXiv: 1004.2251.
8. Alfonso-Faus, A. (2012) arXiv: 1207.0997.
9. Petry, W. (2007) arXiv: 0705.4359.
10. Hartnett, J.G. (2011) arXiv: 1107.2485.
11. Petry, W. (1990) General Relativity and Gravitation, 22, 1045-1065.
12. Petry, W. (1997) Astrophysics and Space Science, 254, 305-317.
13. Petry, W. (2011) arXiv: 1102.1063.
14. Petry, W. (2013) Journal of Modern Physics, 4, 20-25.
15. Petry, W. (1996) Astrophysics and Space Science, 235, 41-57.
16. Petry, W. (2014) Journal of Applied Mathematics and Physics, 2, 50.
17. Petry, W. (2013) Physics Essays, 26, 315.
18. Arp, H.C. (1988) Quasars, Redshifts and Controversies. Cambridge University Press, Cambridge.

CHAPTER 20

The Interacting Generalized Ricci Dark Energy Model in Non-Flat Universe

Masashi Suwa*, **Koji Kobayashi, Hisashi Oshima**

School of Pharmacy, Nihon University, Chiba, Japan

ABSTRACT

We extend our previous analysis and consider the interacting holographic Ricci dark energy (IRDE) model in non-flat universe. We study astrophysical constraints on this model using the recent observations including the type Ia supernovae (SNIa), the baryon acoustic oscillation (BAO), the cosmic microwave background (CMB) anisotropy, and the Hubble parameter. It is shown that the allowed parameter range for the fractional energy density of the curvature is $-0.005 \leq \Omega_{k0} \leq 0.015$ in the presence of the interactions between dark energy and matter. Without the interaction, the flat universe is observationally disfavored in this model.

Keywords: Dark Energy, Ricci, Non-Flat

1. INTRODUCTION

The current astrophysical observations of the Type Ia supernovae (SNIa) [1] [2] , the cosmic microwave background (CMB) [3] -[10] and the large scale structure (LSS) [11] have revealed that the expansion of our universe is accelerated [1] [2] . This indicates that there exists some unknown energy, called dark energy, to realize the accelerated expansion. The simplest interpretation of dark energy is the cosmological constant. However, this model requires an incredible fine-tuning, since the observed cosmological constant is extremely small compared to the fundamental Planck scale $\rho_\Lambda \sim 10^{-120} M_p^4$. Also, this model suffers from the cosmic coincidence problem: why the cosmological constant and matter have comparable energy density today even though their time evolution is so different.

Among various attempts to solve these problems, we focus on the holographic dark energy (HDE) models [12] motivated by the holographic principle of quantum gravity [13] -[16] . Requiring that the total vacuum energy of a system with size L would not exceed the mass of the black hole of the same size, the dark energy density is proposed as

$$\rho_\Lambda = 3c^2 M_p^2 L^{-2} \tag{1}$$

where c is a dimensionless parameter, $M_P = 1/\sqrt{8\pi G}$ is the reduced Planck mass. As for the size L, which is regarded as an IR cutoff, various possibilities are discussed in literatures, such as the Hubble parameter $L^{-1} = H$ [12] , the future event horizon $L = R_h$ [12] , the age of our universe L = T [17] , and the Ricci scalar curvature $L^{-2} = \mathrm{R}$ [18] . With a reasonable choice of L, Equation (1) naturally explains the observed value $\rho_\Lambda \sim M_p^2 H_0^2$ (H_0 is the present Hubble parameter). In our previous work [19] , we studied the Ricci dark energy (RDE) model with $L^{-2} = \mathrm{R}$ by introducing an interaction between dark energy and matter. It was shown that a nonvanishing interaction rate $Q \propto H\rho_\Lambda$ is favored by the observations [19] .

In this paper, we extend our previous analysis, and consider the interacting RDE (IRDE) model in non-flat universe. This paper is organized as follows. In Section 2, we describe the generalized IRDE model in non-flat universe, and obtain analytic expressions for cosmic time evolution. In Section 3, we discuss the observational constraints on this model. We summarize our results in Section 4.

2. THE INTERACTING RICCI DARK ENERGY MODEL

We study the interaction Ricci Dark Energy (IRDE) model in non-flat universe. The Friedmann-Robertson- Walker metric non-flat univrerse is given by

$$\mathrm{d}s^2 = \mathrm{d}t^2 - a^2(t)\left(\frac{\mathrm{d}r^2}{1-kr^2} + r^2\mathrm{d}\theta^2 + r^2\sin^2\theta\mathrm{d}\phi^2\right) \tag{2}$$

where k = 1, 0, −1 for closed, flat, and open geometries. The Friedmann Equation in non-flat univrerse takes the form

$$H^2 = \frac{1}{3M_p^2}\left(\rho_\Lambda + \rho_m + \rho_\gamma + \rho_k\right) \tag{3}$$

where ρ_Λ, ρ_m, ρ_γ and ρ_k represent energy density of dark energy, matter, radiation and curvature, respectively, and $H = \dot{a}/a$ is the Hubble parameter.

We generalize the energy density of the Ricci dark energy as

$$\rho_\Lambda = 3M_p^2\left(\frac{\alpha}{2}\partial_x H^2 + \beta H^2 + \grave{o}ke^{-2x}\right) \tag{4}$$

where α, b and ò are dimensionless parameters and $x = \ln a$. In the case of $\beta = 2\alpha$, $\grave{o} = \alpha$, $k = 0$, this model is reduced to the ordinary RDE model [19] . Moreover, we introduce a phenomenological interaction between dark energy and matter. The energy densities ρ_Λ and ρ_m obey the following Equations [20] -[40]

$$\dot{\rho}_\Lambda + 3H\left(1 + w_\Lambda\right)\rho_\Lambda = -Q \tag{5}$$

$$\dot{\rho}_m + 3H\rho_m = Q \tag{6}$$

We adopt the interaction rate given by

$$Q = \gamma H \rho_{\Lambda} \tag{7}$$

where γ is a dimensionless parameter [19] . To solve H, combining with Equations (4) and (6), the Friedmann Equation (3) is transformed as

$$\frac{\alpha}{2}\frac{d^2H^2}{dx^2} - \left(1 - \frac{3\alpha}{2} - \frac{\alpha\gamma}{2} - \beta\right)\frac{dH^2}{dx} - (3 - 3\beta - \beta\gamma)H^2 - \Omega_{\gamma 0}H_0^2 e^{-4x} - \{1 - \grave{o}(1+\gamma)\}ke^{-2x} = 0 \tag{8}$$

The solution to Equation (8) is given by

$$\frac{H^2}{H_0^2} = A_+ e^{\sigma_+ x} + A_- e^{\sigma_- x} + A_\gamma e^{-4x} + A_k e^{-2x} \tag{9}$$

where

$$\sigma_{\pm} = \frac{(2 - 3\alpha - \alpha\gamma - 2\beta) \pm \sqrt{(2 + 3\alpha - 2\beta)^2 + \alpha\gamma(6\alpha + \alpha\gamma - 4\beta - 4)}}{2\alpha} \tag{10}$$

$\Omega_{\gamma 0} = \rho_{\gamma 0}/\rho_{c0}$, $\Omega_{k0} = -k/H_0^2$ and $\rho_{c0} = 3M_p^2 H_0^2$. When $\sigma_{\pm}$ can be imaginary for sufficiently large α, b and γ, H^2 has oscillatory behavior [19] . The constants A_γ, A_k and $A_{\pm}$ are obtained as

$$A_\gamma = \frac{\Omega_{\gamma 0}}{(2\alpha - \beta)(1 - \gamma) + 1} \tag{11}$$

$$A_k = \frac{1 - (1+\gamma)\grave{o}}{1 + (\alpha - \beta)(1+\gamma)}\Omega_{k0} \tag{12}$$

$$A_{\pm} = \pm\frac{\alpha(\sigma_{\mp} + 4)A_\gamma + \alpha(\sigma_{\mp} + 2)A_k + 2\Omega_{\Lambda 0} + 2\varepsilon\Omega_{k0} - (\alpha\sigma_- + 2\beta)}{\alpha(\sigma_+ - \sigma_-)}. \tag{13}$$

In the case of $\beta = 2\alpha$, $\grave{o} = \alpha$ and $\Omega_{k0} = 0$, Equation (9) reduces to the result obtained in our previous work [19] . Substituting Equation (9) to Equation (4), the Ricci dark energy density is obtained as

$$\rho_\Lambda = \rho_{c0}\left\{\sum_i\left(\frac{\alpha\sigma_i}{2}+\beta\right)A_i e^{\sigma_i x} - \grave{o}\Omega_{k0}e^{-2x}\right\} \tag{14}$$

where $i\in\{+,-,\gamma,k\}$, $\sigma_\gamma=-4,\ \sigma_k=-2$, Likewise, the matter density is obtained as

$$\rho_m = \rho_{c0}\left\{\sum_i\left(1-\frac{\alpha\sigma_i}{2}-\beta\right)A_i e^{\sigma_i x} - \Omega_{\gamma 0}e^{-4x} - (1-\grave{o})\Omega_{k0}e^{-2x}\right\}. \tag{15}$$

To derive the Equation of state parameter w_Λ of the Ricci dark energy, substituting Equation (14) into the following expression:

$$w_\Lambda = -1-\frac{1}{3}\left(\gamma+\frac{1}{\rho_\Lambda}\frac{d\rho_\Lambda}{dx}\right). \tag{16}$$

3. OBSERVATIONAL CONSTRAINTS

In this section, we discuss cosmological constraints on the IRDE model in the non-flat universe $(k\neq 0)$ obtained from SNIa, CMB, BAO and the Hubble parameter observations.

The luminosity distance in the non-flat universe can be written as

$$d_L = \frac{(1+z)}{H_0\sqrt{\Omega_{k0}}}\sinh\left(H_0\sqrt{\Omega_{k0}}\int_0^z\frac{dz'}{H(z')}\right). \tag{17}$$

The SNIa observations measure the distance modulus μ of a supernova and its redshift z. The distance modulus is given by

$$\mu = 5\log_{10}\frac{d_L}{\text{Mpc}}. \tag{18}$$

We use the Union data set of 580 SNIa [41] [42] to obtain limits on the relevant parameters α, γ and $\Omega_{\Lambda 0}$ by minimizing χ^2_{SN} [43] -[46] .

The CMB shift parameter R is given by

$$R=\sqrt{\frac{\Omega_{m0}}{\Omega_{k0}}}\sinh\left(H_0\sqrt{\Omega_{k0}}\int_0^{z_{\mathrm{CMB}}}\frac{dz}{H(z)}\right) \tag{19}$$

where $z_{\mathrm{CMB}}=1089$ is the redshift at recombination, and $\Omega_{m0}=\rho_{m0}/\rho_{c0}$ is the matter fraction at present. We use the value $R=1.725\pm0.018$ obtained from the WMAP9 data [3] -[10] . The CMB constraints are obtained by minimizing χ^2_{CMB} [43] -[46] . The shift parameter gives a complementary bound to the SNIa data $(z\,\hat{}\,2)$, since this parameter involves the large redshift behavior $(z\,\square\,1000)$.

Signatures of the baryon acoustic oscillation (BAO) are provided by the observations of large-scale galaxy clustering. The BAO parameter A is defined by

$$A=\sqrt{\Omega_{m0}}\left(\frac{H_0}{H(z_{\mathrm{BAO}})}\right)^{1/3}\left[\frac{1}{z_{\mathrm{BAO}}\sqrt{\Omega_{k0}}}\sinh\left(H_0\sqrt{\Omega_{k0}}\int_0^{z_{\mathrm{BAO}}}\frac{dz}{H(z)}\right)\right]^{2/3} \tag{20}$$

where $z_{\mathrm{BAO}}=0.35$. We use the measurement of the BAO peak in the distribution of luminous red galaxies (LRGs) observed in SDSS [11] :

$$A=0.469\left(\frac{0.95}{0.98}\right)^{-0.35}\pm0.017. \tag{21}$$

The BAO constraints are obtained by minimizing χ^2_{BAO} [43] -[46] .

The Hubble parameter constraints are given by minimizing

$$\chi^2_H=\sum_i^{28}\left(\frac{H(z_i)-H_{\mathrm{obs}}(z_i)}{\sigma_{H_i}}\right)^2 \tag{22}$$

where σ_{H_i} is the 1σ uncertainty of the observational $H(z)$ data [47] -[52] .

In Figure 1, we plot the Equation of state parameter w_Λ for $\Omega_{k0} = -0.1$ (blue), 0 (red) and 0.1 (green) as a function of the scale factor a. The dotted and solid lines are the results for the case without interaction $(\gamma = 0)$ and with interaction $(\gamma = 0.15)$, respectively. The dark energy parameter is fixed as $\alpha = 0.45$. Though the value $\Omega_{k0} = \pm 0.1$ is too large (see Figure 2), we use these values to visualize the effect of the curvature in the figure. In the case of the curvature $k = -0.1$, the value of w is larger than another case, but the condition for accelerated expansion at present $w_\Lambda < -1/3$ is satisfied for each cases.

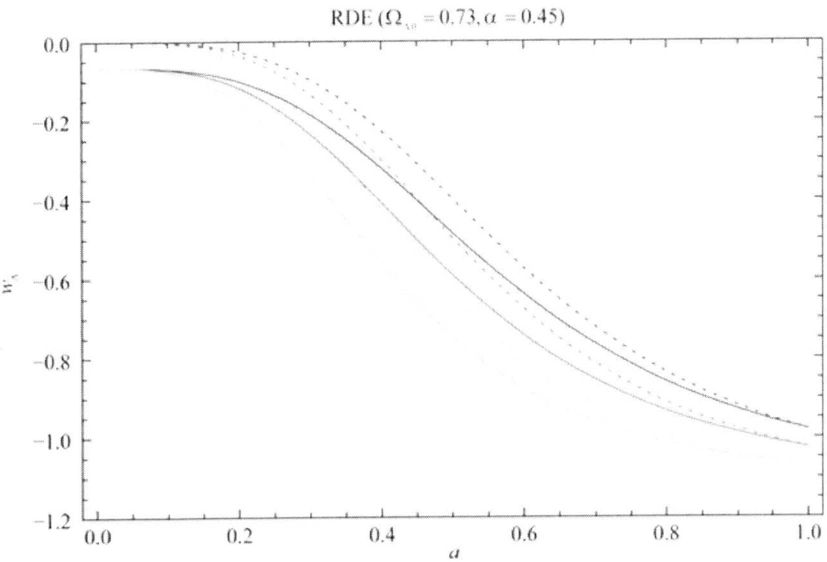

Figure 1. Plot of w_Λ versus a for RDE $(\Omega_\Lambda = 0.73, \alpha = 0.45, \gamma = 0)$. The lines for $k = -0.1$ (blue), 0 (red) and 0.1 (green) in the case without interactions (dotted line) and with interaction (solid line).

In Figure 3, we plot the evolution of the energy density fractions Ω for radiation (green), matter (red) and dark energy (blue) for $\alpha = 0.45$. The dotted and solid lines are the results for the case without interaction $(\gamma = 0)$ and with interaction $(\gamma = 0.15)$, respectively. The panels (a), (b) and (c) corresponds to $\Omega_{k0} = 0$, -0.1 and 0.1, respectively. One can see that the effect of the curvature can be important only for $a \Box 1$. The fractions Ω_Λ and

Ω_m for $\Omega_{k0} = -0.1$ in the panel (b) ($\Omega_{k0} = 0.1$ in the panel (c)) are slightly increased (decreased) around 0.1 峯 a 1 compared to the flat case (a). Around radiation-matter equality, the radiation component Ω_γ is increased by several percents compared to the corresponding result without interaction (dotted line), while the Ω_m is decreased due to the interaction.

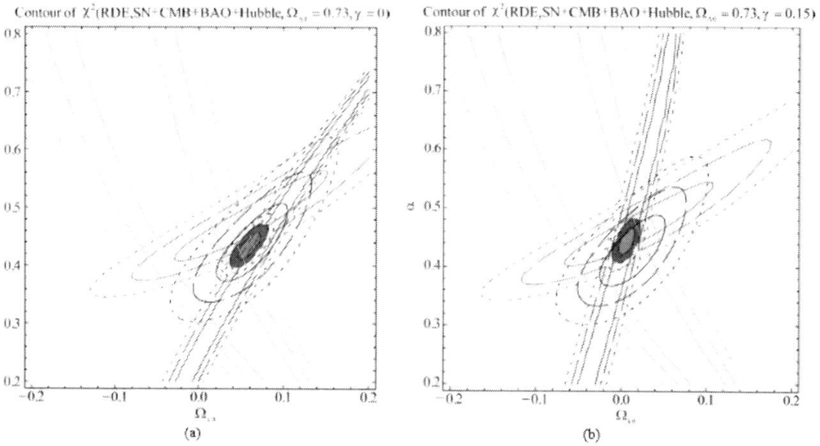

Figure 2. The probability contours for SNIa (red), CMB (blue), BAO (green) and Hubble (black) observations in the (α, Ω_{k0})-plane in the case without interactions $(\gamma = 0)$. The 1σ, 2σ and 3σ contours are drawn with solid, dashed and dotted lines, respectively. The joined constraints using $\chi^2 = \chi^2_{SN} + \chi^2_{CMB} + \chi^2_{BAO} + \chi^2_H$ are shown as shaded contours.

4. CONCLUSION

We have considered the IRDE model in the non-flat universe. We have derived the analytic solutions for the Hubble parameter (9), the dark energy density (14) and matter energy density (15). We have also studied astrophysical constraints on this model using the recent observations including SNIa, BAO, CMB anisotropy, and the Hubble parameter. We have shown that the allowed parameter range for the fractional energy density of the curvature is -0.005 峯 Ω_{k0} 0.015 for $\gamma = 0.15$. The best fit values with 1σ error are $\Omega_{k0} = 0.006 \pm 0.010$ and $\alpha = 0.45 \pm 0.03$ with $\chi^2_{min} = 580$. We have shown that the IRDE model with a small curvature is allowed by observational constraints. Without the interaction, the flat universe is observationally disfavored in this model.

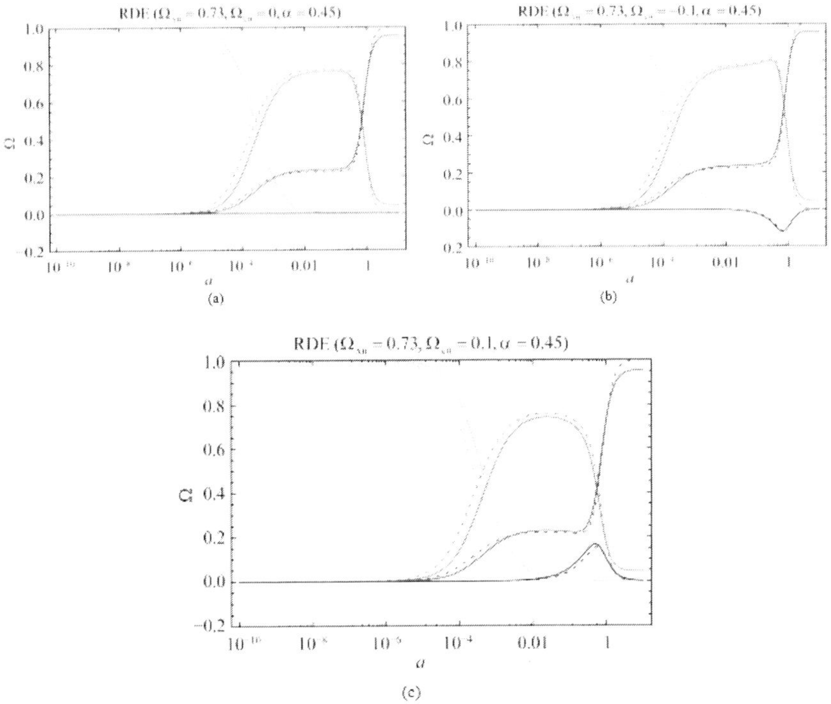

Figure 3. Plot of Ω versus a for RDE $(\Omega_\Lambda = 0.73, \alpha = 0.45, \gamma = 0.15)$. The lines for Radiation (Green), matter (red), dark energy (blue) and curvature (black) in the case without interactions (dotted line) and with interaction (solid line). These figures describe in the case of $\Omega_{k0} = 0$ (a), −0.1 (b) and 0.1 (c).

ACKNOWLEDGEMENTS

Masashi Suwa, Koji Kobayashi, Hisashi Oshima We would like to thank T. Nihei for his valuable discussion, helpful advice and reading the manuscript.

REFERENCES

1. Riess, A.G., et al. (1998) The Astronomical Journal, 116, 1009.
2. Perlmutter, S., et al. (1999) The Astronomical Journal, 517, 565.
3. Bennet, C.L., et al. (2003) The Astronomical Journal, 148, 1.
4. Spergel, D.N., et al. (2003) The Astronomical Journal, 148, 175.
5. Spergel, D.N., et al. (2007) The Astronomical Journal, 170, 377.
6. Page, L., et al. (2007) The Astronomical Journal, 170, 335.
7. Hinshaw, G., et al. (2007) The Astronomical Journal, 170, 263.
8. Komatsu, E., et al. (2009) The Astronomical Journal, 180, 330.
9. Komatsu, E., et al. (2011) The Astronomical Journal, 192, 18.
10. Hinshaw, G., Larson, D., Komatsu, E., Spergel, D.N., Bennett, C.L., Dunkley, J., Nolta, M.R., Halpern, M., et al. arXiv:1212.5226 [astro-ph.CO].
11. Eisenstein, D.J., Zehavi, I., Hogg, D.W., Scoccimarro, R., Blanton, M.R., Nichol, R.C., et al. (2005) Astrophysical Journal, 633, 560-574.
12. Li, M. (2004) Physics Letters B, 603, 1-5.
13. 't Hooft, G. arXiv:gr-qc/9310026.
14. Susskind, L. (1995) Journal of Mathematical Physics, 36, 6377.
15. Maldacena, J.M. (1998) Advances in Theoretical and Mathematical Physics, 2, 231-252.
16. Fischler, W. and Susskind, L. (1998) arXiv:hep-th/9806039.
17. Cai, R.G. (2007) Physics Letters B, 657, 228-231.
18. Gao, C., Wu, F., Chen, X. and Shen, Y.G. (2009) Physical Review D, 79, Article ID: 043511.
19. Suwa, M. and Nihei, T. (2010) Physical Review D, 81, Article ID: 023519.
20. Wang, B., Lin, C.Y. and Abdalla, E. (2006) Physics Letters B, 637, 357-361.
21. Berger, M.S. and Shojaei, H. (2006) Physical Review D, 74, Article ID: 043530.
22. Setare, M.R. (2006) Physics Letters B, 642, 1-4.
23. Setare, M.R. (2007) [hep-th/0701242].
24. Wu, Q., Gong, Y., Wang, A. and Alcaniz, J.S. (2008) Physics Letters B, 659, 34-39.
25. Feng, C., Wang, B., Gong, Y. and Su, R.K. (2007) [arXiv:astro-ph/0706.4033].
26. Karwan, K. (2008) [arXiv:astro-ph/0801.1755].
27. Horvat, R. (2004) Physical Review D, 70, Article ID: 087301.
28. Pavon, D. and Zimdahl, W. (2005) Physics Letters B, 628, 206-210.
29. Guberina, B., Horvat, R. and Nikolic, H. (2007) JCAP, 0701, 012.
30. Zhang, X. and Wu, F.Q. (2005) Physical Review D, 72, Article ID: 043524.
31. Chang, Z., Wu, F.Q. and Zhang, X. (2006) Physics Letters B, 633, 14-18.
32. Zhang, X. and Wu, F.Q. (2007) Physical Review D, 76, Article ID: 023502.
33. Zhang, X. (2005) International Journal of Modern Physics D, 14, 1597.
34. Setare, M.R., Zhang, J. and Zhang, X. (2007) JCAP, 0703, 007.
35. Zhang, J., Zhang, X. and Liu, H. (2008) Physics Letters B, 659, 26-33.
36. Zhang, X. (2007) Physics Letters B, 648, 1-7.
37. Zhang, X. (2006) Physical Review D, 74, Article ID: 103505.
38. Zhang, J., Zhang, X. and Liu, H. (2007) Physics Letters B, 651, 84-88.
39. Ma, Y. and Zhang, X. (2008) Physics Letters B, 661, 239-245.
40. Zhang, J., Zhang, X. and Liu, H. (2007) European Physical Journal C, 52, 693-699.
41. Kowalski, M., Rubin, D., Aldering, G., Agostinho, R.J., Amadon, A., Amanullah, R., et al. (2008) Astrophysical Journal, 686, 749-778. [arXiv:astro-ph/0804.4142]
42. Suzuki, N., Rubin, D., Lidman, C., Aldering, G., Amanullah, R., Barbary, K., et al. (2012) Astrophysical Journal, 746, 85.
43. Nesseris, S. and Perivolaropoulos, L. (2005) Physical Review D, 72, Article ID: 123519.
44. Perivolaropoulos, L. (2005) Physical Review D, 71, Article ID: 063503.
45. Nesseris, S. and Perivolaropoulos, L. (2007) [arXiv:astro-ph/0612653].
46. Pietro, E.D. and Claeskens, J.F. (2003) Monthly Notices of the Royal Astronomical Society, 341, 1299-1310.
47. Simon, J., Verde, L. and Jimenez, R. (2005) Physical Review D, 71, Article ID: 123001.

48. Gaztañaga, E., Cabré, A. and Hui, L. (2009) Monthly Notices of the Royal Astronomical Society, 399, 1663-1680.
49. Stern, D., et al. (2010) JCAP, 1002, 008.
50. Moresco, M., et al. (2012) Journal of Cosmology and Astroparticle Physics, 8, 6.
51. Busca, N.G., Delubac, T., Rich, J., Bailey, S., Font-Ribera, A., Kirkby, D., et al. [arXiv:astro-ph/1211.2616].
52. Farooq, O., Mania, D. and Ratra, B. [arXiv:astrp-ph/1308.0834].

INDEX